Fault-Tolerant Cooperative Control of Unmanned Aerial Vehicles

Ziquan Yu • Youmin Zhang • Bin Jiang •
Chun-Yi Su

Fault-Tolerant Cooperative Control of Unmanned Aerial Vehicles

Springer

Ziquan Yu
College of Automation Engineering
Nanjing University of Aeronautics
and Astronautics
Nanjing, Jiangsu, China

Bin Jiang
College of Automation Engineering
Nanjing University of Aeronautics
and Astronautics
Nanjing, Jiangsu, China

Youmin Zhang
Department of Mechanical, Industrial
and Aerospace Engineering
Concordia University
Montreal, QC, Canada

Chun-Yi Su
Department of Mechanical, Industrial
and Aerospace Engineering
Concordia University
Montreal, QC, Canada

ISBN 978-981-99-7660-7 ISBN 978-981-99-7661-4 (eBook)
https://doi.org/10.1007/978-981-99-7661-4

This Springer imprint is published by the registered company Springer Nature Singapore Pte Ltd.
The registered company address is: 152 Beach Road, #21-01/04 Gateway East, Singapore 189721, Singapore

Paper in this product is recyclable.

To our families for their love and support.

Preface

Over the past few years, unmanned aerial vehicles (UAVs) have been widely used to execute tedious and dangerous tasks, such as remote sensing, search and rescue, surveillance, wireless relay, fire monitoring and detection, agricultural production management. Compared with single UAV, the cooperation of multiple UAVs (multi-UAVs) can significantly increase the task execution efficiency by sharing the information via the communication network. Among these tasks, flight control system (FCS) plays an important role for ensuring the reliability, stability, autonomy, and intelligence of UAV system. It is well known that the UAV dynamics is strongly nonlinear and may simultaneously encounter external disturbances, which remarkably increases the challenge on constructing the FCS. Moreover, the increasing number of sensors, actuators, and system components involved in multi-UAVs makes the formation team vulnerable to faults, which may easily degrade the formation performance or even cause catastrophic accidents if appropriate actions are not adopted in a timely manner. Different from the fault effects encountered by single UAV, which mainly affect the flight performance of the faulty UAV, the fault effects subjected by the UAV formation team will spread via the communication links, since the cooperative control unit embedded in each UAV usually utilizes the neighboring information to generate the control signal. Such a diffusion may destroy the entire formation team if effective control strategies are not activated to attenuate the adverse effects caused by the faults. To solve this difficult problem, the key technology is to design fault-tolerant cooperative control (FTCC) systems against faults by utilizing promising advanced control techniques. Moreover, strongly unknown nonlinearities derived from unknown aerodynamic parameters and external disturbances are also needed to be addressed in the FTCC design for further guaranteeing the flight safety.

The FTCC technique is a promising analytical strategy for improving the operational safety in the fields of aeronautics and astronautics, unmanned systems. However, due to the involvement of communication network, the widely investigated fault-tolerant control (FTC) systems for single UAV cannot be directly extended to handle the faults encountered by the formation team. Up to now, the

FTCC results for multi-UAVs against faults are very rare and there still exist some problems that should be addressed to increase the cooperation safety:

(1) Error constraint requirements are necessary to be constructed and incorporated into the FTCC of multi-UAVs for strictly constraining the formation errors during the faulty phase and post-fault phase.
(2) Input saturation should be considered in the FTCC design since the actuators cannot adjust their signals to arbitrary values to counteract the faults.
(3) The FTCC design for multi-UAVs consisting of multiple leader UAVs and multiple follower UAVs should be investigated to make the FTCC investigation practical in engineering, since the containment control strategies with fault-tolerant capabilities can be used by multiple UAVs to reliably execute tasks in unknown and dangerous environments.
(4) The FTCC strategies can be improved by incorporating the fast fault recovery and compensational capabilities to simultaneously attenuate the adverse effects caused by the strongly unknown nonlinearities, disturbances, and faults.

In view of the aforementioned issues, this monograph focuses on the systematical FTCC design for multi-UAVs against faults, strongly unknown nonlinearities, error constraints, and input saturation. The main contents of this monograph are presented in the following aspects:

(1) For the formation flight of multi-UAVs in a distributed communication network, a prescribed performance-based distributed neural adaptive FTCC scheme is proposed to achieve the longitudinal motions of multi-UAVs under faults and unknown nonlinearities. By transforming the tracking errors into a new set of auxiliary errors, the dynamic surface control (DSC) and minimal learning parameter (MLP) techniques are used to facilitate the FTCC design.
(2) For the neighboring information exchange among multi-UAVs, based on the disturbance observer (DO), a distributed FTCC is developed to handle actuator faults, external disturbances, and input saturation in the longitudinal motions by using the nearest neighbor rule to generate the desired reference for each UAV. To reduce the adverse effects caused by the control input saturation of UAVs, auxiliary dynamic systems are introduced to regulate the control signals when the input signals are saturated for a long time, such that the FTCC capability can be ensured.
(3) To address the FTCC design problem for the longitudinal formation team consisting of multiple leader UAVs and multiple follower UAVs, sliding-mode observer (SMO) and graph theory are first integrated to estimate the unknown references, which are located within the convex hull spanned by the leader UAVs. Then, DO is designed to handle the lumped uncertainty induced by the external disturbances and actuator faults.
(4) To further improve the FTCC capability for the longitudinal formation team consisting of multiple leader UAVs and multiple follower UAVs, a hierarchical FTCC strategy is developed by constructing the distributed finite-time SMOs to estimate the unknown references at the upper layer and finite-time FTC

units to handle the faults at the lower layer. Moreover, the unknown nonlinearities inherent in multi-UAVs, computational burden, and input saturation are simultaneously handled by utilizing neural networks (NNs), MLP, first-order sliding-mode differentiator (FOSMD), and a group of auxiliary systems.

(5) By simultaneously taking the error constraints and finite-time convergence into account, a decentralized finite-time FTCC scheme is investigated to regulate the attitude synchronization among multi-UAVs. By integrating the prescribed performance functions into the synchronization tracking errors, a new set of errors is defined. Based on these transformed errors, a finite-time attitude FTCC scheme is developed by using NNs and finite-time differentiators.

(6) By further considering the directed communications in the distributed communication topology, a decentralized FTCC method is studied to achieve the attitude synchronization of multi-UAVs by using a composite learning algorithm consisting of NNs and DOs. NNs are used to approximate unknown nonlinear terms due to the nonlinearities inherent in UAVs and the loss-of-effectiveness actuator faults. To further compensate for neural approximation errors and bias actuator faults, the DO technique is incorporated into the control scheme to increase the composite approximation capability.

(7) To ensure the cooperative forest fire monitoring of multi-UAVs in the presence of faults during the monitoring mission, a fractional-order (FO) sliding-mode control (SMC) mechanism is utilized to develop the FTCC strategy for multi-UAVs to monitor the elliptical spread of forest fire. By introducing reference systems and sliding-mode differentiators (SMDs), sliding-mode DOs (SMDOs) are developed to estimate the lumped disturbances induced by external disturbances and faults.

This monograph presents systematic and comprehensive descriptions of the FTCC issues for multi-UAVs against faults, external disturbances, strongly unknown nonlinearities, and input saturation. It offers readers with deep understanding and insights on formation system safety and the corresponding design methods. The FTCC methods presented in this monograph can provide guidelines for engineers to improve the safety of aerospace engineering systems and some other related fields. This monograph is suitable for scientists and researchers, aerospace engineers, control engineers, lecturers and teachers, postgraduates, undergraduates in the system and control community, especially those engaged in the field of UAVs and multiple agent systems (MASs).

Nanjing, China — Ziquan Yu
Montreal, QC, Canada — Youmin Zhang
Nanjing, China — Bin Jiang
Montreal, QC, Canada — Chun-Yi Su

Acknowledgments

The contents included in this monograph are the outgrowth and summary of the authors' academic researches to address the FTCC problems for multi-UAVs in the past few years. It was partially supported by National Natural Science Foundation of China (No. 62373188, 62003162, 61833013, 62020106003, 62233009), Natural Science Foundation of Jiangsu Province of China (No. BK20200416, BK20222012), China Postdoctoral Science Foundation (No. 2020TQ0151 and 2020M681590), Science and Technology on Space Intelligent Control Laboratory (No. HTKJ2022KL502015), 111 Project (No. B20007), Industry-University Research Innovation Foundation for the Chinese Ministry of Education (No. 2021ZYA02005), Aeronautical Science Foundation of China (No. 20220007052003, 20200007018001), and Natural Sciences and Engineering Research Council of Canada. The authors greatly appreciate all above financial supports. The authors would like to thank the students Mengna Li, Ruifeng Zhou, Haichuan Yang, Yiwei Xu, Zhongyu Yang, Jiaxu Li, etc. for their contributions and efforts dedicated to this monograph. We appreciate the permission from Springer Nature, Institute of Electrical and Electronics Engineers, Elsevier, SAGE to reuse the results published in the relevant journals, such as *Transactions of the Institute of Measurement and Control* (Chap. 3), *IEEE Transactions on Control Systems Technology* (Chap. 4), *Journal of Intelligent & Robotic Systems* (Chaps. 5 and 9), *IEEE Transactions on Neural Networks and Learning Systems* (Chap. 6), *Journal of the Franklin Institute* (Chap. 7), and *Frontiers of Information Technology & Electronic Engineering* (Chap. 8). In addition, the authors would like to thank the editors in Springer Nature for their help to accomplish the publication of this monograph.

Contents

Acronyms

AFTC	Active Fault-Tolerant Control
BLPF	Butterworth Low-Pass Filter
DO	Disturbance Observer
DSC	Dynamic Surface Control
FCS	Flight Control System
FDD	Fault Detection and Diagnosis
FLS	Fuzzy Logic System
FNN	Fuzzy Neural Network
FO	Fractional-Order
FOSMD	First-Order Sliding Mode Differentiator
FTC	Fault-Tolerant Control
FTCC	Fault-Tolerant Cooperative Control
HGO	High-Gain Observer
HOSMD	High-Order Sliding-Mode Differentiator
IO	Integer-Order
MASs	Multiple Agent Systems
MLP	Minimal Learning Parameter
MLPNN	Minimum Learning Parameter of Neural Network
Multi-UAVs	Multiple Unmanned Aerial Vehicles
NN	Neural Network
PFTC	Passive Fault-Tolerant Control
PID	Proportional-Integration-Derivative
PPC	Prescribed Performance Control
RBFNN	Radial Basis Function Neural Network
RL	Reinforcement Learning
SMC	Sliding-Mode Control
SMD	Sliding-Mode Differentiator
SMDO	Sliding-Mode Disturbance Observer
SMO	Sliding-Mode Observer
UAV	Unmanned Aerial Vehicle
UUB	Uniformly Ultimately Bounded

Chapter 1
Introduction

1.1 Background and Motivations

As an aerial platform, UAV has been widely used to execute forest fire monitoring [88–90], search and rescue [14], powerline/pipeline inspection [32], and reconnaissance and surveillance [9] tasks by equipping task-oriented payloads. Due to the limited coverage and payload capabilities, single UAV usually fails to provide wide-area observation and fine-grained information. To solve this difficult problem, multi-UAVs have been deployed to cooperatively execute the difficult tasks, thus significantly improving the task efficiency. As one of the crucial units to ensure the formation flight safety of multi-UAVs, the cooperative control module embedded in each UAV receives the states from neighboring UAVs and generates the control actions for regulating the cooperative behaviors. Recently, leader-following, behavior-based, virtual structure, collision avoidance, and algebraic graph methods have been used to design effective formation control methods [93]. Moreover, by considering the wake vortices among UAVs in the formation team, close formation control methods are also investigated to keep the relative positions for avoiding the collisions [47, 50, 53, 92]. It should be pointed out that these aforementioned results mainly focus on the development of cooperative control methods for multi-UAVs without the consideration of unexpected situations, such as in-flight faults and disturbances.

As illustrated in Fig. 1.1, the increasing number of actuators, sensors, and components involved in multi-UAVs render the overall formation system susceptible to faults, which may significantly degrade the formation performance and destroy the preset tasks. Different from the faults encountered by the single UAV, which mainly affect the states of the UAV itself, the faults subjected by one UAV within the formation team not only affect the stability of the faulty UAV but also perturb the neighboring UAVs via the information transmissions among UAVs. Therefore, to ensure the formation safety and cooperative operational performance, the FTCC should be developed to stabilize the faulty UAVs and the formation team based on

Z. Yu et al., *Fault-Tolerant Cooperative Control of Unmanned Aerial Vehicles*,
https://doi.org/10.1007/978-981-99-7661-4_1

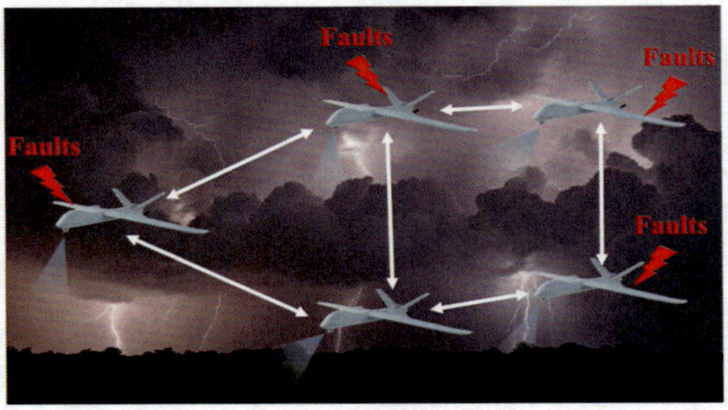

Fig. 1.1 An illustrative example of multi-UAVs against faults

the faulty UAVs' states and the information received from neighboring UAVs, thus significantly increasing the design challenge.

This monograph will present systematical and comprehensive descriptions on the FTCC developments for multi-UAVs against actuator faults. In the following sections, the FTC and FTCC are detailedly analyzed.

1.2 Introduction to Fault-Tolerant Control

As stated in Sect. 1.1, actuator, sensor, and component faults are three typical faults, which are frequently encountered by safety-critical systems [11, 48, 49, 52, 72, 91]. When the actuator faults are subjected by the systems, the faulty actuators will not be able to exactly follow the control signals, which may significantly degrade the system performance if prompt remedies are not activated in a timely manner. With respect to the faulty sensors, the states measured by the faulty sensors will not be the real states, thus providing false feedback information for the closed-loop system. In such a situation, the error convergence will lead to the convergence of the measured states to the expected signals, rather than the convergence of the real states to the expected signals. When the safety-critical systems suffer from component faults, the system structures and dynamics will be changed, making the faulty system prone to crashes. Therefore, it is necessary to handle these commonly encountered actuator, sensor, and component faults for providing reliable operational performance.

According to the detailed analysis in [91], the fault detection and diagnosis (FDD) technique was initially investigated as an independent research area, which was separate from the FTC developments. The FDD module is intended to detect and isolate any system faults as quickly as possible. For the detailed classification of FDD methods, the interested readers can refer to the survey paper [91]. In general, the existing FDD algorithms can be divided into the model-based and data-based methods. With respect to the model-based FDD strategies, the state estimation, parameter estimation, and parity space theories are usually used to facilitate the designs. Regarding the data-based strategies, statistical mechanisms, NNs, fuzzy logic systems (FLSs), pattern recognition, and frequency analysis are the typical design theories. Due to the powerful learning capabilities from massive data, deep learning technique has been innovatively used to develop the FDD units for various systems, leading to numerous innovative FDD strategies [20, 26, 54, 58]. Recently, several survey papers have been reported to show the research progresses of deep learning-based FDD methods [12, 21, 44, 103]. In [10], the reinforcement learning (RL) technique, which is a typical deep learning strategy, was developed to facilitate the intelligent FDD design for rotating machinery. More recently, an intelligent FDD framework was further developed in [67] for rotating machinery by using deep transfer RL.

By integrating the FDD module into the control architecture for dynamically adjusting the control actions against faults, an active FTC (AFTC) mechanism is formulated to stabilize the faulty system in a timely manner [91]. In the AFTC architecture, the FDD module plays an important role in the control system since it provides timely fault diagnosis information to the controller. Therefore, the AFTC design should focus on the reliable FDD module and the corresponding controller design according to the diagnosed fault information from the FDD module. The advantage of AFTC architecture is that some unexpected faults can be attenuated online. The disadvantage of AFTC method is the design complexity for integrating FDD and FTC units. As illustrated in Fig. 1.2, to simplify the FTC design, FDD units are usually removed from the FTC framework, resulting in the passive FTC (PFTC) mechanism [91]. Regarding the PFTC design, the FTC capabilities are mainly obtained by the robustness inherent in the controller. Therefore, the PFTC scheme needs to be constructed and tested offline to verify the fault compensation capability, which is fixed once the PFTC scheme is implemented in engineering systems. The advantage of PFTC architecture is the simple design logic, since the FDD module is no longer needed in the control design. Consider the fact that the

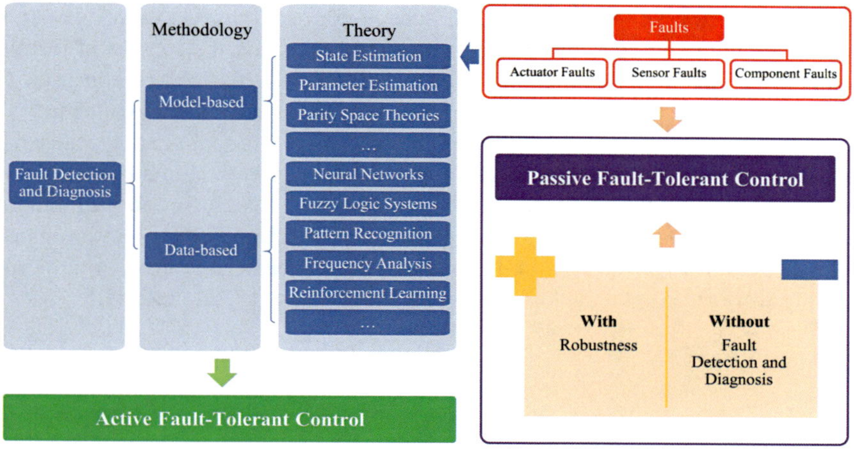

Fig. 1.2 AFTC and PFTC methods

PFTC method mainly relies on the robustness to attenuate the faults, and the obvious disadvantage of PFTC scheme is that the implemented PFTC module cannot tolerate faults, which are not tested offline or exceed the fault thresholds that the PFTC unit can handle. Among various PFTC design theories, sliding-mode control (SMC) has been widely utilized to construct the PFTC schemes, leading to numerous PFTC results for many systems, such as UAVs [45], gasoline engines [2], autonomous electric vehicles [98], and hypersonic vehicle [15].

With respect to the FTC design for single UAV, various methods are explored to construct effective fault handling strategies, such as robust, adaptive, NN, FLS, fuzzy NN (FNN), RL, and Nussbaum techniques [1, 22, 35, 42, 60, 87]. In [25], the authors summarized the relevant FTC results for UAVs by presenting various existing hierarchical or intelligent FTC schemes. In [19], a comprehensive survey was conducted for quadrotor UAVs by introducing the configuration, modeling, identification, collision avoidance, FDD, and FTC methods. In [104], a robust FTC method was developed for underactuated takeoff and landing UAVs subject to thrust and torque faults. By constructing adaptive fault observer and robust item to observe the actuator faults and eliminate the effects of external disturbances, respectively, a backstepping-based robust FTC strategy was investigated in [65] for unmanned autonomous helicopters. Based on the super-twisting algorithm and integral terminal sliding-mode mechanism, a robust adaptive FTC scheme was developed in [43] for quadrotor UAVs to mitigate the actuator faults. To handle the uncertainties induced by the couplings of faults and nonlinearities, fuzzy/neural/fuzzy neural adaptive strategies are skillfully used to construct the FTC schemes. To deal with both additive sensor faults (bias, drift, and loss of accuracy) and multiplicative sensor fault (loss of effectiveness) encountered by a quadrotor UAV, an asymptotic fuzzy adaptive FTC method was developed in [24] by combining the adaptive fuzzy learning technique and Nussbaum strategy. In [56], radial basis function neural networks (RBFNNs) with virtual parameter estimation algorithms were used to design the FTC scheme for quadrotor UAVs. Moreover, by integrating the fuzzy reasoning capabilities from FLSs and the intelligent learning competences from NNs, FNNs have been preliminarily utilized to construct the FTC schemes for UAVs. In [77], the authors incorporated the FNN unit into the FTC architecture for UAV to handle the strongly unknown nonlinearities caused by the faults. By utilizing the strong learning capability from RL, an adaptive model-free FTC scheme was developed in [42] to achieve the tracking control of a highly flexible aircraft. Due to the feature that Nussbaum functions can be used to handle unknown control gains, Nussbaum functions have been widely used in numerous existing works to handle the loss-of-effectiveness faults [18, 22, 23, 84].

1.3 Introduction to Cooperative Control of Multi-UAVs

As stated in Sect. 1.1, to achieve effective formation flight of multi-UAVs, the leader-following, behavior-based, virtual structure, and algebraic graph-based methods are usually utilized to facilitate the cooperative control design [86]. With respect to the leader-following cooperative control architecture, the leader UAV flies along the planned flight path and the follower UAVs track the leader UAV with specified offsets [17, 46, 64]. By weighting the multiple behaviors consisting of formation reconfiguration, formation keeping, trajectory tracking, and collision avoidance, the behavior-based cooperative control strategy is mainly used to handle the conflicting requirements within the formation team of multi-UAVs. In [30], the behavior-based decentralized cooperative control strategy was proposed for multi-UAVs to handle the conflicts among the two tasks that each UAV can fly through predefined waypoints and all UAVs can maintain their distances with respect to their neighboring UAVs. By viewing the formation team as a single entity, the virtual structure-based cooperative control objective can be achieved once the relative geometry relationships among multi-UAVs are satisfied. The distinct feature of the virtual structure-based schemes is the simplicity. However, such a virtual structure method may be difficult to be employed, especially when the formation reconfiguration is frequently utilized to adjust the formation shape [29]. By using the virtual structure, a position synchronization tracking control scheme was developed in [31] for multi-UAVs. Inspired by the explosive research results from MASs, which usually incorporate the algebraic graph into the control design for constructing effective cooperative strategies. In the algebraic graph-based communication network, each node is viewed as a UAV and each edge is used to represent the information transmission route between two UAVs. Recently, the algebraic graph-based strategies have been utilized to facilitate the cooperative control schemes for fixed-wing UAVs, quadrotor UAVs, and unmanned airships [5, 41, 96, 102].

During practical formation flight of multi-UAVs, some unexpected factors can significantly weaken the cooperative performance, such as wind disturbances and wake vortex in close formation flight. In [27], a formation control scheme was developed for multi-UAVs to track a moving target in a windy environment. In [94], a formation flight control algorithm was developed for fixed-wing UAVs by considering wind field. By considering the wake vortex induced by the leading UAV, a safe formation control scheme was developed in [74] for the trailing UAV to handle the additional translational and rotational velocities. By further considering the uncertainties and wake vortex, a distributed close formation flight control algorithm was investigated in [81] for multiple trailing fixed-wing UAVs.

To ensure the flight safety and make the developed cooperative control scheme applicable in engineering, the constraints including input saturation and error constraints should also be explicitly considered in the control design. To handle the input saturation problem, auxiliary dynamic systems are usually constructed to pull back the saturated control signals once the control signals stay at the limits for a while. By using such a strategy, the system stability can be ensured, since the saturated control signals can be pulled back into the unsaturated region in a timely manner. In [61], an auxiliary system was constructed to regulate the DO-based DSC scheme for a transport aircraft, and thus the adverse effects caused by the saturation can be effectively attenuated. To further handle the saturation nonlinearity, a piecewise auxiliary system was developed in [6] to regulate the saturated control signals and facilitate the stability analyses, which were then used and modified in numerous works [40]. By using a smooth hyperbolic tangent function to approximate the asymmetric saturation function and constructing a piecewise auxiliary system to achieve the saturation compensation, a Nussbaum function-based adaptive trajectory control scheme was developed in [100] for a fully actuated surface vehicle. With respect to the formation flight of multi-UAVs in the presence of input saturation, several safe cooperative control schemes have been investigated to reduce the adverse effects of the saturation on the formation system performance. By introducing an auxiliary dynamic system, a distributed cooperative control scheme was developed in [13] for multi-UAVs in the presence of input saturation.

Recently, various strategies have been developed to constrain the system tracking errors, such as barrier Lyapunov functions and prescribed performance functions. Among these methods, prescribed performance control (PPC) has been preliminarily investigated for many unmanned systems. The PPC was first investigated in [4], which mainly transforms the tracking errors into a new set of errors by prescribed performance functions. By using the PPC method, the tracking errors can converge into a small residual set defined by the prescribed performance functions and the convergence rates can be constrained under the predefined values. Motivated by this concept, a few PPC results have been obtained for the flight control of single UAV [51, 55, 59, 95]. Recently, the PPC concept was integrated into the cooperative control scheme to improve the control performance [7, 28, 36, 57]. Cui et al. [8] investigated the distributed output consensus problem for high-order nonlinear MASs in the presence of unknown dead-zone input, and the presented transient, steady performances of synchronization errors were guaranteed. More recently, the PPC mechanisms are also explored to achieve the constrained cooperative control objective for multi-UAVs [3, 63, 66]. In [99], an adaptive distributed SMC strategy was developed for multi-UAVs with prescribed performance.

1.4 Introduction to Fault-Tolerant Cooperative Control of Multi-UAVs

When multi-UAVs are involved within the formation team, numerous actuators, sensors, and components render the overall system susceptible to various faults, significantly threatening the formation flight safety. In general, the existing FTCC

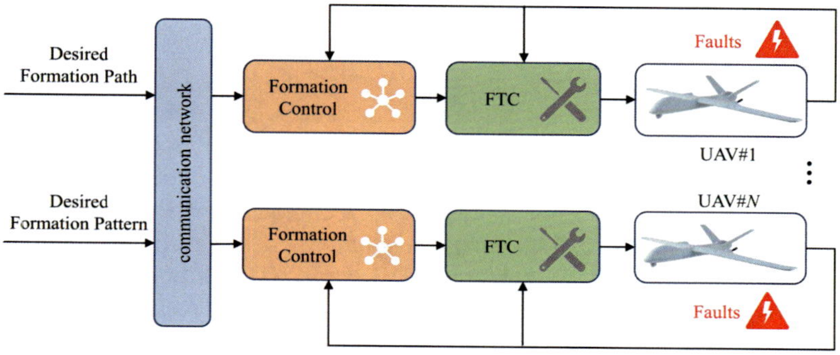

Fig. 1.3 Individual FTCC architecture

strategies can be classified into three types: (1) individual FTCC, (2) leader-following FTCC, and (3) distributed FTCC [86].

1.4.1 Individual Fault-Tolerant Cooperative Control of Multi-UAVs

With respect to the individual FTCC of multi-UAVs, the formation control strategy and the FTC unit are usually developed separately. As illustrated in Fig. 1.3, the formation control module is mainly used to regulate the formation performance according to the predefined formation path and pattern. Based on the references generated from the formation control module, the FTC unit is mainly used for each UAV to tolerate the actuator, sensor, and component faults. Therefore, one can conclude that the formation and FTC units can be independently developed and modified without adjusting the other unit, which is the major advantage of the individual FTCC. However, such a strategy cannot make the formation pattern adaptable to faults. In [37], by designing the leader-following outer-loop formation and inner-loop FTC laws, an FTCC scheme was developed to keep the desired formation flight against potential collision and actuator faults. By using the similar design procedure as [37], the authors in [38] further proposed an FTCC scheme for multi-UAVs by simultaneously considering the actuator faults, obstacle avoidance, and actuator saturation.

Within the individual FTCC architecture, the design strategy consisting of the distributed SMO and FTC unit is usually used to construct the schemes for multi-UAVs. In such an architecture, the distributed SMO is first developed to estimate the desired reference for each follower UAV. Then, the FTC unit is investigated for each follower UAV to steer the output to track the estimated reference for each follower UAV. In [79], a distributed SMO was first investigated for each follower UAV to estimate the desired reference via communications with neighboring UAVs, and then an adaptive neural FTC unit was developed for each UAV to converge to the reference generated by the distributed SMO. By using a distributed SMO similar to that in [79], an FTCC scheme was proposed in [70] for multi-UAVs to simultaneously attenuate the external disturbances, actuator faults, and sensor faults. To concurrently solve the fault and wake vortex issues in close formation flight, an individual FTCC scheme was developed in [74] for nonlinear six-degree-of-freedom follower UAVs by using the DO technique. In the proposed control architecture, the fixed-wing UAV model was first divided into the longitudinal and lateral-directional dynamics, and then the DO-based FTC methods were developed for the longitudinal and lateral-directional subsystems. Recently, by using the similar longitudinal and lateral-directional control architecture presented in [74], an intelligent distributed FTCC scheme was presented in [81] for multiple follower fixed-wing UAVs, which was developed by integrating distributed SMO, DSC, and composite learning algorithms. The distributed SMO was first used for estimating the desired position reference for each follower UAV. Then, the DSC architecture was employed to iteratively design the FTC law for each follower UAV. Moreover, to counteract the in-flight actuator faults and wake vortices, composite learning algorithms were artfully constructed by NNs, DOs, and prediction errors. By considering the formation flight with multiple follower UAVs and multiple leader UAVs against faults, a DO-based individual FTCC strategy was developed in [71] to attenuate the actuator faults, sensor faults, and input saturation. Then, the results in [71] were further extended to the individual FTCC design for multi-UAVs [76], which designed a distributed SMO for estimating the desired reference via communications with neighboring UAVs, DO to estimate the lumped disturbance, and an auxiliary system to handle the input saturation.

1.4.2 Leader-Following Fault-Tolerant Cooperative Control of Multi-UAVs

Regarding the leader-following FTCC for multi-UAVs, extra information flows from the follower UAVs to the leader UAV are generally introduced in the control design, which is illustrated in Fig. 1.4. In such an FTCC structure, the fault information may be sent to the leader UAV, such that the control unit of the leader UAV can adjust its maneuverability to be possibly tracked by the faulty follower UAVs with reduced capabilities. The advantage of this method is that the traditional leader-following control schemes for multi-UAVs can still be used with small modifications. However, it is difficult to handle the completely failed communication links with this method, since there exists only one communication link between each follower UAV and the leader UAV.

By involving a reference generator and finite-time convergence strategy, a leader-following FTCC method was proposed in [69] for multi-UAVs against actuator faults. The reference generator received the state information from both the leader

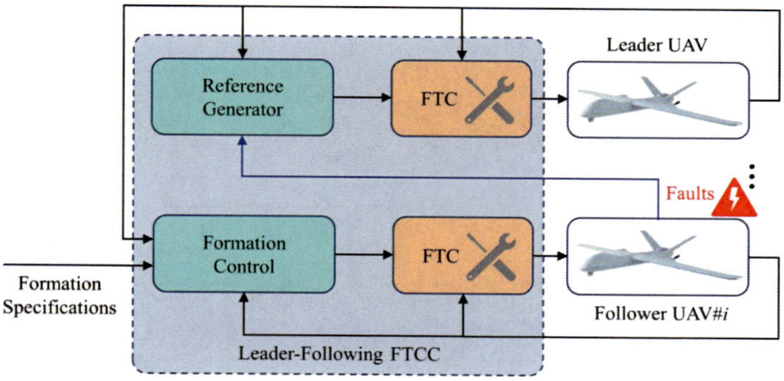

Fig. 1.4 Leader-following FTCC architecture

and follower UAVs. Once the actuator faults were detected in the follower UAVs, the reference generator of the leader UAV modifies its output to adjust the leader UAV's states, such that the faulty follower UAVs with finite-time adaptive FTC units can track the leader UAV. Then, by using a similar FTCC architecture, a leader-following FTCC scheme was further proposed in [33] by incorporating a generator to provide a feasible reference for the leader UAV, a proportional control unit to regulate the outer-loop formation tracking performance, and a sliding-mode FTC unit to achieve the inner-loop attitude tracking performance. Recently, such an FTCC architecture was further researched in [34], and the proportional controller and the estimator-based SMC law were developed for the outer-loop and inner-loop dynamics, respectively. With the outer-loop and inner-loop design architecture, an FTCC scheme was developed in [68] for multi-UAVs by designing an outer-loop formation controller to ensure the formation stability and an SMO-based inner-loop attitude controller to track the signals generated by the outer-loop controller.

1.4.3 Distributed Fault-Tolerant Cooperative Control of Multi-UAVs

For the distributed FTCC of multi-UAVs, the distributed communication network is usually involved into the FTCC design. Different from the individual FTCC and leader-following FTCC, the distributed FTCC integrates the formation control and FTC objectives into a unified framework, such that the formation of multi-UAVs can be adaptable to faulty UAVs. Figure 1.5 illustrates the distributed FTCC structure for multi-UAVs. Moreover, the distributed FTCC has high communication robustness against communication link faults, since each UAV uses the information from neighboring UAVs to decide its behavior. If some communication links among UAVs are attacked by the hackers or become faulty, the overall formation stability can still be maintained by using the remaining healthy communication links.

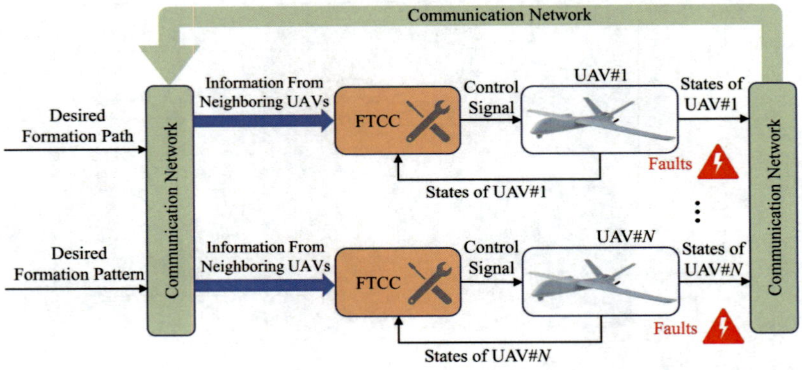

Fig. 1.5 Distributed FTCC architecture

According to the number of leader UAVs, the distributed FTCC of multi-UAVs can be classified into the following three types: (1) distributed FTCC without a leader UAV, (2) distributed FTCC with only one leader UAV, and (3) distributed FTCC with multiple leader UAVs. The distributed FTCC without a leader UAV can be viewed as the consensus control with the integration of FTC capability. The distributed FTCC with only one leader UAV can be viewed as the distributed tracking control of multi-UAVs in the presence of faults. In this framework, only a portion of follower UAVs has direct access to the leader UAV, and the communication network for follower UAVs is connected or strongly connected. There exists at least one path from the leader UAV to each follower UAV. The distributed FTCC with multiple leader UAVs is also named as the fault-tolerant containment control, which ensures that all follower UAVs including faulty and healthy UAVs can converge into the convex hull spanned by multiple leader UAVs.

In [63], a distributed FTCC method was proposed for multiple helicopters to achieve the attitude synchronization against actuator faults. By utilizing the updating mechanism from adaptive laws, a distributed FTCC scheme was developed in [62] for multiple heterogeneous linear UAVs. By using the nearest neighboring rule, a distributed FTCC scheme was developed in [73] for attenuating actuator faults. In the proposed FTCC scheme, the desired reference was first generated for each follower UAV in a distributed communication network by averaging the outputs from neighboring UAVs, and then DOs were constructed for estimating the lumped disturbances induced by the actuator faults. Recently, an FTCC scheme was developed for multiple octorotor UAVs in the presence of uncertainties and faults, such that all UAVs can be steered to achieve a desired formation flight in a distributed communication network [39]. By incorporating command filters into the backstepping design architecture and estimating the upper bounds of faults and external disturbances, a distributed robust FTCC scheme was developed in [101] for multi-UAVs. To further increase the robustness against actuator faults and improve the formation stability, a distributed finite-time FTCC scheme was investigated in

[97] by designing distributed finite-time position, attitude, and propeller FTC units for the position, attitude, and propeller loops of multiple quadrotors, respectively. In [16], the authors further investigated the FTCC of multi-UAVs for forest fire monitoring. A distributed robust formation control unit was first developed to achieve the desired formation shape during the forest fire search stage. Once a forest fire was detected, another distributed formation control unit was activated to distribute UAVs around the fire perimeter. If some UAVs failed to continue their tasks due to faults, the task reassignment strategy was adopted to switch the tasks of faulty UAVs to healthy UAVs, leading to an FTCC scheme.

When multi-UAVs cooperatively execute a task, all UAVs within a distributed communication network may need to track their individual references in a synchronized manner. In such a situation, the synchronization and individual tracking performance are both important for the formation team against actuator faults. Moreover, the synchronization performance should be prioritized especially when all UAVs cooperatively carry a heavy and costly object against actuator faults. To solve this difficult FTCC design problem, the synchronization tracking errors were constructed in [75, 78, 80, 83] by combining the individual tracking errors and synchronization errors. In [78], a decentralized neural adaptive FTCC method was proposed to achieve the attitude synchronization tracking control objective. In [75], prescribed performance functions were first introduced into the synchronization tracking errors, and then a composite learning-based decentralized FTCC scheme was developed for the formation team. Within the proposed composite learning architecture, NNs were utilized for approximating the unknown nonlinear functions related to the inherent nonlinearities and loss-of-effectiveness actuator faults, and DOs were utilized to compensate for the NN approximation errors and bias actuator faults. By continuing to impose the prescribed performance functions into the synchronized tracking errors, the authors in [80] further considered the decentralized finite-time adaptive FTCC design for multi-UAVs to rapidly react to actuator faults. In the proposed finite-time FTCC, prescribed performance functions, NNs, and finite-time differentiators were combined to facilitate the control design, and the MLPNN was adopted to reduce the computational burden. Recently, fractional-order (FO) calculus was introduced into the synchronization tracking error-based FTCC design to improve the control performance for multi-UAVs. In [83], wind effects were explicitly considered in the formation team and a composite learning algorithm was developed for each UAV to estimate the uncertain terms induced by the actuator faults and wind effects. By utilizing the backstepping control architecture and the high-order sliding-mode differentiator (HOSMD), the overall control signal was developed at the last step. The authors in [82] investigated the FTCC design with the integration of FO sliding-mode surfaces, nonlinear DOs, intelligent fuzzy wavelet NNs, synchronization errors, and robust control signals, such that an intelligent FO FTCC scheme was developed for multi-UAVs. More recently, a fault-tolerant time-varying elliptical formation control scheme was proposed in [85] for multiple fixed-wing UAVs to cooperatively monitor a forest fire with the consideration of actuator faults. In the proposed method, FO calculus and sliding-mode DO are used to facilitate the FTCC design.

1.5 Organization of the Monograph

This monograph consists of 10 chapters. Chapter 1 presents the background and motivations, as well as the descriptions of FTC and FTCC. Chapter 2 presents the UAV dynamics, which will be used for the subsequent chapters.

Chapters 3, 4, 5, and 6 are about the FTCC design for longitudinal motions of multi-UAVs against faults. In particular, Chap. 3 investigates the prescribed performance-based distributed neural adaptive FTCC design for achieving the longitudinal motions of multi-UAVs under faults and unknown nonlinearities. To further handle the compact effects of actuator faults, external disturbances, and input saturation, a DO-based distributed FTCC is developed in Chap. 4 by using the nearest neighbor rule to generate the desired reference for each UAV in the longitudinal formation team. Consider the situation that multiple leader UAVs are involved into the formation team, and Chap. 5 integrates the SMO and graph theory to estimate the unknown references and develops a DO for each UAV to handle the lumped uncertainty including external disturbances and actuator faults, leading to the distributed FTCC scheme. Furthermore, to increase the fault compensation speed for improving the FTCC capability to ensure the longitudinal motion safety of UAVs against faults, a hierarchical FTCC strategy is developed by constructing the distributed finite-time SMOs to estimate the unknown references at the upper layer and finite-time FTC units to handle the faults at the lower layer.

Chapters 7 and 8 are about the FTCC design for attitude motions of multi-UAVs against faults. In particular, Chap. 7 simultaneously incorporates the error constraint requirements and finite-time convergence characteristics into the FTCC design to achieve the attitude synchronization among multi-UAVs, such that an enhanced FTCC strategy is obtained for the formation team. Considering the fact that unidirectional communications are widely used in the cooperative task execution of multi-UAVs, Chap. 8 presents a distributed FTCC method to achieve the attitude synchronization of multi-UAVs by using the composite learning algorithm consisting of NNs and DOs.

Chapter 9 considers the FTCC design for the outer-loop position motions of multi-UAVs against faults and develops an FTCC strategy for multi-UAVs to monitor the elliptical spread of forest fire in the presence of faults. Chapter 10 concludes this monograph and gives some open problems and challenges faced by the FTCC design of multi-UAVs.

Therefore, this monograph presents the FTCC design from longitudinal motions to attitude motions and to outer-loop position motions of multi-UAVs.

References

1. A. Abbaspour, K.K. Yen, P. Forouzannezhad, A. Sargolzaei, A neural adaptive approach for active fault-tolerant control design in UAV. IEEE Trans. Syst. Man Cybern. -Syst. **50**(9), 3401–3411 (2018)

2. A.A. Amin, K. Mahmoodul Hasan, Robust passive fault tolerant control for air fuel ratio control of internal combustion gasoline engine for sensor and actuator faults. IEEE J. Res. **69**, 1–16 (2021)
3. B.H. An, B. Wang, H.J. Fan, L. Liu, H. Hu, Y.J. Wang, Fully distributed prescribed performance formation control for UAVs with unknown maneuver of leader. Aerosp. Sci. Technol. **130**, 107886 (2022)
4. C.P. Bechlioulis, G.A. Rovithakis, Robust adaptive control of feedback linearizable MIMO nonlinear systems with prescribed performance. IEEE Trans. Autom. Control **53**(9), 2090–2099 (2008)
5. M. Campion, P. Ranganathan, S. Faruque, UAV swarm communication and control architectures: a review. J. Unmanned Veh. Syst. **7**(2), 93–106 (2018)
6. M. Chen, S.Z.S. Ge, B.B. Ren, Adaptive tracking control of uncertain MIMO nonlinear systems with input constraints. Automatica **47**(3), 452–465 (2011)
7. G. Chen, Y. Zhao, Distributed adaptive output-feedback tracking control of non-affine multi-agent systems with prescribed performance. J. Frankl. Inst. **355**(13), 6087–6110 (2018)
8. G.Z. Cui, S.Y. Xu, Q. Ma, Y.M. Li, Z.Q. Zhang, Prescribed performance distributed consensus control for nonlinear multi-agent systems with unknown dead-zone input. Int. J. Control **91**(5), 1053–1065 (2018)
9. C. Deng, S.W. Wang, Z. Huang, Z.F. Tan, J.Y. Liu, Unmanned aerial vehicles for power line inspection: a cooperative way in platforms and communications. J. Commun. **9**, 687–692 (2014)
10. Y. Ding, L. Ma, J. Ma, M.L. Suo, L.F. Tao, Y.J. Cheng, C. Lu, Intelligent fault diagnosis for rotating machinery using deep Q-network based health state classification: a deep reinforcement learning approach. Adv. Eng. Inform. **42**, 100977 (2019)
11. Q. Dong, Q. Zong, B.L. Tian, C.F. Zhang, W.J. Liu, Adaptive disturbance observer-based finite-time continuous fault-tolerant control for reentry RLV. Int. J. Robust Nonlinear Control **27**, 4275–4295 (2017)
12. L.X. Duan, M.Y. Xie, J.J. Wang, T.B. Bai, Deep learning enabled intelligent fault diagnosis: overview and applications. J. Intell. Fuzzy Syst. **35**(5), 5771–5784 (2018)
13. H.B. Duan, Y. Yuan, Z.G. Zeng, Distributed cooperative control of multiple UAVs in the presence of actuator faults and input constraints. IEEE T. Circuits-II **69**(11), 4463–4467 (2022)
14. D. Erdos, A. Erdos, S.E. Watkins, An experimental UAV system for search and rescue challenge. IEEE Trans. Aerosp. Electron. Syst. **28**(5), 32–37 (2013)
15. Z.F. Gao, B. Jiang, P. Shi, J.Y. Liu, Y.F. Xu, Passive fault-tolerant control design for near-space hypersonic vehicle dynamical system. Circ. Syst. Signal Pr. **31**, 565–581 (2012)
16. K.A. Ghamry, Y.M. Zhang, Fault-tolerant cooperative control of multiple UAVs for forest fire detection and tracking mission, in *The 3rd Conference on Control and Fault-Tolerant Systems (SysTol)*, Barcelona, Spain (2016)
17. Y. Gu, B. Seanor, G. Campa, M.R. Napolitano, L. Rowe, S. Gururajan, S. Wan, Design and flight testing evaluation of formation control laws. IEEE Trans. Control Syst. Technol. **14**(6), 1105–1112 (2006)
18. W. Hao, B. Xian, Nonlinear adaptive fault-tolerant control for a quadrotor UAV based on immersion and invariance methodology. Nonlinear Dyn. **90**, 2813–2826 (2017)
19. S. Hassan, A. Ali, Y. Rafic, A survey on quadrotors: Configurations, modeling and identification, control, collision avoidance, fault diagnosis and tolerant control. IEEE Aero. El. Sys. Mag. **33**, 14–33 (2018)
20. M. He, D. He, Deep learning based approach for bearing fault diagnosis. IEEE Trans. Ind. Appl. **53**(3), 3057–3065 (2017)
21. D.T. Hoang, H.J. Kang, A survey on deep learning based bearing fault diagnosis. Neurocomputing **335**, 327–335 (2019)
22. C.F. Hu, X.P. Zhou, B.H. Sun, W.J. Liu, Q. Zong, Nussbaum-based fuzzy adaptive nonlinear fault-tolerant control for hypersonic vehicles with diverse actuator faults. Aerosp. Sci. Technol. **71**, 432–440 (2017)

23. C.F. Hu, Z.L. Zhang, X.P. Zhou, B.H. Sun, N. Wang, Fuzzy adaptive fault-tolerant controller design based on Nussbaum for nonlinear quadrotor UAV system with input saturation. Int. J. Syst. Control Inf. Pr. **3**(1), 1–25 (2019)
24. C.F. Hu, Z.L. Zhang, X.P. Zhou, N. Wang, Command filter-based fuzzy adaptive nonlinear sensor-fault tolerant control for a quadrotor unmanned aerial vehicle. Trans. Inst. Meas. Control **42**(2), 198–213 (2020)
25. I. Sadeghzadeh, Y.M. Zhang, *A review on Fault Tolerant Control for Unmanned Aerial Vehicles (UAVs), AIAA InfoTech@Aerospace*, St. Louis, Missouri, USA (2011)
26. F. Jia, Y.G. Lei, L. Guo, J. Lin, S.B. Xing, A neural network constructed by deep learning technique and its application to intelligent fault diagnosis of machines. Neurocomputing **272**, 619–628 (2018)
27. J.B. Jia, X. Chen, W.Z. Wang, K.L. Wu, M.Y. Xie, Distributed observer-based finite-time control of moving target tracking for UAV formation. ISA Trans. **140**, 1–17 (2023). https://doi.org/10.1016/j.isatra.2023.06.017
28. Y.B. Jiang, Z.X. Liu, Z.Q. Chen, Robust distributed formation control with prescribed performance for nonlinear multi-agent systems subjected to compound disturbances. Int. J. Adapt. Control Signal Process. (2023). https://doi.org/10.1002/acs.3649
29. M.A. Kamel, X. Yu, Y.M. Zhang, Formation control and coordination of multiple unmanned ground vehicles in normal and faulty situations: a review. Annu. Rev. Control **49**, 128–144 (2020)
30. S. Kim, Y. Kim, Optimum design of three-dimensional behavioural decentralized controller for UAV formation flight. Eng. Optimiz. **41**(3), 199–224 (2009)
31. N.H. Li, H.H. Liu, Formation UAV flight control using virtual structure and motion synchronization, in *American Control Conference*, Seattle, WA, USA (2008)
32. Z. Li, Y. Liu, R. Hayward, J. Zhang, J. Cai, Knowledge-based power line detection for UAV surveillance and inspection systems, in *International Conference Image and Vision Computing*, Christchurch, New Zealand (2008)
33. P. Li, X. Yu, X.Y. Peng, Z.Q. Zheng, Y.M. Zhang, Fault-tolerant cooperative control for multiple UAVs based on sliding mode techniques. Sci. China Inf. Sci. **60**, 1–13 (2017)
34. P. Li, X. Yu, Y.M. Zhang, Fault-tolerant cooperative control for multiple UAVs based on UDE and model following SMC, in *IFAC Symposium on Fault Detection, Supervision and Safety for Technical Processes*, Warsaw, Poland (2018)
35. D.J. Li, P. Yang, Z.X. Liu, Z.X. Wang, Z.Q. Zhang, Fault-tolerant aircraft control based on self-constructing fuzzy neural network for quadcopter. Int. J. Autom. Tech. **15**(1), 109–122 (2021)
36. Z.B. Lin, Z. Liu, C.Y. Su, Y.N. Wang, C.P. Chen, Y. Zhang, Adaptive fuzzy prescribed performance output-feedback cooperative control for uncertain nonlinear multi-agent systems. IEEE Trans. Fuzzy Syst. (2023). https://doi.org/10.1109/TFUZZ.2023.3285649
37. Z.X. Liu, X. Yu, C. Yuan, Y.M. Zhang, Leader-follower formation control of unmanned aerial vehicles with fault tolerant and collision avoidance capabilities, in *International Conference on Unmanned Aircraft Systems (ICUAS)*, Denver, CO, USA (2015)
38. Z.X. Liu, C. Yuan, X. Yu, Y.M. Zhang, Fault-tolerant formation control of unmanned aerial vehicles in the presence of actuator faults and obstacles. Unmanned Syst. **4**(03), 197–211 (2016)
39. D.Y. Liu, H. Liu, J.X. Xi, Fully distributed adaptive fault-tolerant formation control for octorotors subject to multiple actuator faults. Aerosp. Sci. Technol. **108**, 106366 (2021)
40. Y. Liu, X.Q. Yao, W. Zhao, Distributed neural-based fault-tolerant control of multiple flexible manipulators with input saturations. Automatica **156**, 111202 (2023)
41. F.F. Lizzio, E. Capello, G. Guglieri, A review of consensus-based multi-agent UAV implementations. J. Intell. Robot. Syst. **106**(2), 43 (2022)
42. J.J. Ma, C. Peng, Adaptive model-free fault-tolerant control based on integral reinforcement learning for a highly flexible aircraft with actuator faults. Aerosp. Sci. Technol. **119**, 107204 (2021)

43. S. Mallavalli, A. Fekih, Adaptive fault tolerant control design for actuator fault mitigation in quadrotor UAVs, in *IEEE Conference on Control Technology and Applications*, Copenhagen, Denmark (2018)
44. M. Mansouri, M. Trabelsi, H. Nounou, M. Nounou, Deep learning-based fault diagnosis of photovoltaic systems: a comprehensive review and enhancement prospects. IEEE Access **9**, 126286–126306 (2021)
45. A.R. Merheb, H. Noura, F. Bateman, Passive fault tolerant control of quadrotor UAV using regular and cascaded sliding mode control, in *Conference on Control and Fault-Tolerant Systems (SysTol)*, Nice, France (2013)
46. T.S. No, Y. Kim, M.J. Tahk, G.E. Jeon, Cascade-type guidance law design for multiple-UAV formation keeping. Aerosp. Sci. Technol. **15**(6), 431–439 (2011)
47. M. Pachter, J.J. D'Azzo, A.W. Proud, Tight formation flight control. J. Guidance Control Dyn. **24**(2), 246–254 (2001)
48. L.G. Qin, X. He, D.H. Zhou, A survey of fault diagnosis for swarm systems. Syst. Sci. Control Eng. **2**, 13–23 (2014)
49. L.G. Qin, X. He, D.H. Zhou, Fault-tolerant cooperative output regulation for multi-vehicle systems with sensor faults. Int. J. Control **90**(10), 2227–2248 (2017)
50. C. Riceand, Y. Gu, H.Y. Chao, T. Larrabee, S. Gururajan, M. Napolitano, T. Mandal, M. Rhudy, Autonomous close formation flight control with fixed wing and quadrotor test beds. Int. J. Aerosp. Eng. **2016**, 15 (2016)
51. K. Sasaki, Z.J. Yang, Disturbance observer-based control of UAVs with prescribed performance. Int. J. Syst. Sci. **51**(5), 939–957 (2020)
52. P. Sequeira, P.A. Antonio, A fault detection and isolation scheme for formation control of fixed-wing UAVs, in *International Conference on Unmanned Aircraft Systems (ICUAS)* (2017)
53. J.J. Shan, H.T. Liu, Close-formation flight control with motion synchronization. J. Guidance Control Dyn. **28**(6), 1316–1320 (2005)
54. S.Y. Shao, W.J. Sun, R.Q. Yan, P. Wang, R.X. Gao, A deep learning approach for fault diagnosis of induction motors in manufacturing. Chin. J. Mech. Eng. **30**(6), 1347–1356 (2017)
55. Z.P. Shen, F. Li, X.M. Cao, C. Guo, Prescribed performance dynamic surface control for trajectory tracking of quadrotor UAV with uncertainties and input constraints. Int. J. Control **94**(11), 2945–2955 (2021)
56. Y.D. Song, L. He, D. Zhang, J.Y. Qian, J. Fu, Neuroadaptive fault-tolerant control of quadrotor UAVs: a more affordable solution. IEEE Trans. Neural Netw. Learn. Syst. **30**(7), 1975–1983 (2018)
57. C.J. Stamouli, C.P. Bechlioulis, K.J. Kyriakopoulos, Multi-agent formation control based on distributed estimation with prescribed performance. IEEE Robot. Autom. Let. **5**(2), 2929–2934 (2020)
58. Y.L. Wang, Z.F. Pan, X.F. Yuan, C.H. Yang, W.H. Gui, A novel deep learning based fault diagnosis approach for chemical process with extended deep belief network. ISA Trans. **96**, 457–467 (2020)
59. Z.H. Wu, J.K. Ni, W. Qian, X.H. Bu, B.J. Liu, Composite prescribed performance control of small unmanned aerial vehicles using modified nonlinear disturbance observer. ISA Trans. **116**, 30–45 (2021)
60. B. Xian, W. Hao, Nonlinear robust fault-tolerant control of the tilt trirotor UAV under rear servo's stuck fault: theory and experiments. IEEE Trans. Ind. Inform. **15**(4), 2158–2166 (2018)
61. B. Xu, Disturbance observer-based dynamic surface control of transport aircraft with continuous heavy cargo airdrop. IEEE Trans. Syst. Man Cybern. Syst. **47**(1), 161–170 (2016)
62. H.L. Yang, B. Jiang, H. Yang, Fault tolerant cooperative control for heterogeneous multiple UAVs, in *IEEE-CSAA Guidance, Navigation and Control Conference*, Xiamen, China (2018)
63. H.L. Yang, B. Jiang, H. Yang, H. Liu, Synchronization of multiple 3-DOF helicopters under actuator faults and saturations with prescribed performance. ISA Trans. **75**, 118–126 (2018)

64. J. Yang, X.M. Wang, S. Baldi, S. Singh, S. Farì, A software-in-the-loop implementation of adaptive formation control for fixed-wing UAVs. IEEE-CAA J. Automatica **6**(5), 1230–1239 (2019)
65. K. Yan, M. Chen, Q.X. Wu, K. Lu, Robust attitude fault-tolerant control for unmanned autonomous helicopter with flapping dynamics and actuator faults. Trans. I. Meas. Control **41**(5), 1266–1277 (2019)
66. K.B. Yang, W.H. Dong, Y.Y. Tong, L. He, Leader-follower formation consensus of quadrotor UAVs based on prescribed performance adaptive constrained backstepping control. Int. J. Control Autom. **20**(10), 3138–3154 (2022)
67. D.G. Yang, H.R. Karimi, M. Pawelczyk, A new intelligent fault diagnosis framework for rotating machinery based on deep transfer reinforcement learning. Control Eng. Pract. **134**, 105475 (2023)
68. L. Yin, J.W. Liu, P. Yang, J.P. Shi, Sliding mode observer-based active fault tolerant control for UAVs formation. IOP Confer. Ser. Mater. Sci. Eng. **452**, 042068 (2018)
69. X. Yu, Z.X. Liu, Y.M. Zhang, Fault-tolerant formation control of multiple UAVs in the presence of actuator faults. Int. J. Robust Nonlinear Control **26**(12), 2668–2685 (2016)
70. Z.Q. Yu, Y.H. Qu, Y.M. Zhang, Y.T. Zhang, Distributed adaptive fault-tolerant cooperative control for multi-UAVs against actuator and sensor faults, in *International Design Engineering Technical Conferences and Computers and Information in Engineering Conference*, Ohio, USA (2017)
71. Z.Q. Yu, Y.M. Zhang, Y.H. Qu, Y.T. Zhang, Distributed fault-tolerant containment control for multi-UAVs with actuator and sensor faults, in *International Conference on Unmanned Aircraft Systems (ICUAS)*, Miami, FL, USA (2017)
72. X. Yu, Y. Fu, P. Li, Y.M. Zhang, Fault-tolerant aircraft control based on self-constructing fuzzy neural networks and multivariable SMC under actuator faults. IEEE Trans. Fuzzy Syst. **26**, 2324–2335 (2018)
73. Z.Q. Yu, Y.H. Qu, Y.M. Zhang, Distributed fault-tolerant cooperative control for multi-UAVs under actuator fault and input saturation. IEEE Trans. Control Syst. Technol. **27**(6), 2417–2429 (2018)
74. Z.Q. Yu, Y.H. Qu, Y.M. Zhang, Safe control of trailing UAV in close formation flight against actuator fault and wake vortex effect. Aerosp. Sci. Technol. **77**, 189–205 (2018)
75. Z.Q. Yu, Z.X. Liu, Y.M. Zhang, Y.H. Qu, C.Y. Su, Decentralized fault-tolerant cooperative control of multiple UAVs with prescribed attitude synchronization tracking performance under directed communication topology. Front. Inform. Technol. Elect. Eng. **20**(5), 685–700 (2019)
76. Z.Q. Yu, Y.H. Qu, Y.M. Zhang, Fault-tolerant containment control of multiple unmanned aerial vehicles based on distributed sliding-mode observer. J. Intell. Robot. Syst. **93**, 163–177 (2019)
77. Z.Q. Yu, Y.M. Zhang, Z.X. Liu, Y.H. Qu, C.Y. Su, Distributed adaptive fractional-order fault-tolerant cooperative control of networked unmanned aerial vehicles via fuzzy neural networks. IET Contr. Theory Appl. **13**(17), 2917–2929 (2019)
78. Z.Q. Yu, Y.M. Zhang, Y.H. Qu, C.Y. Su, Y.J. Ma, B. Jiang, Decentralized adaptive fault-tolerant cooperative control of multi-UAVs under actuator faults and directed communication topology, in *Asian Control Conference (ASCC)*, Kitakyushu, Japan (2019)
79. Z.Q. Yu, Y.M. Zhang, Y.H. Qu, C.Y. Su, Y.T. Zhang, Z.W. Xing, Fault-tolerant adaptive neural control of multi-UAVs against actuator faults, in *International Conference on Unmanned Aircraft Systems (ICUAS)*, Atlanta, GA, USA (2019)
80. Z.Q. Yu, Y.M. Zhang, Z.X. Liu, Y.H. Qu, C.Y. Su, B. Jiang, Decentralized finite-time adaptive fault-tolerant synchronization tracking control for multiple UAVs with prescribed performance. J. Frankl. Inst. **357**(16), 11830–11862 (2020)
81. Z.Q. Yu, Y.M. Zhang, B. Jiang, X. Yu, J. Fu, Y. Jin, T.Y. Chai, Distributed adaptive fault-tolerant close formation flight control of multiple trailing fixed-wing UAVs. ISA Trans. **106**, 181–199 (2020)

82. Z.Q. Yu, Y.M. Zhang, B. Jiang, J. Fu, Y. Jin, T.Y. Chai, Composite adaptive disturbance observer-based decentralized fractional-order fault-tolerant control of networked UAVs. IEEE Trans. Syst. Man Cybern. Syst. **52**(2), 799–813 (2020)
83. Z.Q. Yu, Y.M. Zhang, B. Jiang, C.Y. Su, J. Fu, Y. Jin, T.Y. Chai, Decentralized fractional-order backstepping fault-tolerant control of multi-UAVs against actuator faults and wind effects. Aerosp. Sci. Technol. **104**, 105939 (2020)
84. Z.Q. Yu, Y.M. Zhang, B. Jiang, C.Y. Su, J. Fu, Y. Jin, T.Y. Chai, Nussbaum-based finite-time fractional-order backstepping fault-tolerant flight control of fixed-wing UAV against input saturation with hardware-in-the-loop validation. Mech. Syst. Signal Pr. **153**, 107406 (2021)
85. Z.Q. Yu, Y.M. Zhang, B. Jiang, X. Yu, Fault-tolerant time-varying elliptical formation control of multiple fixed-wing UAVs for cooperative forest fire monitoring. J. Intell. Robot. Syst. **101**, 1–15 (2021)
86. Z.Q. Yu, Y.M. Zhang, B. Jiang, F. Jun, J. Ying, A review on fault-tolerant cooperative control of multiple unmanned aerial vehicles. Chin. J. Aeronaut. **35**(1), 1–18 (2022)
87. Y.J. Yu, J. Guo, M. Chadli, Z.R. Xiang, Distributed adaptive fuzzy formation control of uncertain multiple unmanned aerial vehicles with actuator faults and switching topologies. IEEE Trans. Fuzzy Syst. **31**(3), 919–929 (2022)
88. C. Yuan, Y.M. Zhang, Z.X. Liu, A survey on technologies for automatic forest fire monitoring, detection, and fighting using unmanned aerial vehicles and remote sensing techniques. Can. J. For. Res. **45**, 783–792 (2015)
89. C. Yuan, Z.X. Liu, Y.M. Zhang, Aerial images-based forest fire detection for firefighting using optical remote sensing techniques and unmanned aerial vehicles. J. Intell. Robot. Syst. **88**, 635–654 (2017)
90. C. Yuan, Z.X. Liu, Y.M. Zhang, Learning-based smoke detection for unmanned aerial vehicles applied to forest fire surveillance. J. Intell. Robot. Syst. **93**, 337–349 (2019)
91. Y.M. Zhang, J. Jiang, Bibliographical review on reconfigurable fault-tolerant control systems. Annu. Rev. Control **32**, 229–252 (2008)
92. Q.R. Zhang, H.H. Liu, Robust design of close formation flight control via uncertainty and disturbance estimator, in *AIAA Guidance, Navigation, and Control Conference*, San Diego, CA, USA (2016)
93. Y.M. Zhang, H. Mehrjerdi, A survey on multiple unmanned vehicles formation control and coordination: Normal and fault situations, in *International Conference on Unmanned Aircraft Systems (ICUAS)*, Atlanta, GA, USA (2013)
94. J.L. Zhang, J.G. Yan, A novel control approach for flight-stability of fixed-wing UAV formation with wind field. IEEE Syst. J. **15**(2), 2098–2108 (2020)
95. Y. Zhang, S.H. Wang, B. Chang, W.H. Wu, Adaptive constrained backstepping controller with prescribed performance methodology for carrier-based UAV. Aerosp. Sci. Technol. **92**, 55–65 (2019)
96. P. Zhang, Q.B. Wang, J.W. Tang, D.P. Duan, Agent-based stratosphere airship cooperative control for earth-observing system. Int. J. Robust Nonlinear Control **32**(8), 4966–4979 (2022)
97. X.Y. Zhao, Q. Zong, B.L. Tian, D.D. Wang, M. You, Finite-time fault-tolerant formation control for multiquadrotor systems with actuator fault. Int. J. Robust Nonlinear Control **28**(17), 5386–5405 (2018)
98. J. Zhao, X.W. Wang, Z.C. Liang, W.F. Li, X.B. Wang, P.K. Wong, Adaptive event-based robust passive fault tolerant control for nonlinear lateral stability of autonomous electric vehicles with asynchronous constraints. ISA Trans. **127**, 310–323 (2022)
99. X.Y. Zhao, B.L. Tian, M. You, L. Ma, Adaptive distributed sliding mode control for multiple unmanned aerial vehicles with prescribed performance. IEEE Trans. Veh. Technol. **71**(11), 11480–11490 (2022)
100. Z.W. Zheng, Y.T. Huang, L.H. Xie, B. Zhu, Adaptive trajectory tracking control of a fully actuated surface vessel with asymmetrically constrained input and output. IEEE Trans. Control Syst. Technol. **26**(5), 1851–1859 (2017)
101. Z. Zheng, M.S. Qian, P. Li, H. Yi, Distributed adaptive control for UAV formation with input saturation and actuator fault. IEEE Access **7**, 144638–144647 (2019)

102. Y.K. Zhou, B. Rao, W. Wang, UAV swarm intelligence: recent advances and future trends. IEEE Access **8**, 183856–183878 (2020)
103. Z.Q. Zhu, Y.B. Lei, G.Q. Qi, Y. Chai, N. Mazur, Y.Y. An, X.H. Huang, A review of the application of deep learning in intelligent fault diagnosis of rotating machinery. Measurement **206**, 112346 (2022)
104. Y. Zou, K.W. Xia, Robust fault-tolerant control for underactuated takeoff and landing UAVs. IEEE Trans. Aerosp. Electron. Syst. **56**(5), 3545–3555 (2020)

Chapter 2
Fixed-Wing UAV Model

This chapter introduces the fixed-wing UAV model for the subsequent FTCC design. Since this monograph is mainly focused on the FTCC design for fixed-wing UAVs and there are numerous monographs and literatures containing UAV modeling procedures, the fixed-wing UAV model is directly introduced in this chapter. For more details about the modeling procedure of fixed-wing UAV, the interested readers can refer to the monographs and literatures [1, 3, 9].

2.1 Fixed-Wing UAV Dynamics

With respect to the fixed-wing UAV illustrated in Fig. 2.1, the aileron, elevator, rudder, and engine actuators are usually involved to steer the UAV for keeping the formation flight. The dynamics of the ith fixed-wing UAV is expressed as [2, 4–7]

$$\begin{cases} \dot{x}_i = V_i \cos\gamma_i \cos\chi_i \\ \dot{y}_i = V_i \cos\gamma_i \sin\chi_i \\ \dot{z}_i = V_i \sin\gamma_i \end{cases} \tag{2.1}$$

$$\begin{cases} \dot{V}_i = (-D_i + T_i \cos\alpha_i \cos\beta_i)/m_i - g \sin\gamma_i \\ \dot{\chi}_i = (L_i \sin\mu_i + Y_i \cos\mu_i)/(m_i V_i \cos\gamma_i) + T_i \sin\alpha_i \sin\mu_i/(m_i V_i \cos\gamma_i) \\ \qquad -T_i \cos\alpha_i \sin\beta_i \cos\mu_i/(m_i V_i \cos\gamma_i) \\ \dot{\gamma}_i = (L_i \cos\mu_i - Y_i \sin\mu_i)/(m_i V_i) + T_i \cos\alpha_i \sin\beta_i \sin\mu_i/(m_i V_i) \\ \qquad +T_i \sin\alpha_i \cos\mu_i/(m_i V_i) - g \cos\gamma_i / V_i \end{cases} \tag{2.2}$$

Z. Yu et al., *Fault-Tolerant Cooperative Control of Unmanned Aerial Vehicles*,
https://doi.org/10.1007/978-981-99-7661-4_2

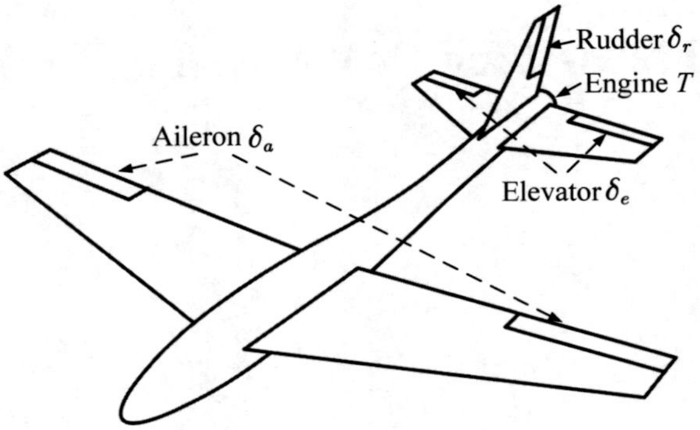

Fig. 2.1 An illustrative example of fixed-wing UAV

$$\begin{cases} \dot{\alpha}_i = q_i - \tan\beta_i(p_i\cos\alpha_i + r_i\sin\alpha_i) \\ \qquad -(\dot{\chi}_i\cos\gamma_i\sin\mu_i + \dot{\gamma}_i\cos\mu_i)/\cos\beta_i \\ \dot{\beta}_i = p_i\sin\alpha_i - r_i\cos\alpha_i + \dot{\chi}_i\cos\gamma_i\cos\mu_i - \dot{\gamma}_i\sin\mu_i \\ \dot{\mu}_i = (p_i\cos\alpha_i + r_i\sin\alpha_i)/\cos\beta_i + \dot{\chi}_i(\sin\gamma_i + \cos\gamma_i\sin\mu_i\tan\beta_i) \\ \qquad +\dot{\gamma}_i\cos\mu_i\tan\beta_i \end{cases} \tag{2.3}$$

$$\begin{cases} \dot{p}_i = (c_{i1}r_i + c_{i2}p_i)q_i + c_{i3}\mathcal{L}_i + c_{i4}\mathcal{N}_i \\ \dot{q}_i = c_{i5}p_ir_i - c_{i6}(p_i^2 - r_i^2) + c_{i7}\mathcal{M}_i \\ \dot{r}_i = (c_{i8}p_i - c_{i2}r_i)q_i + c_{i4}\mathcal{L}_i + c_{i9}\mathcal{N}_i \end{cases} \tag{2.4}$$

where the subscript i denotes the ith UAV in the formation team. x_i, y_i, and z_i denote the spatial positions of the ith UAV; V_i, χ_i, and γ_i are the velocity, heading angle, and flight path angle, respectively; μ_i, α_i, and β_i denote the bank angle, angle of attack, and sideslip angle, respectively; p_i, q_i, and r_i represent the angular rates; m_i and g are the mass and gravity constant, respectively; $c_{i1}, c_{i2}, \ldots, c_{i9}$ are the inertial terms; T_i, L_i, D_i, and Y_i are the thrust, lift, drag, and side forces, respectively; $\mathcal{L}_i$, $\mathcal{M}_i$, and $\mathcal{N}_i$ denote the roll, pitch, and yaw moments, respectively. The expressions of $c_{i1}, c_{i2}, \ldots, c_{i9}$ are given by Yu et al. [8]

$$\begin{cases} c_{i1} = \frac{(I_y - I_z)I_z - I_{xz}^2}{I_xI_z - I_{xz}^2},\ c_{i2} = \frac{I_{xz}(I_x - I_y + I_z)}{I_xI_z - I_{xz}^2},\ c_{i3} = \frac{I_z}{I_xI_z - I_{xz}^2} \\ c_{i4} = \frac{I_{xz}}{I_xI_z - I_{xz}^2},\ c_{i5} = \frac{I_z - I_x}{I_y},\ c_{i6} = \frac{I_{xz}}{I_y} \\ c_{i7} = \frac{1}{I_y},\ c_{i8} = \frac{I_x(I_x - I_y) + I_{xz}^2}{I_xI_z - I_{xz}^2},\ c_{i9} = \frac{I_x}{I_xI_z - I_{xz}^2} \end{cases} \tag{2.5}$$

where I_x, I_y, and I_z are the moments of inertia. I_{xz} is the product of inertia.

The expressions of T_i, L_i, D_i, Y_i, $\mathcal{L}_i$, $\mathcal{M}_i$, and $\mathcal{N}_i$ are calculated as

$$\begin{cases} T_i = T_{i\max}\delta_{T_i}, \ L_i = \bar{q}_i s_i C_{iL}, \ D_i = \bar{q}_i s_i C_{iD}, \ Y_i = \bar{q}_i s_i C_{iY} \\ \mathcal{L}_i = \bar{q}_i s_i b_i C_{il}, \ \mathcal{M}_i = \bar{q}_i s_i c_i C_{im}, \ \mathcal{N}_i = \bar{q}_i s_i b_i C_{in} \end{cases} \tag{2.6}$$

where $T_{i\max}$ and δ_{T_i} are the maximum engine thrust and instantaneous thrust throttle setting, respectively. δ_{T_i} acts as the thrust actuator input. $\bar{q}_i = \frac{1}{2}\rho V_i^2$ is the dynamic pressure and ρ is the air density. s_i, b_i, and c_i represent the wing area, wing span, and mean aerodynamic chord, respectively. The aerodynamic expressions C_{iL}, C_{iD}, C_{iY}, C_{il}, C_{im}, and C_{in} are given by

$$\begin{cases} C_{iL} = C_{iL0} + C_{iL\alpha}\alpha_i + C_{iL\delta_e}\delta_{ie} \\ C_{iD} = C_{iD0} + C_{iD\alpha}\alpha_i + C_{iD\alpha_2}\alpha_i^2 \\ C_{iY} = C_{iY0} + C_{iY\beta}\beta_i \\ C_{il} = C_{il0} + C_{il\beta}\beta_i + C_{il\delta_a}\delta_{ia} + C_{il\delta_r}\delta_{ir} + \frac{C_{ilp} b_i p_i}{2V_i} + \frac{C_{ilr} b_i r_i}{2V_i} \\ C_{im} = C_{im0} + C_{im\alpha}\alpha_i + C_{im\delta_e}\delta_{ie} + \frac{C_{imq} C_i q_i}{2V_i} \\ C_{in} = C_{in0} + C_{in\beta}\beta_i + C_{in\delta_a}\delta_{ia} + C_{in\delta_r}\delta_{ir} + \frac{C_{inp} b_i p_i}{2V_i} + \frac{C_{inr} b_i r_i}{2V_i} \end{cases} \tag{2.7}$$

where δ_{ia}, δ_{ie}, and δ_{ir} are the aileron, elevator, and rudder actuator inputs, respectively. C_{iL0}, $C_{iL\alpha}$, $C_{iL\delta_e}$, C_{iD0}, $C_{iD\alpha}$, $C_{iD\alpha_2}$, C_{iY0}, $C_{iY\beta}$, C_{il0}, $C_{il\beta}$, $C_{il\delta_a}$, $C_{il\delta_r}$, C_{ilp}, C_{ilr}, C_{im0}, $C_{im\alpha}$, $C_{im\delta_e}$, C_{imq}, C_{in0}, $C_{in\beta}$, $C_{in\delta_a}$, $C_{in\delta_r}$, C_{inp}, and C_{inr} are the aerodynamic parameters. The effect of $C_{iL\delta_e}\delta_{ie}$ on the lift force is usually small and can be ignored.

2.2 Fixed-Wing UAV Longitudinal Model

Based on the fixed-wing UAV model (2.1)–(2.4), the longitudinal dynamics can be extracted by only considering the longitudinal motion. As shown in Fig. 2.2, the longitudinal model is expressed as [5]

$$\dot{V}_i = -\frac{D_i}{m_i} + \frac{T_i \cos\alpha_i}{m_i} - g\sin\gamma_i \tag{2.8}$$

$$\dot{z}_i = V_i \sin\gamma_i \tag{2.9}$$

$$\dot{\gamma}_i = \frac{L_i - m_i g \cos\gamma_i + T_i \sin\alpha_i}{m_i V_i} \tag{2.10}$$

$$\dot{\theta}_i = q_i \tag{2.11}$$

$$\dot{q}_i = \frac{\mathcal{M}_i}{I_{iy}} \tag{2.12}$$

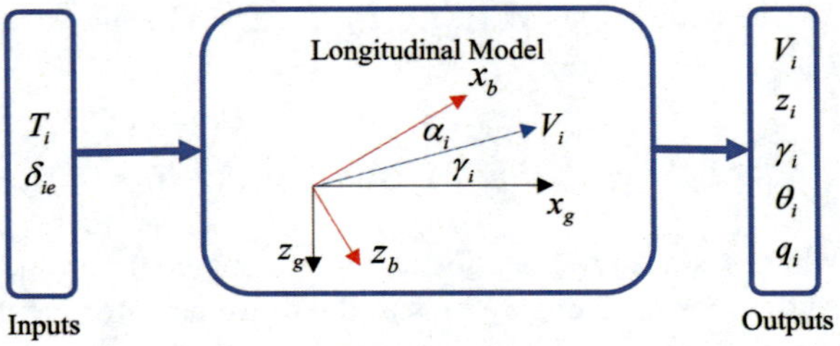

Fig. 2.2 The longitudinal motion of the ith fixed-wing UAV

where the relationship $\theta_i = \alpha_i + \gamma_i$ is used and θ_i is the pitch angle. Five states V_i, z_i, γ_i, θ_i, q_i and two control inputs T_i, δ_{ie} are usually involved in the longitudinal motion. $x_g y_g z_g$, $x_b y_b z_b$, $x_w y_w z_w$, and $x_k y_k z_k$ are the inertial reference frame, body-fixed coordinate frame, wind coordinate frame, and flight path coordinate frame, respectively.

2.3 Fixed-Wing UAV Inner-Loop Attitude Model

As shown in Fig. 2.3, by solely considering the inner-loop attitude motion of the ith fixed-wing UAV, the attitude kinematics can be directly described as

$$\begin{cases} \dot{\alpha}_i = q_i - \tan\beta_i (p_i \cos\alpha_i + r_i \sin\alpha_i) \\ \qquad -(\dot{\chi}_i \cos\gamma_i \sin\mu_i + \dot{\gamma}_i \cos\mu_i)/\cos\beta_i \\ \dot{\beta}_i = p_i \sin\alpha_i - r_i \cos\alpha_i + \dot{\chi}_i \cos\gamma_i \cos\mu_i - \dot{\gamma}_i \sin\mu_i \\ \dot{\mu}_i = (p_i \cos\alpha_i + r_i \sin\alpha_i)/\cos\beta_i + \dot{\chi}_i (\sin\gamma_i + \cos\gamma_i \sin\mu_i \tan\beta_i) \\ \qquad +\dot{\gamma}_i \cos\mu_i \tan\beta_i \end{cases} \tag{2.13}$$

The attitude dynamics can be derived as

$$\begin{cases} \dot{p}_i = (c_{i1} r_i + c_{i2} p_i) q_i + c_{i3} \mathcal{L}_i + c_{i4} \mathcal{N}_i \\ \dot{q}_i = c_{i5} p_i r_i - c_{i6} (p_i^2 - r_i^2) + c_{i7} \mathcal{M}_i \\ \dot{r}_i = (c_{i8} p_i - c_{i2} r_i) q_i + c_{i4} \mathcal{L}_i + c_{i9} \mathcal{N}_i \end{cases} \tag{2.14}$$

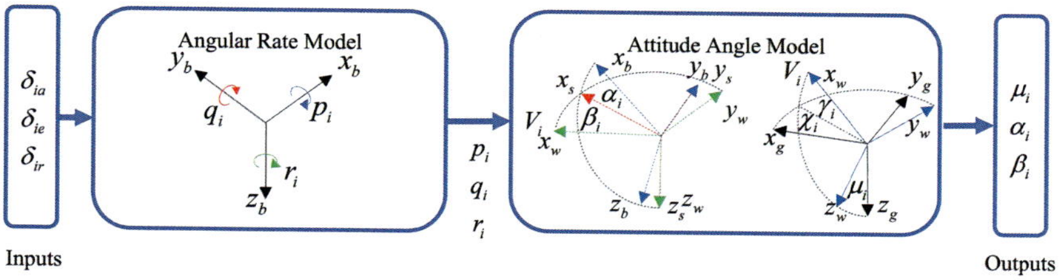

Fig. 2.3 The inner-loop attitude motion of the ith fixed-wing UAV

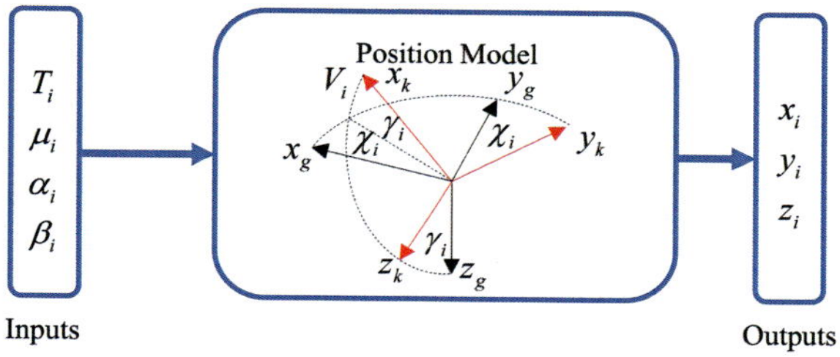

Fig. 2.4 The outer-loop position motion of the ith fixed-wing UAV

2.4 Fixed-Wing UAV Outer-Loop Position Model

As illustrated in Fig. 2.4, by solely considering the outer-loop position motion of the ith fixed-wing UAV, the position model of the ith fixed-wing UAV can be expressed as

$$\begin{cases} \dot{x}_i = V_i \cos \gamma_i \cos \chi_i \\ \dot{y}_i = V_i \cos \gamma_i \sin \chi_i \\ \dot{z}_i = V_i \sin \gamma_i \end{cases} \tag{2.15}$$

$$\begin{cases} \dot{V}_i = (-D_i + T_i \cos \alpha_i \cos \beta_i)/m_i - g \sin \gamma_i \\ \dot{\chi}_i = (L_i \sin \mu_i + Y_i \cos \mu_i)/(m_i V_i \cos \gamma_i) + T_i \sin \alpha_i \sin \mu_i/(m_i V_i \cos \gamma_i) \\ \qquad -T_i \cos \alpha_i \sin \beta_i \cos \mu_i/(m_i V_i \cos \gamma_i) \\ \dot{\gamma}_i = (L_i \cos \mu_i - Y_i \sin \mu_i)/(m_i V_i) + T_i \cos \alpha_i \sin \beta_i \sin \mu_i/(m_i V_i) \\ \qquad +T_i \sin \alpha_i \cos \mu_i/(m_i V_i) - g \cos \gamma_i / V_i \end{cases} \tag{2.16}$$

2.5 Conclusions

This chapter has directly presented the fixed-wing UAV model and the corresponding longitudinal, inner-loop attitude, and outer-loop position models, which will be used to facilitate the FTCC design for multi-UAVs in the subsequent chapters.

References

1. G.J.J. Ducard, *Fault-Tolerant Flight Control and Guidance Systems: Practical Methods for Small Unmanned Aerial Vehicles* (Springer, Berlin, 2009)
2. F. Gavilan, J.A. Acosta, R. Vazquez, Control of the longitudinal flight dynamics of an UAV using adaptive backstepping, in *IFAC World Congress*, Milano (2011)
3. T. Lee, Y. Kim, Nonlinear adaptive flight control using backstepping and neural networks controller. J. Guid. Control Dyn. **24**(4), 675–682 (2001)
4. W. Lin, Distributed UAV formation control using differential game approach. Aerosp. Sci. Technol. **35**, 54–62 (2014)
5. C. Liu, W.H. Chen, Disturbance rejection flight control for small fixed-wing unmanned aerial vehicles. J. Guid. Control Dyn. **39**(12), 2810–2819 (2016)
6. P.K. Menon, G.D. Sweriduk, B. Sridhar, Optimal strategies for free-flight air traffic conflict resolution. J. Guid. Control Dyn. **22**(2), 202–211 (1999)
7. L. Sonneveldt, E.R. Van Oort, Q. Chu, J. Mulder, Nonlinear adaptive trajectory control applied to an F-16 model. J. Guid. Control Dyn. **32**(1), 25–39 (2009)
8. Z.Q. Yu, Y.H. Qu, Y.M. Zhang, Safe control of trailing UAV in close formation flight against actuator fault and wake vortex effect. Aerosp. Sci. Technol. **77**, 189–205 (2018)
9. J.M. Zhang, Q. Li, N. Cheng, B. Liang, Adaptive dynamic surface control for unmanned aerial vehicles based on attractive manifolds. J. Guid. Control Dyn. **36**(6), 1776–1783 (2013)

Chapter 3
Distributed FTCC of Multi-UAVs with Prescribed Performance

3.1 Introduction

As a branch of cooperative control method, distributed control has better robustness and less communication requirement than traditional centralized control approaches [17]. In general, distributed control problem can be divided into three types: (1) regulation problem, (2) tracking control problem with one leader and multiple followers, and (3) containment control problem with multiple leaders and multiple followers [14]. Recently, distributed control strategy has been investigated to regulate the formation flight of multiple vehicles [11]. By considering the fixed and directed topologies, the distributed formation containment control problem was addressed in [7], in which the leader UAVs can exchange information with each other to maintain a prespecified formation and different protocols were used to control the leader UAVs and follower UAVs. When numerous UAVs are flying within a distributed communication network, the number of actuators in the formation team significantly increases, which raises the fault probability. Therefore, more investigations on the distributed FTCC schemes should be conducted to achieve safe formation of multi-UAVs.

Inspired by the PPC concept, numerous studies have been presented in [1–4, 8]. Recently, the PPC concept was further integrated into the distributed control scheme to improve the cooperative performance. In [15], an NN observer-based consensus control scheme was investigated for unknown nonlinear MASs with prescribed performance and input quantization, meanwhile the directed communications are considered in the control design. Within the directed communication network, a distributed PPC method was developed in [20] for consensus tracking of nonlinear semi-strict feedback systems in the presence of mismatched nonlinear uncertainties, external disturbances, and uncertain nonlinear virtual control coefficients of the subsystems. By considering unknown time-varying delayed nonlinear interaction faults, a decentralized adaptive FTC with prescribed performance bounds was proposed for nonlinear large-scale systems [21]. Then, by combining the PPC method,

Z. Yu et al., *Fault-Tolerant Cooperative Control of Unmanned Aerial Vehicles*,
https://doi.org/10.1007/978-981-99-7661-4_3

the distributed containment control problem of networked nonlinear systems under a directed graph was investigated in [22]. Although there have been significant investigations on the FTC design for single UAV with prescribed performance [16, 18], very few researches have considered the distributed FTCC for multi-UAVs with prescribed performance, which needs further investigations.

Motivated by the above discussions, this chapter investigates the difficult distributed FTCC problem for multi-UAVs with prescribed performance. Since external disturbances and input saturation are often encountered by nonlinear systems [6, 19], to make the study practical, input saturation and external disturbances are simultaneously considered in this chapter. By using the distributed SMO, the leader UAV's trajectory is first estimated by each follower UAV. Then, by using the PPC method, the tracking errors of follower UAVs with respect to the estimated references are transformed into a new set of errors. Based on the transformed error variables, a set of distributed PPC laws is finally constructed for multi-UAVs by the combination of NN and DSC techniques. Compared to other existing results, the main contributions of this chapter are listed as follows:

(1) Compared with the centralized control strategy [12], which involved massive information exchange, this chapter presents a distributed control scheme for multi-UAVs. In the distributed communication network, each UAV only needs to exchange information with its neighboring UAVs, thus significantly reducing the communication requirement.
(2) Different from [10, 11], which investigated the distributed control for multi-UAVs, this chapter further considers the prescribed performance requirement of distributed tracking control scheme. Moreover, the distributed SMO and NN are used to facilitate the controller design. Furthermore, both MLPNN and DSC techniques are utilized to reduce the computational burden, which makes the control scheme easy to achieve. It should be stressed that input saturation, actuator faults, and external disturbances are simultaneously addressed in this chapter.

The rest of this chapter is organized as follows. The preliminaries and problem statement are introduced in Sect. 3.2. Then, distributed SMO, tracking error transformation with prescribed performance function, distributed FTCC design, and stability analysis are given in Sect. 3.3. Subsequently, simulation results are presented in Sect. 3.4 to show the effectiveness of the proposed control scheme. Finally, conclusions are drawn in Sect. 3.5.

(*Remark: The main control schemes and contents of this chapter are from the published journal paper "Z. Q. Yu, Y. M. Zhang, Y. H. Qu. Prescribed performance-based distributed fault-tolerant cooperative control for multi-UAVs. Transactions of the Institute of Measurement and Control. 2019, 41(4): 975–989." The authors appreciate the permission from SAGE to reuse the results published in the relevant journal.*)

3.2 Preliminaries and Problem Statement

3.2.1 UAV Longitudinal Dynamics with Faults

Consider a group of UAVs including N follower UAVs and one leader UAV. By recalling the longitudinal model (2.8)–(2.12) and the elevator faults consisting of loss-of-effectiveness and bias faults, the longitudinal dynamics of the ith faulty fixed-wing UAV is expressed as [9]

$$\dot{V}_i = -\frac{D_i}{m_i} + \frac{T_i \cos\alpha_i}{m_i} - g\sin\gamma_i + d_{i1} \tag{3.1}$$

$$\dot{z}_i = V_i \sin\gamma_i \tag{3.2}$$

$$\dot{\gamma}_i = \frac{L_i - m_i g\cos\gamma_i + T_i \sin\alpha_i}{m_i V_i} + d_{i2} \tag{3.3}$$

$$\dot{\theta}_i = q_i \tag{3.4}$$

$$\dot{q}_i = \frac{\mathcal{M}_i}{I_{iy}} + d_{i3} \tag{3.5}$$

where d_{i1}, d_{i2}, and d_{i3} are external disturbances. $\mathcal{M}_i = \bar{q}_i s_i c_i C_{im}$ and $C_{im} = C_{im0} + C_{im\alpha}\alpha_i + C_{imq}\frac{c_i}{2V_i}q_i + C_{im\delta_e}\delta_{ie}$. The elevator actuator fault model is given by [23]

$$\delta_{ie} = \rho_i \delta_{ie0} + u_{if}, \quad i = 1, 2, \ldots, N \tag{3.6}$$

where δ_{ie} and δ_{ie0} are the applied control signal and the desired control signal to be constructed, respectively, and $\boldsymbol{\delta}_{e0} = [\delta_{1e0}, \delta_{2e0}, \ldots, \delta_{Ne0}]^T$ is the elevator deflection vector. $0 < \rho_i \leq 1$ is the remaining actuator effectiveness factor and u_{if} is the bounded bias fault signal of the ith follower UAV.

The representation (3.6) consists of the normal and faulty cases: (1) when $\rho_i = 1$ and $u_{if} = 0$, it means that there is no fault for the ith UAV's elevator, (2) when $0 < \rho_i < 1$ and $u_{if} = 0$, the ith UAV's elevator is encountered by the loss of effectiveness fault, (3) when $\rho_i = 1$ and $u_{if} \neq 0$, the corresponding elevator is in the bias fault mode, and (4) the stuck fault occurs in the ith UAV's elevator if $\rho_i = 0$ and $u_{if} \neq 0$.

By using the functional decomposition, the dynamics of each follower UAV can be divided into the velocity subsystem and the attitude subsystem. For the velocity subsystem, the proportional-integration-derivative (PID) controller is used to regulate the velocity tracking error of each follower UAV. Regarding the attitude subsystem, a set of distributed FTCC laws δ_{ie0}, $i = 1, 2, \ldots, N$, is designed for multi-UAVs to track the desired altitude reference z_0 from the leader UAV even in the presence of actuator faults, input saturation, and external disturbances, while all altitude tracking errors with respect to the estimated altitude references can be confined within the predefined performance bounds.

3.2.2 Dynamics Transformation

By using functional decomposition, the longitudinal dynamics of the ith follower UAV can be divided into the velocity subsystem and the attitude subsystem. For the velocity subsystem (3.1), the PID controller is used as

$$T_i = k_{11}e_{i1} + k_{12}\int e_{i1}dt + k_{13}\dot{e}_{i1}, \quad i = 1, 2, \ldots, N \tag{3.7}$$

where $k_{11} > 0$, $k_{12} > 0$, and $k_{13} > 0$ are the design parameters. $e_{i1} = V_0 - V_i$ is the velocity tracking error of the ith follower UAV with respect to the velocity reference from the leader UAV and $\boldsymbol{e}_1 = [e_{11}, e_{21}, \ldots, e_{N1}]^T$, $\boldsymbol{T} = [T_1, T_2, \ldots, T_N]^T$, and $\boldsymbol{V} = [V_1, V_2, \ldots, V_N]^T$ are the corresponding vectors. In this chapter, the leader UAV is used to generate references for follower UAVs.

Consider the actuator fault model (3.6), and the attitude subsystem is formulated as

$$\dot{z}_i = V_i\gamma_i \tag{3.8}$$

$$\dot{\gamma}_i = f_{i\gamma}(V_i, \gamma_i, \theta_i) + \theta_i + d_{i2} \tag{3.9}$$

$$\dot{\theta}_i = q_i \tag{3.10}$$

$$\dot{q}_i = f_{iq}(V_i, \gamma_i, \theta_i, q_i, u_i) + u_i + d_{i3} \tag{3.11}$$

$$y_i = z_i, \ i = 1, 2, \ldots, N$$

where y_i is the output. $u_i = -\delta_{ie0}$, $\boldsymbol{z} = [z_1, z_2, \ldots, z_N]^T$, $\boldsymbol{\gamma} = [\gamma_1, \gamma_2, \ldots, \gamma_N]^T$, $\boldsymbol{\theta} = [\theta_1, \theta_2, \ldots, \theta_N]^T$, $\boldsymbol{\alpha} = \boldsymbol{\theta} - \boldsymbol{\gamma} = [\alpha_1, \alpha_2, \ldots, \alpha_N]^T$, $\boldsymbol{q} = [q_1, q_2, \ldots, q_N]^T$, and the unknown nonlinear functions $f_{i\gamma}(\cdot)$ and $f_{iq}(\cdot)$ are given by

$$f_{i\gamma}(V_i, \gamma_i, \theta_i) = \frac{L_i - m_i g_i \cos\gamma_i + T_i \sin\alpha_i}{m_i V_i} - \theta_i \tag{3.12}$$

$$f_{iq}(V_i, \gamma_i, \theta_i, q_i, u_i) = \frac{\mathcal{M}_{if}}{I_{iy}} + \delta_{ie0} \tag{3.13}$$

$$\mathcal{M}_{if} = \bar{q}_i s_i c_i \left[C_{im0} + C_{im\alpha}\alpha_i + C_{imq}\frac{c_i}{2V_i}q_i + C_{im\delta_e}(\rho_i\delta_{ie0} + u_{if}) \right] \tag{3.14}$$

Since the cruise flight phase is considered in this chapter, $\sin\gamma_i \approx \gamma_i$ is used in (3.8). It should be noted that if θ_i and u_i are used as the virtual control signal and the control signal in the DSC design, respectively, the algebraic loops will be introduced into the closed-loop system, which makes the scheme not applicable in practical applications. To solve the problem, two Butterworth low-pass filters (BLPFs) are used as [25]

$$\begin{aligned} \theta_{if} &= B_{i1}(s)\theta_i \approx \theta_i \\ u_{if} &= B_{i2}(s)u_i \approx u_i \end{aligned}$$

where $B_{i1}(s)$ and $B_{i2}(s)$ are BLPFs and θ_{if} and u_{if} are filtered signals.

Therefore, one has

$$f_{i\gamma}(V_i, \gamma_i, \theta_i) = f_{i\gamma}(V_i, \gamma_i, \theta_{if}) + \epsilon_{i1} \tag{3.15}$$

$$f_{iq}(V_i, \gamma_i, \theta_i, q_i, u_i) = f_{iq}(V_i, \gamma_i, \theta_i, q_i, u_{if}) + \epsilon_{i2} \tag{3.16}$$

where ϵ_{i1} and ϵ_{i2} are filter errors, bounded by $|\epsilon_{i1}| \le \bar{\epsilon}_{i1}$, $|\epsilon_{i2}| \le \bar{\epsilon}_{i2}$.

3.2.3 Basic Graph Theory

In this chapter, the topology of information flows among follower UAVs is described by an undirected and fixed graph $\mathcal{G} = (\Upsilon, \mathcal{E}, \mathcal{A})$, where $\Upsilon = \{r_1, r_2, \ldots, r_N\}$ is the set of follower UAVs, $\mathcal{E} \subseteq \Upsilon \times \Upsilon$ is the set of edges, and $\mathcal{A} = [a_{ij}] \in R^{N\times N}$ is the weighted adjacency matrix with nonnegative elements, where $a_{ii} = 0$ and $a_{ij} = a_{ji}$. An edge in $\mathcal{G}$ is denoted by an unordered pair (i, j) and $(i, j) \in \mathcal{E}$ if and only if there is information exchange between the ith follower UAV and the jth follower UAV. a_{ij} denotes the communication quality between the ith follower UAV and the jth follower UAV, i.e., $(i, j) \in E \Leftrightarrow a_{ij} > 0$. For any two follower UAVs i and j, if there exists a path between them, then $\mathcal{G}$ is called a connected graph.

Let $\mathcal{D} = \text{diag}\{D_1, D_2, \ldots, D_N\}$ represent the degree matrix of graph $\mathcal{G}$, where $D_i = \sum_{j=1}^{N} a_{ij}$. The Laplacian matrix $\mathcal{L}$ of graph $\mathcal{G}$ is defined as $\mathcal{L} = \mathcal{D} - \mathcal{A}$. Next, consider another graph $\bar{\mathcal{G}}$, which consists of N follower UAVs and one leader UAV. Define the adjacency matrix of the leader UAV as $\mathcal{B} = \text{diag}\{b_1, b_2, \ldots, b_N\}$, where $b_i > 0$ if the ith follower UAV has access to the leader UAV; otherwise, $b_i = 0$, $i = 1, 2, \ldots, N$.

Lemma 3.1 *If the graph $\bar{\mathcal{G}}$ is connected, then the matrix $\mathcal{L} + \mathcal{B}$ is symmetric and positive [24].*

3.2.4 Radial Basis Function Neural Network

The RBFNN technique will be used to approximate an unknown continuous function $f(\boldsymbol{Z})$ defined on the compact set Ω in this chapter. In [13], the authors have proven that the RBFNN can approximate any continuous function $f(\mathbf{Z})$ to arbitrary accuracy λ, which is given by

$$f(\mathbf{Z}) = \boldsymbol{\phi}^{*T}\boldsymbol{\varphi}(\mathbf{Z}) + \lambda \tag{3.17}$$

where $\mathbf{Z} \in \Omega \subset R^l$ is the input vector with l being the input dimension of RBFNN. $\boldsymbol{\phi}^* = [\phi_1^*, \phi_2^*, \ldots, \phi_r^*]^T \in R^r$ is the ideal weight vector, and r is the number of RBFNN nodes. λ is the approximation error, satisfying $|\lambda| \leq \lambda_m$. $\boldsymbol{\varphi}(\mathbf{Z}) = [\varphi_1, \varphi_2, \ldots, \varphi_r]^T \in R^r$ denotes the basis function vector.

The ideal weight vector is defined as

$$\boldsymbol{\phi}^* = \arg\min_{\boldsymbol{\phi}\in R^r}\{\sup_{\mathbf{Z}\in\Omega}|f(\mathbf{Z}) - \boldsymbol{\phi}^T\varphi(\mathbf{Z})|\} \tag{3.18}$$

where $\boldsymbol{\phi} \in R^r$ is the weight vector.

The basis function φ_i is chosen as the commonly used Gaussian function, which is expressed as

$$\varphi_i = \exp\left[-\frac{(\mathbf{Z} - \mathbf{C}_i)^T(\mathbf{Z} - \mathbf{C}_i)}{n_i^2}\right], \; i = 1, 2, \ldots, r \tag{3.19}$$

where $\mathbf{C}_i$ and n_i denote the center and the width of the ith neural cell, respectively.

3.2.5 Control Objective

The control objective of this chapter is to design a set of distributed FTCC laws for multi-UAVs, such that the follower UAVs can track the leader UAV even when actuator faults, input saturation, and external disturbances are simultaneously encountered by a portion of follower UAVs, meanwhile the tracking errors are confined within the predefined performance bounds.

3.3 Main Results

In this section, a distributed neural adaptive FTCC scheme is proposed based on the DSC design architecture and the RBFNN technique. The control scheme contains three parts: the distributed SMO design, the tracking error transformation, and the distributed PPC design. The trajectory of the leader UAV is first estimated by each UAV in a distributed manner with the proposed distributed SMO. Then, to achieve the PPC of altitude tracking errors of follower UAVs with respect to the estimated altitude references, the tracking errors are transformed into a new set of errors. Finally, by using the transformed error variables, a distributed FTCC scheme is proposed. To handle the input saturation, auxiliary dynamic signals are introduced into the control scheme to attenuate the adverse effect of input saturation. The detailed design process is described as follows.

3.3.1 Distributed Sliding-Mode Observer

In this section, the distributed SMO is employed to estimate the altitude reference of leader UAV based on graph theory. The distributed SMO is expressed as

$$\begin{aligned}\dot{z}_{id} =& -\beta_1\left[\sum_{j=1}^{N} a_{ij}(z_{id}-z_{jd}) + b_i(z_{id}-z_0)\right]\\ &-\beta_2\text{sign}\left[\sum_{j=1}^{N} a_{ij}(z_{id}-z_{jd}) + b_i(z_{id}-z_0)\right]\end{aligned} \tag{3.20}$$

where z_{id} is the altitude reference estimated by the ith follower UAV. z_0 is the altitude trajectory of the leader UAV. β_1 and β_2 are positive design parameters. $\text{sign}(\cdot)$ is the sign function and $z_d = [z_{1d}, z_{2d}, \ldots, z_{Nd}]^T$.

Lemma 3.2 *If the distributed SMO is constructed as (3.20) to estimate the altitude trajectory of the leader UAV for each follower UAV, then the estimated altitude references z_{id}, $i = 1, 2, \ldots, N$, can converge to z_0 by properly choosing β_1 and β_2.*

Proof Define the estimation error as $e_{si} = z_{id} - z_0$, $i = 1, 2, \ldots, N$, then one has

$$\begin{aligned}\dot{e}_{si} =& -\beta_1\left(\sum_{j=1}^{N} l_{ij}e_{sj} + b_i e_{si}\right)\\ &-\beta_2\text{sign}\left(\sum_{j=1}^{N} l_{ij}e_{s_j} + b_i e_{si}\right) - \dot{z}_0\end{aligned} \tag{3.21}$$

where l_{ij} represents the element of the Laplacian matrix $\mathcal{L}$.

Choose the Lyapunov candidate function as

$$V_s = \frac{1}{2}\boldsymbol{e}_s^T \mathcal{P} \boldsymbol{e}_s \tag{3.22}$$

where $\boldsymbol{e}_s = [e_{s1}, e_{s2}, \ldots, e_{sN}]^T$, $\mathcal{P} = \mathcal{L} + \mathcal{B}$. From Lemma 3.1, one can see that matrix $\mathcal{P}$ is positive definite.

Taking the time derivative of (3.22) yields

$$\begin{aligned}\dot{V}_s &= \boldsymbol{e}_s^T \mathcal{P}\left[-\beta_1 \mathcal{P}\boldsymbol{e}_s - \beta_2 \text{sign}(\mathcal{P}\boldsymbol{e}_s) - \boldsymbol{Q}_0\right] \\ &= -\beta_1 \boldsymbol{e}_s^T \mathcal{P}\mathcal{P}\boldsymbol{e}_s - \beta_2 \boldsymbol{e}_s^T \mathcal{P}\text{sign}(\mathcal{P}\boldsymbol{e}_s) - \boldsymbol{e}_s^T \mathcal{P}\boldsymbol{Q}_0 \\ &\le -\frac{2\beta_1 \lambda_{\min}^2(\mathcal{P})}{\lambda_{\max}(\mathcal{P})} V_s - \beta_2 ||\boldsymbol{e}_s^T \mathcal{P}||_1 + z_{0d} ||\boldsymbol{e}_s^T \mathcal{P}||_1 \\ &\le -\frac{2\beta_1 \lambda_{\min}^2(\mathcal{P})}{\lambda_{\max}(\mathcal{P})} V_s - (\beta_2 - z_{0d}) ||\boldsymbol{e}_s^T \mathcal{P}||_1 \\ &\le -\frac{2\beta_1 \lambda_{\min}^2(\mathcal{P})}{\lambda_{\max}(\mathcal{P})} V_s - (\beta_2 - z_{0d}) ||\boldsymbol{e}_s^T \mathcal{P}||_2 \\ &\le -\frac{2\beta_1 \lambda_{\min}^2(\mathcal{P})}{\lambda_{\max}(\mathcal{P})} V_s - \frac{2^{1/2}(\beta_2 - z_{0d})\lambda_{\min}(\mathcal{P})}{\lambda_{\max}^{1/2}(\mathcal{P})} V_s^{1/2}\end{aligned} \tag{3.23}$$

where $\boldsymbol{Q}_0 = [\dot{z}_0, \dot{z}_0, \ldots, \dot{z}_0]^T \in R^{N\times 1}$ and $|\dot{z}_0| \le z_{0d}$. $\lambda_{\max}(\mathcal{P})$ and $\lambda_{\min}(\mathcal{P})$ are the maximum and minimum eigenvalues of $\mathcal{P}$.

According to the finite-time stability theory, if $\beta_1 > 0$, $\beta_2 > z_{0d}$, then $\boldsymbol{e}_s \to 0$ in finite time T_0, which is given by

$$T_0 \le \frac{\lambda_{\max}(\mathcal{P})}{\beta_1 \lambda_{\min}^2(\mathcal{P})} \ln\left(\frac{2\beta_1 \lambda_{\min}^2(\mathcal{P}) V_s^{1/2}(0)}{2^{1/2}(\beta_2 - z_{0d})\lambda_{\min}(\mathcal{P})\lambda_{\max}^{1/2}(\mathcal{P})} + 1\right) \tag{3.24}$$

□

Remark 3.1 From (3.24), it is observed that the convergence time T_0 decreases when the design parameters β_2 increase. Since the distributed SMO (3.20) can be embedded in each follower UAV to estimate the altitude trajectory of leader UAV and the estimation errors can be regulated to zero in finite time, then the PPC method will be mainly used to design the controller for each follower UAV to track the estimated altitude reference z_{id} with guaranteed performance.

3.3.2 Tracking Error Transformation with Prescribed Performance Function

The PPC design means that the tracking error is confined within the predefined residual set and the convergence rate cannot exceed the predefined value. The prescribed performance can be achieved if the altitude tracking error $e_{i2} = z_i - z_{id}$ strictly evolves within a predefined decaying bounds, defined by

$$-\underline{k}_{i1}\varepsilon_{i1}(t) < e_{i2}(t) < \overline{k}_{i1}\varepsilon_{i1}(t),\ i = 1, 2, \ldots, N \tag{3.25}$$

where $\underline{k}_{i1} > 0$ and $\overline{k}_{i1} > 0$ are design parameters. The performance function $\varepsilon_{i1}(t)$ is a strictly positive decreasing smooth function with $\lim_{t\to\infty}\varepsilon_{i1}(t) = \varepsilon_{i1\infty}$. $\varepsilon_{i1\infty}$ represents the maximum allowable value of $e_{i2}(t)$ at steady state. In this chapter, the performance function is chosen as $\varepsilon_{i1}(t) = (\varepsilon_{i10} - \varepsilon_{i1\infty})e^{-\eta_{i1}t} + \varepsilon_{i1\infty}$, where ε_{i10}, $\varepsilon_{i1\infty}$, and η_{i1} are positive constants and $\varepsilon_{i10} > \varepsilon_{i1\infty}$. $\varepsilon_{i10} = \varepsilon_{i1}(0)$ is selected such that $-\underline{k}_{i1}\varepsilon_{i1}(0) < e_{i2}(0) < \overline{k}_{i1}\varepsilon_{i1}(0)$. η_{i1} is the decreasing rate, which represents the required convergence speed of $e_{i2}(t)$.

To facilitate the error constrained controller design, (3.25) can be transformed into an equality form, which is given by

$$e_{i2} = \varepsilon_{i1}(t)\Gamma(E_{i2}) \tag{3.26}$$

where E_{i2} is the transformed error. $\Gamma(E_{i2})$ is the error transformation function, expressed as

$$\Gamma(E_{i2}) = \frac{\overline{k}_{i1}e^{E_{i2}+\nu_i} - \underline{k}_{i1}e^{-E_{i2}-\nu_i}}{e^{E_{i2}+\nu_i} + e^{-E_{i2}-\nu_i}} \tag{3.27}$$

where $\nu_i = \frac{1}{2}\ln\frac{\underline{k}_{i1}}{\overline{k}_{i1}}$.

The transformation function $\Gamma(E_{i2})$ has the following properties: (1) $\Gamma(0) = 0$, (2) $-\underline{k}_{i1} \le \Gamma(E_{i2}) \le \overline{k}_{i1}$, and (3) $\lim_{E_{i2}\to+\infty}\Gamma(E_{i2}) = \overline{k}_{i1}$, $\lim_{E_{i2}\to-\infty}\Gamma(E_{i2}) = -\underline{k}_{i1}$. Therefore, e_{i2} will be confined within the bounds $\left[-\underline{k}_{i1}\varepsilon_{i1}(t),\ \overline{k}_{i1}\varepsilon_{i1}(t)\right]$ if the transformed error E_{i2} is uniformly ultimately bounded (UUB). With the definition of $\Gamma(E_{i2})$, (3.26) can be used to describe (3.25). Then, the transformed error E_{i2} is given by

$$\begin{aligned} E_{i2} &= \Gamma^{-1}\left(\frac{e_{i2}}{\varepsilon_{i1}}\right) \\ &= \frac{1}{2}\ln\frac{\overline{k}_{i1}\underline{k}_{i1} + \overline{k}_{i1}\sigma_{i1}}{\overline{k}_{i1}\underline{k}_{i1} - \underline{k}_{i1}\sigma_{i1}} \end{aligned} \tag{3.28}$$

where $\sigma_{i1} = \frac{e_{i2}}{\varepsilon_{i1}}$.

Taking the time derivative of (3.28) yields

$$\begin{aligned} \dot{E}_{i2} &= \frac{1}{2\varepsilon_{i1}}\left(\frac{1}{\underline{k}_{i1} + \sigma_{i1}} + \frac{1}{\overline{k}_{i1} - \sigma_{i1}}\right)\left(\dot{e}_{i2} - \frac{e_{i2}\dot{\varepsilon}_{i1}}{\varepsilon_{i1}}\right) \\ &= \xi_{i1}\left(\dot{e}_{i2} - \frac{e_{i2}\dot{\varepsilon}_{i1}}{\varepsilon_{i1}}\right) \end{aligned} \tag{3.29}$$

where $\xi_{i1} = \frac{1}{2\varepsilon_{i1}}\left(\frac{1}{\underline{k}_{i1}+\sigma_{i1}} + \frac{1}{\overline{k}_{i1}-\sigma_{i1}}\right)$.

3.3.3 Prescribed Performance-Based Distributed FTCC Design

To achieve the FTCC design for the altitudes of multi-UAVs with guaranteed prescribed performance, the DSC architecture with a first-order filter is adopted to facilitate the control design, which contains four steps.

Step 1 Design the intermediate control signal $\bar{\gamma}_{id}$ as

$$\bar{\gamma}_{id} = \frac{1}{V_i}\left[-\left(k_{21} - \frac{\dot{\xi}_{i1}}{2\xi_{i1}^2}\right)E_{i2} + \dot{z}_{id} + \frac{e_{i2}\dot{\varepsilon}_{i1}}{\varepsilon_{i1}}\right] \tag{3.30}$$

where $k_{21} > 0$ is the design parameter.

Then, pass the intermediate control signal $\bar{\gamma}_{id}$ into a first-order filter with time constant τ_{i1} to obtain the filtered virtual control signal γ_{id}.

$$\tau_{i1}\dot{\gamma}_{id} + \gamma_{id} = \bar{\gamma}_{id},\ \gamma_{id}(0) = \bar{\gamma}_{id}(0) \tag{3.31}$$

where $\tau_{i1} > 0$ is a design parameter.

Remark 3.2 Compared with the direct computation of $\dot{\bar{\gamma}}_{id}$ in the traditional backstepping architecture, $\dot{\gamma}_{id} = (\bar{\gamma}_{id} - \gamma_{id})/\tau_{i1}$ is used in this chapter, which replaces the differential operation with an algebraic operation.

Define the filter error as $\eta_{i1} = \gamma_{id} - \bar{\gamma}_{id}$, then one has

$$\dot{\eta}_{i1} = -\frac{\eta_{i1}}{\tau_{i1}} + M_{i1} \tag{3.32}$$

where $M_{i1} = -\dot{\bar{\gamma}}_{id}$.

Choose the Lyapunov candidate function as

$$L_{i1} = \frac{1}{2\bar{V}_i^2\xi_{i1}}E_{i2}^2 + \frac{1}{2}\eta_{i1}^2 \tag{3.33}$$

where $\bar{V}_i$ is the maximum velocity of the ith follower UAV. The introduction of $\bar{V}_i$ is only used for the stability analysis.

Taking the time derivatives of (3.33) along the trajectories of (3.8), (3.29), and (3.32) gives

$$\begin{aligned}\dot{L}_{i1} =& -\frac{\dot{\xi}_{i1}}{2\bar{V}_i^2\xi_{i1}^2}E_{i2}^2 + \frac{1}{\bar{V}_i^2\xi_{i1}}E_{i2}\dot{E}_{i2} + \eta_{i1}\dot{\eta}_{i1}\\ =& -\frac{\dot{\xi}_{i1}}{2\bar{V}_i^2\xi_{i1}^2}E_{i2}^2 + \frac{1}{\bar{V}_i^2\xi_{i1}}E_{i2}\xi_{i1}\left(\dot{e}_{i2} - \frac{e_{i2}\dot{\varepsilon}_{i1}}{\varepsilon_{i1}}\right) + \eta_{i1}\dot{\eta}_{i1}\end{aligned}$$

$$
\begin{aligned}
&= -\frac{\dot{\xi}_{i1}}{2\bar{V}_i^2\xi_{i1}^2}E_{i2}^2 + \eta_{i1}\dot{\eta}_{i1} \\
&\quad + \frac{1}{\bar{V}_i^2}E_{i2}\left(V_i\bar{\gamma}_{id} + V_i\eta_{i1} + V_iE_{i3} - \dot{z}_{id} - \frac{e_{i2}\dot{\varepsilon}_{i1}}{\varepsilon_{i1}}\right) \\
&= -\frac{k_{21}}{\bar{V}_i^2}E_{i2}^2 + \frac{1}{\bar{V}_i^2}E_{i2}V_i\eta_{i1} + \frac{1}{\bar{V}_i^2}E_{i2}V_iE_{i3} + \eta_{i1}\dot{\eta}_{i1}
\end{aligned}
\tag{3.34}
$$

where $E_{i3} = \gamma_i - \gamma_{id}$ is the flight path angle tracking error of the ith follower UAV.

By using Young's inequality, one has

$$
\frac{1}{\bar{V}_i^2}E_{i2}V_i\eta_{i1} \leq \frac{E_{i2}^2}{2w_{i11}^2\bar{V}_i^2} + \frac{w_{i11}^2\eta_{i1}^2}{2} \tag{3.35}
$$

$$
\frac{1}{\bar{V}_i^2}E_{i2}V_iE_{i3} \leq \frac{E_{i2}^2}{2w_{i12}^2\bar{V}_i^2} + \frac{w_{i12}^2E_{i3}^2}{2} \tag{3.36}
$$

where w_{i11} and w_{i12} are positive parameters.

Then, (3.35) has the following form:

$$
\begin{aligned}
\dot{L}_{i1} \leq &-\frac{1}{\bar{V}_i^2}\left(k_{21} - \frac{1}{2w_{i11}^2} - \frac{1}{2w_{i12}^2}\right)E_{i2}^2 + \frac{w_{i12}^2}{2}E_{i3}^2 \\
&-\left(\frac{1}{\tau_{i1}} - \frac{1}{2} - \frac{w_{i11}^2}{2}\right)\eta_{i1}^2 + \frac{M_{i1}^2}{2}
\end{aligned}
\tag{3.37}
$$

Step 2 Design the intermediate control signal $\bar{\theta}_{id}$ and the adaptive law as

$$
\bar{\theta}_{id} = -\left(k_{31} + \frac{\hat{\Phi}_{i\gamma}\boldsymbol{\varphi}_{i\gamma}^T\boldsymbol{\varphi}_{i\gamma}}{2w_{i21}^2} + \frac{1}{2w_{i22}^2} + \frac{1}{2w_{i23}^2}\right)E_{i3} + \dot{\gamma}_{id} \tag{3.38}
$$

$$
\dot{\hat{\Phi}}_{i\gamma} = g_{i11}\left(\frac{E_{i3}^2\varphi_{i\gamma}^T\boldsymbol{\varphi}_{i\gamma}}{2w_{i21}^2} - w_{i24}\hat{\Phi}_{i\gamma}\right) \tag{3.39}
$$

where k_{31}, w_{i21}, w_{i22}, w_{i23}, w_{i24}, and g_{i11} are positive design parameters. $\hat{\Phi}_{i\gamma}$ is the estimation of $\Phi_{i\gamma}$. $\Phi_{i\gamma}$ and $\boldsymbol{\varphi}_{i\gamma}$ will be defined later.

Pass the intermediate control signal $\bar{\theta}_{id}$ into a first-order filter with time constant τ_{i2} to obtain the filtered virtual control signal θ_{id}.

$$
\tau_{i2}\dot{\theta}_{id} + \theta_{id} = \bar{\theta}_{id}, \ \theta_{id}(0) = \bar{\theta}_{id}(0) \tag{3.40}
$$

where $\tau_{i2} > 0$ is a design parameter.

Define the filter error as $\eta_{i2} = \theta_{id} - \bar{\theta}_{id}$, then one has

$$\dot{\eta}_{i2} = -\frac{\eta_{i2}}{\tau_{i2}} + M_{i2} \tag{3.41}$$

where $M_{i2} = -\dot{\bar{\theta}}_{id}$.

Choose the Lyapunov candidate function as

$$L_{i2} = \frac{1}{2}E_{i3}^2 + \frac{1}{2g_{i11}}\tilde{\Phi}_{i\gamma}^2 + \frac{1}{2}\eta_{i2}^2 \tag{3.42}$$

where $\tilde{\Phi}_{i\gamma} = \Phi_{i\gamma} - \hat{\Phi}_{i\gamma}$ is the estimation error.

Then, taking the time derivative of (3.42) yields

$$\begin{aligned}
\dot{L}_{i2} &= E_{i3}\dot{E}_{i3} + \frac{1}{g_{i11}}\tilde{\Phi}_{i\gamma}\dot{\tilde{\Phi}}_{i\gamma} + \eta_{i2}\dot{\eta}_{i2} \\
&= E_{i3}[f_{i\gamma}(V_i, \gamma_i, \theta_i) + \theta_i + d_{i2} - \dot{\gamma}_{id}] \\
&+ \frac{1}{g_{i11}}\tilde{\Phi}_{i\gamma}\dot{\tilde{\Phi}}_{i\gamma} + \eta_{i2}\dot{\eta}_{i2} \\
&= E_{i3}[f_{i\gamma}(V_i, \gamma_i, \theta_{if}) + \epsilon_{i1} + \theta_i + d_{i2} - \dot{\gamma}_{id}] \\
&+ \frac{1}{g_{i11}}\tilde{\Phi}_{i\gamma}\dot{\tilde{\Phi}}_{i\gamma} + \eta_{i2}\dot{\eta}_{i2}
\end{aligned} \tag{3.43}$$

The RBFNN is used to approximate $f_{i\gamma}(V_i, \gamma_i, \theta_{if})$, which is given by

$$f_{i\gamma}(V_i, \gamma_i, \theta_{if}) = \boldsymbol{\phi}_{i\gamma}^{*T}\boldsymbol{\varphi}_{i\gamma}(V_i, \gamma_i, \theta_{if}) + \lambda_{i11} \tag{3.44}$$

For brevity, $(\cdot)$ of $\boldsymbol{\varphi}_{i\gamma}(V_i, \gamma_i, \theta_{if})$ is omitted in the subsequent analysis.

Define $\lambda_{i12} = f_{i\gamma}(V_i, \gamma_i, \theta_i) - f_{i\gamma}(V_i, \gamma_i, \theta_{if}) + \lambda_{i11}$, and by using Young's inequality, one has

$$E_{i3}\boldsymbol{\phi}_{i\gamma}^{*T}\boldsymbol{\varphi}_{i\gamma} \le \frac{E_{i3}^2\Phi_{i\gamma}\boldsymbol{\varphi}_{i\gamma}^T\boldsymbol{\varphi}_{i\gamma}}{2w_{i21}^2} + \frac{w_{i21}^2}{2} \tag{3.45}$$

$$E_{i3}\lambda_{i12} \le \frac{E_{i3}^2}{2w_{i22}^2} + \frac{w_{i22}^2\lambda_{i12m}^2}{2} \tag{3.46}$$

$$E_{i3}d_{i2} \le \frac{E_{i3}^2}{2w_{i23}^2} + \frac{w_{i23}^2 d_{i2m}^2}{2} \tag{3.47}$$

where $\Phi_{i\gamma} = \boldsymbol{\phi}_{i\gamma}^{*T}\boldsymbol{\phi}_{i\gamma}^*$, $\boldsymbol{\Phi}_\gamma = [\Phi_{1\gamma}, \Phi_{2\gamma}, \ldots, \Phi_{N\gamma}]^T$, $|\lambda_{i12}| \le \lambda_{i12m}$, $|d_{i2}| \le d_{i2m}$.

Then, one has

$$\begin{aligned}
\dot{L}_{i2} &= E_{i3}(\boldsymbol{\phi}_{i\gamma}^{*T}\boldsymbol{\varphi}_{i\gamma} + \lambda_{i12} + \theta_i + d_{i2} - \dot{\gamma}_{id}) + \frac{1}{g_{i11}}\tilde{\Phi}_{i\gamma}\dot{\tilde{\Phi}}_{i\gamma} + \eta_{i2}\dot{\eta}_{i2} \\
&\le E_{i3}\left(\frac{E_{i3}}{2w_{i22}^2} + \frac{E_{i3}\Phi_{i\gamma}\boldsymbol{\varphi}_{i\gamma}^T\boldsymbol{\varphi}_{i\gamma}}{2w_{i21}^2} + \frac{E_{i3}}{2w_{i23}^2} + \bar{\theta}_{id} + \eta_{i2}\right) + E_{i3}E_{i4} \\
&\quad - E_{i3}\dot{\gamma}_{id} + \frac{w_{i21}^2}{2} + \frac{w_{i22}^2\lambda_{i12m}^2}{2} + \frac{w_{i23}^2 d_{i2m}^2}{2} + \frac{1}{g_{i11}}\tilde{\Phi}_{i\gamma}\dot{\tilde{\Phi}}_{i\gamma} + \eta_{i2}\dot{\eta}_{i2} \\
&\le -\left(k_{31} - \frac{1}{2w_{i25}^2} - \frac{1}{2w_{i26}^2}\right)E_{i3}^2 + \frac{w_{i26}^2}{2}E_{i4}^2 - \left(w_{i24} - \frac{1}{2}\right)\tilde{\Phi}_{i\gamma}^2 \\
&\quad - \left(\frac{1}{\tau_{i2}} - \frac{w_{i25}^2}{2} - \frac{1}{2}\right)\eta_{i2}^2 + \frac{w_{i21}^2}{2} + \frac{w_{i22}^2\lambda_{i12m}^2}{2} + \frac{w_{i23}^2 d_{i1m}^2}{2} \\
&\quad + \frac{M_{i2}^2}{2} + \frac{w_{i24}^2}{2}\Phi_{i\gamma}^2
\end{aligned} \tag{3.48}$$

where $E_{i4} = \theta_i - \theta_{id}$ is the pitch angle tracking error. The inequalities $E_{i3}\eta_{i2} \le E_{i3}^2/(2w_{i25}^2) + w_{i25}^2\eta_{i2}^2/2$, $E_{i3}E_{i4} \le E_{i3}^2/(2w_{i26}^2) + w_{i26}^2E_{i4}^2/2$ are also used in (3.48), where w_{i25} and w_{i26} are positive parameters.

Step 3 Design the intermediate control signal $\bar{q}_{id}$ as

$$\bar{q}_{id} = -k_{41}E_{i4} + \dot{\theta}_{id} \tag{3.49}$$

where k_{41} is a positive design parameter.

Pass the intermediate control signal $\bar{q}_{id}$ into a first-order filter with time constant τ_{i3} to obtain the filtered virtual control signal q_{id}.

$$\tau_{i3}\dot{q}_{id} + q_{id} = \bar{q}_{id},\ q_{id}(0) = \bar{q}_{id}(0) \tag{3.50}$$

where $\tau_{i3} > 0$ is a design parameter.

Define the filter error as $\eta_{i3} = q_{id} - \bar{q}_{id}$, then one has

$$\dot{\eta}_{i3} = -\frac{\eta_{i3}}{\tau_{i3}} + M_{i3} \tag{3.51}$$

where $M_{i3} = -\dot{\bar{q}}_{id}$.

Choose the Lyapunov candidate function as

$$L_{i3} = \frac{1}{2}E_{i4}^2 + \frac{1}{2}\eta_{i3}^2 \tag{3.52}$$

Then, taking the time derivative of (3.52) yields

$$\begin{aligned}\dot{L}_{i3} &= E_{i4}(-k_{41}E_{i4} + \eta_{i3} + E_{i5}) + \eta_{i3}\dot{\eta}_{i3} \\ &= -k_{41}E_{i4}^2 + E_{i4}\eta_{i3} + E_{i4}E_{i5} + \eta_{i3}\dot{\eta}_{i3} \\ &\leq -\left(k_{41} - \frac{1}{2w_{i31}^2} - \frac{1}{2w_{i32}^2}\right)E_{i4}^2 + \frac{w_{i32}^2 E_{i5}^2}{2} \\ &\quad - \left(\frac{1}{\tau_{i3}} - \frac{w_{i31}^2}{2} - \frac{1}{2}\right)\eta_{i3}^2 + \frac{M_{i3}^2}{2}\end{aligned} \tag{3.53}$$

where $E_{i5} = q_i - q_{id}$ is the pitch rate tracking error and w_{i31} and w_{i32} are positive parameters.

Step 4 Design the desired control input signal u_{id} and adaptive law as

$$u_{id} = -\left(k_{51} + \frac{\hat{\Phi}_{iq}\varphi_{iq}^T\varphi_{iq}}{2w_{i41}^2} + \frac{1}{2w_{i42}^2} + \frac{1}{2w_{i43}^2}\right)E_{i5} + \dot{q}_{id} + k_{52}\zeta_i \tag{3.54}$$

$$\dot{\hat{\Phi}}_{iq} = g_{i21}\left(\frac{E_{i5}^2\varphi_{iq}^T\varphi_{iq}}{2w_{i41}^2} - w_{i44}\hat{\Phi}_{iq}\right) \tag{3.55}$$

where k_{51}, k_{52}, w_{i41}, w_{i42}, w_{i43}, w_{i44}, and g_{i21} are positive design parameters. ζ_i is the auxiliary signal related to the input saturation. $\hat{\Phi}_{iq}$ is the estimation of Φ_{iq}. ζ_i, Φ_{iq}, and φ_{iq} will be defined later.

In practical engineering applications, the control input often encounters input saturation nonlinearities. Hence, the applied control signal is expressed as

$$u_i = \text{sat}(u_{id}) = \begin{cases} u_{i\max}, & u_{id} \geq u_{i\max} \\ u_{id}, & u_{i\min} < u_{id} < u_{i\max} \\ u_{i\min}, & u_{id} \leq u_{i\min} \end{cases} \tag{3.56}$$

where $i = 1, 2, \ldots, N$. $u_{i\min} = -\delta_{ie0\max}$, $u_{i\max} = -\delta_{ie0\min}$. $\delta_{ie0\max}$ and $\delta_{ie0\min}$ are the maximum and minimum elevator deflection angles, respectively.

To handle the input saturation, an auxiliary dynamic system is designed as [5]

$$\dot{\zeta}_i = \begin{cases} -k_{52}\zeta_i - \frac{|E_{i5}\Delta u_i| + 0.5\Delta u_i^2}{|\zeta_i|^2}\zeta_i + \Delta u_i & |\zeta_i| \geq \mu_i \\ 0 & |\zeta_i| < \mu_i \end{cases} \tag{3.57}$$

where $\Delta u_i = u_i - u_{id}$.

When $|\zeta_i| \geq \mu_i$, one has

$$\zeta_i \dot{\zeta}_i \leq -k_{52}\zeta_i^2 - |E_{i5}\Delta u_i| + \frac{\zeta_i^2}{2} \tag{3.58}$$

When $|\zeta_i| < \mu_i$, one has

$$\zeta_i \dot{\zeta}_i = 0 \tag{3.59}$$

Remark 3.3 Note that in (3.57), $\dot{\zeta}_i = -k_{52}\zeta_i$ if $\Delta u_i = 0$, which means that ζ_i converges into the region $|\zeta_i| < \mu_i$. If the input saturation occurs, i.e., $\Delta u_i \neq 0$, when $|\zeta_i| < \mu_i$, one can reset ζ_i to $|\zeta_i| \geq \mu_i$. Then, the auxiliary system (3.57) can be reactivated until $\Delta u_i = 0$ and $|\zeta_i| < \mu_i$ are regained.

Choose the Lyapunov candidate function as

$$L_{i4} = \frac{1}{2}E_{i5}^2 + \frac{1}{2g_{i21}}\tilde{\Phi}_{iq}^2 + \frac{1}{2}\zeta_i^2 \tag{3.60}$$

where $\tilde{\Phi}_{iq} = \Phi_{iq} - \hat{\Phi}_{iq}$ is the estimation error.

Then, taking the time derivative of (3.60) yields

$$\begin{aligned}
\dot{L}_{i4} &= E_{i5}\dot{E}_{i5} + \frac{1}{g_{i21}}\tilde{\Phi}_{iq}\dot{\tilde{\Phi}}_{iq} + \zeta_i\dot{\zeta}_i \\
&= E_{i5}\left[f_{iq}(V_i, \gamma_i, \theta_i, q_i, u_i) + u_i + d_{i3} - \dot{q}_{id}\right] \\
&\quad + \frac{1}{g_{i21}}\tilde{\Phi}_{iq}\dot{\tilde{\Phi}}_{iq} + \zeta_i\dot{\zeta}_i \\
&= E_{i5}\left[f_{iq}(V_i, \gamma_i, \theta_i, q_i, u_{if}) + \epsilon_{i2} + u_i + d_{i3} - \dot{q}_{id}\right] \\
&\quad + \frac{1}{g_{i21}}\tilde{\Phi}_{iq}\dot{\tilde{\Phi}}_{iq} + \zeta_i\dot{\zeta}_i
\end{aligned} \tag{3.61}$$

The RBFNN is used to approximate $f_{iq}(V_i, \gamma_i, \theta_i, q_i, u_{if})$, which is given by

$$f_{iq}(V_i, \gamma_i, \theta_i, q_i, u_{if}) = \boldsymbol{\phi}_{iq}^{*T}\boldsymbol{\varphi}_{iq}(V_i, \gamma_i, \theta_i, q_i, u_{if}) + \lambda_{i21} \tag{3.62}$$

In the subsequent analysis, $\boldsymbol{\varphi}_{iq}(V_i, \gamma_i, \theta_i, q_i, u_{if}) \rightarrow \boldsymbol{\varphi}_{iq}$ is used. Define $\lambda_{i22} = \epsilon_{i2} + \lambda_{i21}$, and by using Young's inequality, one has

$$E_{i5}\boldsymbol{\phi}_{iq}^{*T}\boldsymbol{\varphi}_{iq} \leq \frac{E_{i5}^2\Phi_{iq}\boldsymbol{\varphi}_{iq}^T\boldsymbol{\varphi}_{iq}}{2w_{i41}^2} + \frac{w_{i41}^2}{2} \tag{3.63}$$

$$E_{i5}\lambda_{i22} \leq \frac{E_{i5}^2}{2w_{i42}^2} + \frac{w_{i42}^2\lambda_{i22m}^2}{2} \tag{3.64}$$

$$E_{i5}d_{i3} \le \frac{E_{i5}^2}{2w_{i43}^2} + \frac{w_{i43}^2 d_{i3m}^2}{2} \tag{3.65}$$

where $\Phi_{iq} = \boldsymbol{\phi}_{iq}^{*T}\boldsymbol{\phi}_{iq}^{*}$, $\boldsymbol{\Phi}_q = [\Phi_{1q}, \Phi_{2q}, \dots \Phi_{Nq}]^T$, $|\lambda_{i22}| \le \lambda_{i22m}$, $|d_{i3}| \le d_{i3m}$.
Then, one has

$$\begin{aligned}
\dot{L}_{i4} \le\ & E_{i5}\left(\frac{E_{i5}\Phi_{iq}\boldsymbol{\varphi}_{iq}^T\boldsymbol{\varphi}_{iq}}{2w_{i41}^2} + \frac{E_{i5}}{2w_{i42}^2} + \frac{E_{i5}}{2w_{i43}^2} + u_i - \dot{q}_{id}\right) \\
& + \frac{1}{g_{i21}}\tilde{\Phi}_{iq}\dot{\hat{\Phi}}_{iq} + \zeta_i\dot{\zeta}_i + \frac{w_{i41}^2}{2} + \frac{w_{i42}^2\lambda_{i22m}^2}{2} + \frac{w_{i43}^2 d_{i3m}^2}{2} \\
\le\ & E_{i5}(-k_{51}E_{i5} + \Delta u_i + k_{52}\zeta_i) + w_{i44}\tilde{\Phi}_{iq}\hat{\Phi}_{iq} + \zeta_i\dot{\zeta}_i \\
& + \frac{w_{i41}^2}{2} + \frac{w_{i42}^2\lambda_{i22m}^2}{2} + \frac{w_{i43}^2 d_{i3m}^2}{2}
\end{aligned} \tag{3.66}$$

When $|\zeta_i| \ge \mu_i$, (3.66) is given by

$$\begin{aligned}
\dot{L}_{i4} \le & -\left(k_{51} - \frac{k_{52}}{2}\right)E_{i5}^2 - \left(\frac{k_{52}}{2} - \frac{1}{2}\right)\zeta_i^2 - \left(w_{i44} - \frac{1}{2}\right)\tilde{\Phi}_{iq}^2 \\
& + \frac{w_{i44}^2\Phi_{iq}^2}{2} + \frac{w_{i41}^2}{2} + \frac{w_{i42}^2\lambda_{i22m}^2}{2} + \frac{w_{i43}^2 d_{i3m}^2}{2}
\end{aligned} \tag{3.67}$$

When $|\zeta_i| < \mu_i$, (3.66) is derived as

$$\begin{aligned}
\dot{L}_{i4} \le & -\left(k_{51} - \frac{1}{2} - \frac{k_{52}}{2}\right)E_{i5}^2 - \left(w_{i44} - \frac{1}{2}\right)\tilde{\Phi}_{iq}^2 \\
& + \frac{k_{52}}{2}\mu_i^2 + \frac{w_{i44}^2}{2}\Phi_{iq}^2 + \frac{\Delta u_i^2}{2} \\
& + \frac{w_{i41}^2}{2} + \frac{w_{i42}^2\lambda_{i22m}^2}{2} + \frac{w_{i43}^2 d_{i3m}^2}{2}
\end{aligned} \tag{3.68}$$

3.3.4 Stability Analysis

Theorem 3.1 *Consider a group of N follower UAVs under actuator faults (3.6) and one leader UAV, and the dynamics of each follower UAV is given by (3.1)–(3.5). If the distributed SMO is designed as (3.20), the error transformation is conducted as (3.25), the control laws are designed as (3.30), (3.38), (3.49), and (3.54), the adaptive laws are designed as (3.39) and (3.55), and the auxiliary system is designed as (3.57), then all signals of the closed-loop system are UUB and all* *follower UAVs can track the leader UAV. Furthermore, the prescribed tracking performance (3.25) can be guaranteed.*

Proof To analyze the stability of the closed-loop system, the Lyapunov candidate function can be chosen as

$$L_i = L_{i1} + L_{i2} + L_{i3} + L_{i4} \tag{3.69}$$

Taking the time derivative of L_i along the trajectories of (3.37), (3.48), and (3.53) and synthesizing (3.67) and (3.68), then one has

$$\begin{aligned}\dot{L}_i \le & -\frac{1}{\bar{V}_i^2}\left(k_{21} - \frac{1}{2w_{i11}^2} - \frac{1}{2w_{i12}^2}\right)E_{i2}^2 + \frac{w_{i12}^2}{2}E_{i3}^2 - \left(\frac{1}{\tau_{i1}} - \frac{1}{2} - \frac{w_{i11}^2}{2}\right)\eta_{i1}^2 \\ & -\left(k_{31} - \frac{1}{2w_{i25}^2} - \frac{1}{2w_{i26}^2}\right)E_{i3}^2 + \frac{w_{i26}^2}{2}E_{i4}^2 \\ & -\left(w_{i24} - \frac{1}{2}\right)\tilde{\Phi}_{i\gamma}^2 - \left(\frac{k_{52}}{2} - \frac{1}{2}\right)\zeta_i^2 \\ & -\left(\frac{1}{\tau_{i2}} - \frac{w_{i25}^2}{2} - \frac{1}{2}\right)\eta_{i2}^2 - \left(k_{41} - \frac{1}{2w_{i31}^2} - \frac{1}{2w_{i32}^2}\right)E_{i4}^2 + \frac{w_{i32}^2 E_{i5}^2}{2} \\ & -\left(\frac{1}{\tau_{i3}} - \frac{w_{i31}^2}{2} - \frac{1}{2}\right)\eta_{i3}^2 - \left(k_{51} - \frac{1}{2} - \frac{k_{52}}{2}\right)E_{i5}^2 - \left(w_{i44} - \frac{1}{2}\right)\tilde{\Phi}_{iq}^2 + \varsigma_i\end{aligned} \tag{3.70}$$

where $\varsigma_i = \frac{M_{i1m}^2}{2} + \frac{w_{i21}^2}{2} + \frac{w_{i22}^2\lambda_{i12m}^2}{2} + \frac{w_{i23}^2 d_{i1m}^2}{2} + \frac{M_{i2m}^2}{2} + \frac{w_{i24}^2}{2}\Phi_{i\gamma}^2 + \frac{M_{i3m}^2}{2} + \frac{k_{52}}{2}\mu_i^2 + \frac{w_{i44}^2}{2}\Phi_{iq}^2 + \frac{\Delta u_i^2}{2} + \frac{w_{i41}^2}{2} + \frac{w_{i42}^2\lambda_{i22m}^2}{2} + \frac{w_{i43}^2 d_{i3m}^2}{2}$, and $|M_{i1}| \le M_{i1m}$, $|M_{i2}| \le M_{i2m}$, $|M_{i3}| \le M_{i3m}$.

The corresponding design parameters k_{21}, k_{31}, k_{41}, k_{51}, w_{i24}, τ_{i1}, τ_{i2}, and τ_{i3} should be chosen such that

$$\begin{cases} k_{21} > \frac{1}{2w_{i11}^2} + \frac{1}{2w_{i12}^2}, k_{31} > \frac{1}{2w_{i25}^2} + \frac{1}{2w_{i26}^2} + \frac{w_{i12}^2}{2}, \\ k_{41} > \frac{1}{2w_{i31}^2} + \frac{1}{2w_{i32}^2} + \frac{w_{i26}^2}{2}, \\ k_{51} > \frac{1}{2} + \frac{k_{52}}{2} + \frac{w_{i32}^2}{2}, w_{i24} > \frac{1}{2}, w_{i44} > \frac{1}{2}, \frac{1}{\tau_{i1}} > \frac{1}{2} + \frac{w_{i11}^2}{2}, \\ \frac{1}{\tau_{i2}} > \frac{1}{2} + \frac{w_{i25}^2}{2}, \frac{1}{\tau_{i3}} > \frac{1}{2} + \frac{w_{i31}^2}{2} \end{cases} \tag{3.71}$$

Define

$$\begin{cases} \Omega_{E_{i2}} = \left\{ |E_{i2}| \le \bar{V}_i \sqrt{\frac{\varsigma_i}{k_{21} - \frac{1}{2w_{i11}^2} - \frac{1}{2w_{i12}^2}}} \right\}, \\ \Omega_{E_{i3}} = \left\{ |E_{i3}| \le \sqrt{\frac{\varsigma_i}{k_{31} - \frac{1}{2w_{i25}^2} - \frac{1}{2w_{i26}^2} - \frac{w_{i12}^2}{2}}} \right\} \\ \Omega_{E_{i4}} = \left\{ |E_{i4}| \le \sqrt{\frac{\varsigma_i}{k_{41} - \frac{1}{2w_{i31}^2} - \frac{1}{2w_{i32}^2} - \frac{w_{i26}^2}{2}}} \right\}, \\ \Omega_{E_{i5}} = \left\{ |E_{i5}| \le \sqrt{\frac{\varsigma_i}{k_{51} - \frac{1}{2} - \frac{k_{52}}{2} - \frac{w_{i32}^2}{2}}} \right\} \\ \Omega_{\tilde{\Phi}_{i\gamma}} = \left\{ |\tilde{\Phi}_{i\gamma}| \le \sqrt{\frac{\varsigma_i}{w_{i24} - \frac{1}{2}}} \right\}, \Omega_{\tilde{\Phi}_{iq}} = \left\{ |\tilde{\Phi}_{iq}| \le \sqrt{\frac{\varsigma_i}{w_{i44} - \frac{1}{2}}} \right\} \\ \Omega_{\eta_{i1}} = \left\{ |\eta_{i1}| \le \sqrt{\frac{\varsigma_i}{\frac{1}{\tau_{i1}} - \frac{1}{2} - \frac{w_{i11}^2}{2}}} \right\}, \Omega_{\eta_{i2}} = \left\{ |\eta_{i2}| \le \sqrt{\frac{\varsigma_i}{\frac{1}{\tau_{i2}} - \frac{1}{2} - \frac{w_{i25}^2}{2}}} \right\} \\ \Omega_{\eta_{i3}} = \left\{ |\eta_{i3}| \le \sqrt{\frac{\varsigma_i}{\frac{1}{\tau_{i3}} - \frac{1}{2} - \frac{w_{i31}^2}{2}}} \right\}, \Omega_{\zeta_i} = \left\{ |\zeta_i| \le \sqrt{\frac{\varsigma_i}{\frac{k_{52}}{2} - \frac{1}{2}}} \right\} \end{cases} \tag{3.72}$$

Then, $\dot{L}_i$ will be negative if $E_{i2} \notin \Omega_{E_{i2}}$, $E_{i3} \notin \Omega_{E_{i3}}$, $E_{i4} \notin \Omega_{E_{i4}}$, $E_{i5} \notin \Omega_{E_{i5}}$, $\tilde{\Phi}_{i\gamma} \notin \Omega_{\tilde{\Phi}_{i\gamma}}$, $\tilde{\Phi}_{iq} \notin \Omega_{\tilde{\Phi}_{iq}}$, $\eta_{i1} \notin \Omega_{\eta_{i1}}$, $\eta_{i2} \notin \Omega_{\eta_{i2}}$, $\eta_{i3} \notin \Omega_{\eta_{i3}}$, and $\zeta_i \notin \Omega_{\zeta_i}$. Therefore, the errors E_{i2}, E_{i3}, E_{i4}, E_{i5}, $\tilde{\Phi}_{i\gamma}$, $\tilde{\Phi}_{iq}$, η_{i1}, η_{i2}, η_{i3}, and ζ_i are bounded. Based on the tracking error transformation (3.26), the error e_{i2} of the ith follower UAV is confined within the prescribed performance bound $[-\underline{k}_{i1}\varepsilon_{i1}, \ \overline{k}_{i1}\varepsilon_{i1}]$ when E_{i2} is bounded. By recalling Lemma 3.2, one can conclude that all follower UAVs can track the leader UAV and the tracking errors $e_{si} + e_{i2}$ are bounded. This ends the proof. □

Remark 3.4 From (3.71) and (3.72), it can be seen that the errors E_{i2}, E_{i3}, E_{i4}, E_{i5}, $\tilde{\Phi}_{i\gamma}$, $\tilde{\Phi}_{iq}$, η_{i1}, η_{i2}, and η_{i3} can be made very small by increasing k_{21}, k_{31}, k_{41}, k_{51}, w_{i24}, and w_{44} and decreasing τ_{i1}, τ_{i2}, τ_{i3}, and ς_i. Based on this recipe for parameter selection, the control parameters are chosen by trial and error until a good performance is obtained.

3.4 Simulation Results

In this section, a group of three follower UAVs and one leader UAV is considered, and simulations are conducted to verify the effectiveness of the proposed distributed FTCC scheme. The aerodynamic coefficients and model parameters are shown in Table 3.1 and the input saturation limits are presented in Table 3.2. The topology is illustrated in Fig. 3.1 and the weighted adjacency matrices $\mathcal{A}$ and $\mathcal{B}$ are chosen as

$$\mathcal{A} = \begin{bmatrix} 0 & 0.3 & 0.5 \\ 0.3 & 0 & 0.4 \\ 0.5 & 0.4 & 0 \end{bmatrix}, \ \mathcal{B} = \text{diag}(0.7, 0, 0) \tag{3.73}$$

In the simulation, the initial conditions are set as $V_i(0) = 30\,\text{m/s}$, $z_i(0) = 1000\,\text{m}$, $\gamma_i(0) = 0$ rad, $\theta_i(0) = 0.0322\,\text{rad}$, and $q_i(0) = 0\,\text{rad/s}$, $i = 1, 2, 3$. The design parameters for the control scheme are given in Table 3.3. The performance function $\varepsilon_{i1}(t)$ is chosen as $\varepsilon_{i1}(t) = 1.55e^{-0.3t} + 0.45$. The initial values of the distributed SMO (3.20) are set as $z_d(0) = [z_{1d}(0), z_{2d}(0), z_{3d}(0)] =$

Table 3.1 Parameters and coefficients of the ith follower UAV

Coefficient	Value	Unit	Coefficient	Value	Unit
s_i	1.463	m^2	$C_{iD_{\alpha^2}}$	1.0778	–
c_i	0.451	m	C_{iL_0}	0.2153	–
m_i	25	kg	C_{iL_α}	4.6333	rad^{-1}
g	9.8	m.s^{-2}	$C_{iL_{\delta_e}}$	0.2865	rad^{-1}
ρ	1.205	kg.m^{-3}	C_{im_0}	−0.0954	–
I_{iy}	6.765	kg.m^2	C_{im_α}	−2.8206	rad^{-1}
C_{iD_0}	0.0225	–	$C_{im_{\delta_e}}$	−0.9455	rad^{-1}
C_{iD_α}	0.1002	rad^{-1}	C_{im_q}	−13.8189	$\text{rad}^{-1}\cdot\text{s}$

Table 3.2 Input saturation limits of the ith follower UAV

Lower limit	Upper limit
$T_{i\min} = 0\,\text{N}$	$T_{i\max} = 100\,\text{N}$
$\delta_{ie0\min} = -0.87\,\text{rad}$	$\delta_{ie0\max} = 0.87\,\text{rad}$

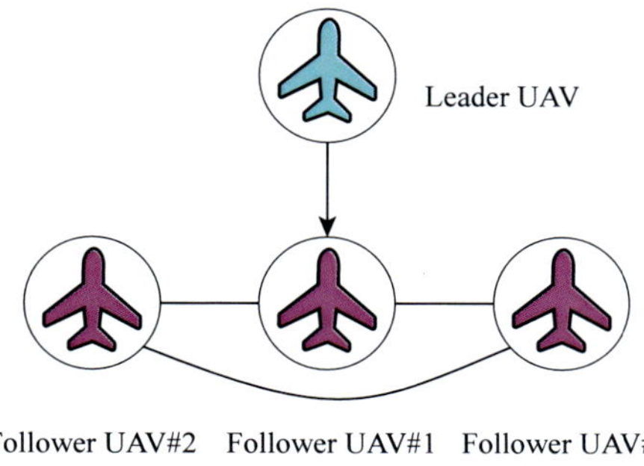

Fig. 3.1 Communication topology

Table 3.3 Design parameters

Parameter	Value	Parameter	Value	Parameter	Value
k_{11}	186	k_{12}	45	k_{13}	0.8
β_1	8	β_2	4	$\underline{k}_{i1}$	1
$\overline{k}_{i1}$	1	k_{21}	1.5	τ_{i1}	0.05
k_{31}	2.72	w_{i21}	1.06	w_{i22}	2.5
w_{i23}	4	g_{i11}	18	w_{i24}	0.06
τ_{i2}	0.05	k_{41}	3.15	τ_{i3}	0.05
k_{51}	15	w_{i41}	2	w_{i42}	2.71
w_{i43}	4.06	g_{i21}	25	w_{i44}	0.02
k_{52}	3	μ_i	0.01		

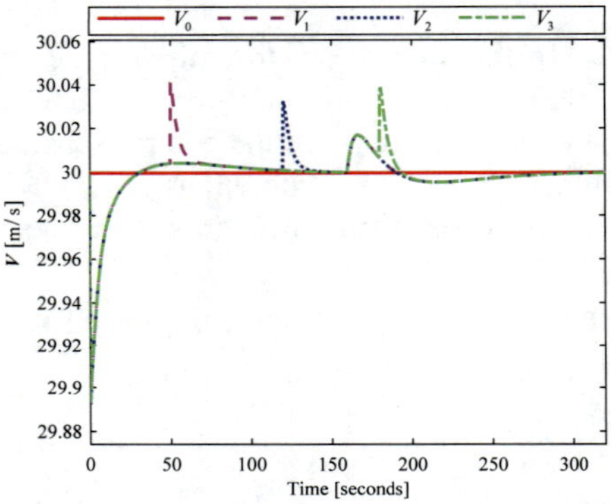

Fig. 3.2 Velocities of UAV#1–UAV#3

[1000, 1000, 1000] m. The initial values of the adaptive laws (3.39) and (3.55) are set as $\hat{\Phi}_{i\gamma}(0) = 0$ and $\hat{\Phi}_{iq}(0) = 0$. To verify the effectiveness of the proposed control scheme, the external disturbances and actuator faults encountered by three follower UAVs are as follows: UAV#1 suffers from the faults $\rho_1 = 0.7$ and $u_{1f} = 0.3$ at $t = 30$ s and the disturbances $d_{11} = 0.2$, $d_{12} = 0.1$, and $d_{13} = 0.1$ at $t = 50$ s; UAV#2 suffers from the faults $\rho_2 = 0.75$ and $u_{2f} = 0.3$ and disturbances $d_{21} = 0.2$, $d_{22} = 0.2$, and $d_{23} = 0.2$ at $t = 120$ s; and UAV#3 suffers from the disturbances $d_{31} = 0.2$, $d_{32} = 0.3$, and $d_{33} = 0.25$ at $t = 180$ s and the faults $\rho_3 = 0.65$ and $u_{3f} = 0.2$ at $t = 250$ s. The external disturbance signals are introduced into the system through a first-order filter $1/(0.1s + 1)$. These actuator faults are chosen based on the fault model (3.6) and the external disturbance signals are chosen to verify the disturbance attenuation performance. Since the sign function is used in (3.20), the chattering phenomenon may be induced due to the discontinuous design. To reduce the chattering, a smooth hyperbolic tangent function $\tanh(\cdot/\epsilon)$ is utilized to replace $\text{sign}(\cdot)$, where $\epsilon = 0.1$. In the simulation, both the engine dynamics and the elevator actuator dynamics are assumed to be a first-order system $20/(s + 20)$. The altitude trajectory of the leader UAV is generated by filtering a step signal, which changes from 1000 m to 1100 m at $t = 0$ s and then changes from 1100 m to 1000 m at $t = 160$ s. The filter is chosen as $0.04^2/(s^2 + 0.072s + 0.04^2)$.

The velocities of UAV#1-UAV#3 are shown in Fig. 3.2 and the velocity tracking errors are depicted in Fig. 3.3. It can be observed from Figs. 3.2 and 3.3 that the

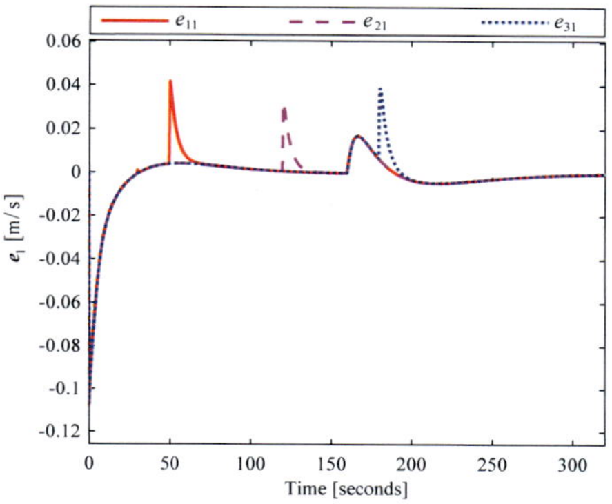

Fig. 3.3 Velocity tracking errors of UAV#1–UAV#3

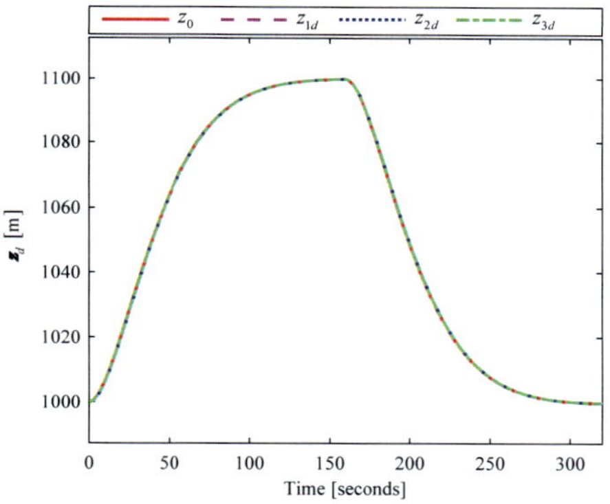

Fig. 3.4 Estimated altitude references

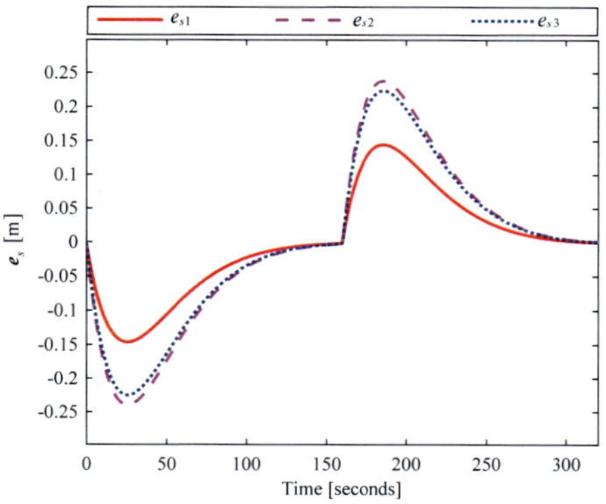

Fig. 3.5 Altitude reference estimation errors of UAV#1–UAV#3

velocities of all follower UAVs can track the velocity reference very well and the tracking errors are bounded. As can be seen from Figs. 3.4 and 3.5, the altitude references estimated by follower UAVs converge to the altitude trajectory of the leader UAV and the estimation errors under the developed SMO are very small. Figure 3.6 shows the altitude responses of follower UAV#1–UAV#3, which are bounded under the supervision of the proposed control scheme. Figure 3.7 illustrates the altitude tracking errors, which are strictly confined within the prescribed bounds.

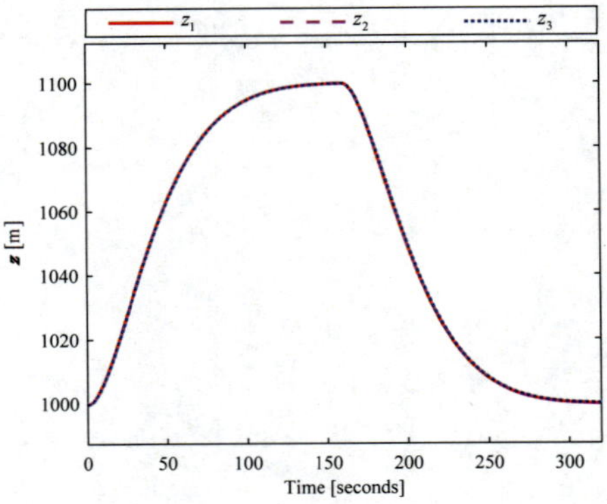

Fig. 3.6 Altitudes of follower UAV#1–UAV#3

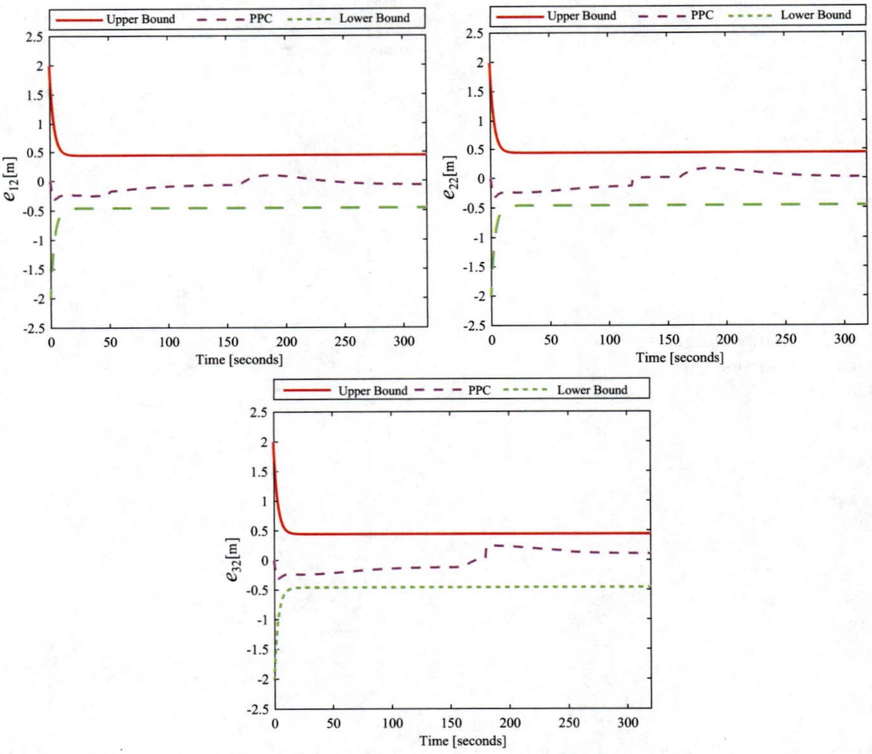

Fig. 3.7 Altitude tracking errors of UAV#1–UAV#3

Figures 3.8 and 3.9 show the states of UAV#1–UAV#3 and it is observed that these signals are UUB. From Fig. 3.10, it can be found that under the proposed distributed neural adaptive control scheme with prescribed performance bounds,

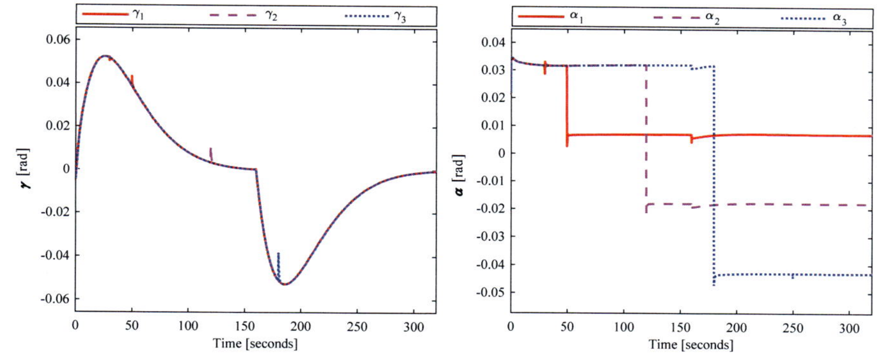

Fig. 3.8 Flight path angles and angles of attack of UAV#1–UAV#3

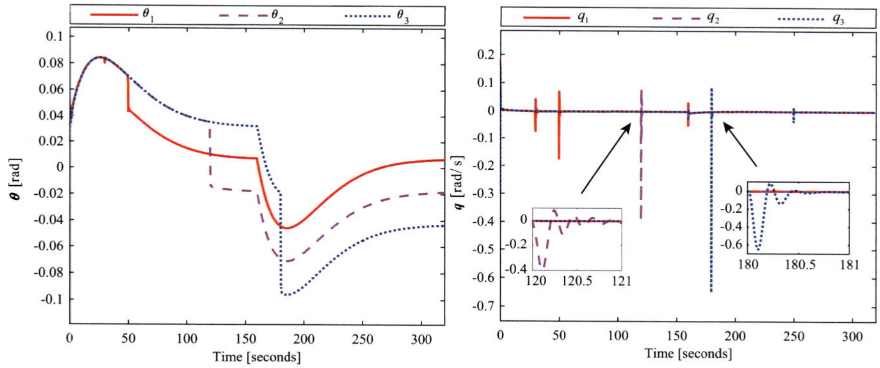

Fig. 3.9 Pitch angles and pitch rates of UAV#1–UAV#3

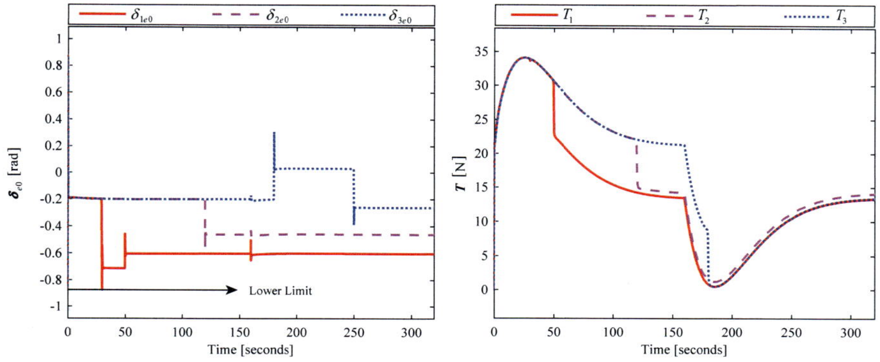

Fig. 3.10 Control inputs of UAV#1–UAV#3

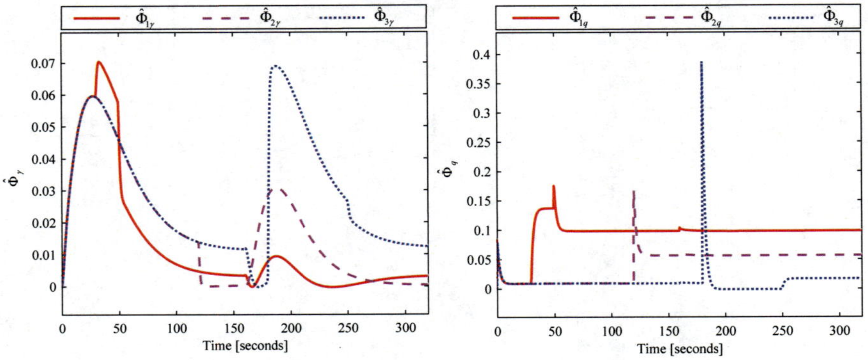

Fig. 3.11 Estimated $\hat{\Phi}_\gamma$ and $\hat{\Phi}_q$

the thrust inputs and elevator deflections react to the external disturbances and internal faults in a timely manner. When the actuator fault is encountered by UAV#1 at $t = 30$ s, the elevator reaches its lower limit. Then, the auxiliary dynamic system (3.57) starts working to pull the saturated signal back to the unsaturated region. The estimated values $\hat{\Phi}_\gamma$ and $\hat{\Phi}_q$ are shown in Fig. 3.11, which are bounded under the adaptive laws (3.39) and (3.55).

3.5 Conclusion

This chapter presents a distributed neural adaptive FTCC scheme for the longitudinal dynamics of multi-UAVs in the presence of external disturbances, actuator faults, and input saturation. With the help of distributed SMO, the altitude trajectory of leader UAV is first estimated by each follower UAV. Then, based on the PPC method, the tracking errors of follower UAVs with respect to the estimated altitude references are transformed into a new set of errors. Finally, with the transformed error variables, the PPC scheme has been designed. The transient and steady performances of the tracking errors with respect to the estimated references have been guaranteed. In the proposed FTCC scheme, the unknown nonlinear functions are approximated by the RBFNN and only one parameter needs to be updated at each step. Furthermore, the adverse effects of input saturation have been attenuated by introducing an auxiliary dynamic system. Simulations have revealed that the tracking errors with respect to the estimated references are confined in the prescribed bounds and all follower UAVs can track the trajectory from the leader.

References

1. C.P. Bechlioulis, G.A. Rovithakis, A low-complexity global approximation-free control scheme with prescribed performance for unknown pure feedback systems. Automatica **50**(4), 1217–1226 (2014)
2. X.W. Bu, Guaranteeing prescribed output tracking performance for air-breathing hypersonic vehicles via non-affine back-stepping control design. Nonlinear Dyn. **91**(1), 525–538 (2018)
3. X.W. Bu, Y. Xiao, K. Wang, A prescribed performance control approach guaranteeing small overshoot for air-breathing hypersonic vehicles via neural approximation. Aerosp. Sci. Technol. **71**, 485–498 (2017)
4. X.W. Bu, G.J. He, K. Wang, Tracking control of air-breathing hypersonic vehicles with non-affine dynamics via improved neural back-stepping design. ISA Trans. **75**, 88–100 (2018)
5. M. Chen, S.S. Ge, B.B. Ren, Adaptive tracking control of uncertain MIMO nonlinear systems with input constraints. Automatica **47**(3), 452–465 (2011)
6. T.T. Gao, J.S. Huang, Y. Zhou, Q.G. Wang, Finite-time consensus control of second-order nonlinear systems with input saturation. Trans. I. Meas. Control **38**(11), 1381–1391 (2016)
7. T. Han, M. Chi, Z.H. Guan, B. Hu, J.W. Xiao, Y.H. Huang, Distributed three-dimensional formation containment control of multiple unmanned aerial vehicle systems. Asian J. Control **19**(3), 1103–1113 (2017)
8. Y.M. Li, S.C. Tong, L. Liu, G. Feng, Adaptive output-feedback control design with prescribed performance for switched nonlinear systems. Automatica **80**, 225–231 (2017)
9. C.J. Liu, W.H. Chen, Disturbance rejection flight control for small fixed-wing unmanned aerial vehicles. J. Guid. Control Dyn. **39**(12), 2810–2819 (2016)
10. Q. Luo, H.B. Duan, Distributed UAV flocking control based on homing pigeon hierarchical strategies. Aerosp. Sci. Technol. **70**, 257–264 (2017)
11. H.X. Qiu, H.B. Duan, Multiple UAV distributed close formation control based on in-flight leadership hierarchies of pigeon flocks. Aerosp. Sci. Technol. **70**, 471–486 (2017)
12. G. Sánchez-Ayala, V. Centeno, J. Thorp, Gain scheduling with classification trees for robust centralized control of PSSs. IEEE Trans. Power Syst. **31**(3), 1933–1942 (2016)
13. R.M. Sanner, J.J. Slotine, Gaussian networks for direct adaptive control. IEEE Trans. Neural Netw. **3**(6), 837–863 (1992)
14. Q. Shen, B. Jiang, P. Shi, J. Zhao, Cooperative adaptive fuzzy tracking control for networked unknown nonlinear multiagent systems with time-varying actuator faults. IEEE Trans. Fuzzy Syst. **22**(3), 494–504 (2014)
15. Z.Q. Shi, C. Zhou, J. Guo, Neural network observer based consensus control of unknown nonlinear multi-agent systems with prescribed performance and input quantization. Int. J. Control Autom. **19**(5), 1944–1952 (2021)
16. W. Wang, C.Y. Wen, Adaptive actuator failure compensation control of uncertain nonlinear systems with guaranteed transient performance. Automatica **46**(12), 2082–2091 (2010)
17. X.Y. Wang, S.H. Li, P. Shi, Distributed finite-time containment control for double-integrator multiagent systems. IEEE T. Cybern. **44**(9), 1518–1528 (2014)
18. Z.H. Wu, J.C. Lu, J.P. Shi, Y. Liu, Q. Zhou, Robust adaptive neural control of morphing aircraft with prescribed performance. Math. Probl. Eng. **2017**, 1401427 (2017)
19. B. Xiao, Q.C. Dong, D. Ye, L. Liu, X. Huo, A general tracking control framework for uncertain systems with exponential convergence performance. IEEE-ASME Trans. Mechatron. **23**(1), 111–120 (2018)
20. Z.J. Yang, Distributed prescribed performance control for consensus output tracking of nonlinear semi-strict feedback systems using finite-time disturbance observers. Int. J. Syst. Sci. **50**(5), 989–1005 (2019)
21. S.J. Yoo, Neural-network-based decentralized fault-tolerant control for a class of nonlinear large-scale systems with unknown time-delayed interaction faults. J. Frankl. Inst. **351**(3), 1615–1629 (2014)

22. S.J. Yoo, A low-complexity design for distributed containment control of networked pure-feedback systems and its application to fault-tolerant control. Int. J. Robust Nonlinear Control **27**(3), 363–379 (2017)
23. X. Yu, P. Li, Y.M. Zhang, The design of fixed-time observer and finite-time fault-tolerant control for hypersonic gliding vehicles. IEEE Trans. Ind. Electron. **65**(5), 4135–4144 (2018)
24. A.M. Zou, K.D. Kumar, Robust attitude coordination control for spacecraft formation flying under actuator failures. J. Guid. Control Dyn. **35**(4), 1247–1255 (2012)
25. A.M. Zou, Z.G. Hou, M. Tan, Adaptive control of a class of nonlinear pure-feedback systems using fuzzy backstepping approach. IEEE Trans. Fuzzy Syst. **16**(4), 886–897 (2008)

Chapter 4
Distributed FTCC of Multi-UAVs Under Actuator Fault and Input Saturation

4.1 Introduction

Research on the formation flight of UAVs has attracted tremendous attention in control community [11, 26, 43]. As mentioned in [15, 37, 40] and analyzed in Chap. 1, multi-UAVs can provide high efficiency in many fields, such as forest fire monitoring, detection and fighting [37], and surveillance [22, 24]. As a kind of cooperative control strategies, distributed control schemes have been investigated for significantly reducing the communication cost and enhancing the robustness against communication failures in multi-UAVs [1, 18]. In [19], a distributed cooperative controller was designed for multi-UAVs by using differential game theory. With the aid of the state observer and the virtual structure, the output-feedback formation control problem of multi-UAVs without linear and angular velocity measurements was investigated in [12]. With respect to the distributed control of multi-UAVs, the researches are mainly conducted on multi-UAVs with first-order or second-order linear/nonlinear dynamics, and the results on distributed control of multi-UAVs with high-order nonlinear dynamics are scarce.

In FCSs, failure to counteract in-flight faults may degrade flight performance or even cause catastrophic consequences, and malfunctions in system components of each UAV not only have adverse impacts on the safety of the individual UAV, but also induce the instability of the overall networked UAVs. Therefore, fault-tolerant capability is significant for the safe formation flight of multi-UAVs [38]. To react to faults with prompt actions, many effective FTC methods have been proposed to guarantee the formation flight system stability [4, 9, 16, 20, 30, 35, 36, 38]. In [44], a distributed attitude coordination control scheme was proposed for the formation flight of spacecrafts in the presence of external disturbances. This result was then promoted to [45], where a robust adaptive fault-tolerant attitude coordination control scheme was proposed for the formation flight of spacecrafts against actuator faults. The time-varying attitude reference for each spacecraft was constructed with neighboring spacecrafts' information. Recently, in [41], by using neighboring

Z. Yu et al., *Fault-Tolerant Cooperative Control of Unmanned Aerial Vehicles*,
https://doi.org/10.1007/978-981-99-7661-4_4

synchronization, a finite-time control algorithm was designed by using adaptive mechanism, and all follower spacecrafts can track the reference generated from the leader in finite time. However, the information exchange via the distributed communication network and the high-order strongly nonlinear dynamics of UAV make the distributed FTCC design complex, and the aforementioned results cannot be directly extended to the distributed FTCC of multi-UAVs. Therefore, more new and effective distributed FTCC schemes should be further developed.

In practical engineering applications, input saturation and external disturbances can also cause system instability and performance degradation [2, 7, 39, 42]. In [5], a piecewise auxiliary system was introduced to deal with the non-symmetric input constraints for a class of uncertain multi-input and multi-output (MIMO) nonlinear systems. Another auxiliary signal was constructed in [32] to handle the input saturation based on the error between the desired control input and the saturated control input. It should be emphasized that although numerous efforts have been made in the aforementioned works, very few literatures have addressed the input saturation and external disturbances encountered by multi-UAVs in a distributed communication network. However, input saturation and external disturbances are inevitable in practical formation flight of multi-UAVs, which may cause instability to multi-UAVs once such issues are not handled in a timely manner.

In this chapter, a distributed FTCC scheme is designed for a group of UAVs by simultaneously considering the distributed communication network, actuator faults, input saturation, and external disturbances. The DO technique is utilized to compensate for the lumped uncertainties caused by the actuator faults and external disturbances. Auxiliary dynamic systems are introduced as compensatory terms to reduce the influence caused by the input saturation. Furthermore, in order to attenuate the computational complexity in the traditional backstepping design procedure, the DSC technique with the first-order filter is incorporated to facilitate the controller design. The main contributions of this chapter are as follows:

(1) This chapter proposes an FTCC scheme for multi-UAVs to enhance the flight safety for the overall formation team. In [35], a leader-following FTCC scheme was presented for multi-UAVs. In such a scheme, all follower UAVs only communicate with the leader UAV, and there is no information exchange among follower UAVs, which may lead to instability if a portion of follower UAVs loses the communication with the leader UAV. Moreover, collisions among UAVs may be caused if the data links between the faulty UAVs and the leader UAV are broken in the formation team. To increase the flight safety and reduce the communication cost, this chapter considers the FTCC design problem in a distributed communication network, which is more complex than the FTC design for a single UAV or the FTCC design for multi-UAVs with the leader-following architecture.

(2) Compared with the cooperative control schemes for multi-UAVs, such as [12, 14, 19, 21, 34], which mainly handled the distributed formation issues, this chapter presents a solution for the integrated design to simultaneously address the actuator faults, input saturation, and external disturbances.

(3) In contrast to the existing works on the longitudinal motion of UAV [27, 33], which gave the flight path angle command directly for the altitude dynamics, and the Lyapunov stability analyses are solely conducted for the pitch rate, pitch angle, and flight path angle dynamics. In this chapter, such a restriction is removed by introducing the velocity maximum in the Lyapunov stability analyses. With this method, the stability analyses including pitch rate, pitch angle, flight path angle, and altitude dynamics are carried out as a whole. Furthermore, by integrating nearest neighbor rule to generate reference, DO technique for lumped uncertainties estimations and auxiliary dynamic systems for the compensation of input saturation, a distributed FTCC scheme is proposed to achieve the safe formation flight.

The rest of this chapter is organized as follows. Section 4.2 describes the UAV longitudinal dynamics, actuator fault model, basic graph theory, and control objective. Then, the main results of this chapter are given in Sect. 4.3, which include the controller design and stability analyses. Numerical simulations are performed in Sect. 4.4 to illustrate the effectiveness of the proposed distributed FTCC. Finally, concluding remarks are drawn in Sect. 4.5.

(Remark: The main control schemes and contents of this chapter are from the published journal paper "Z. Q. Yu, Y. H. Qu, and Y. M. Zhang. Distributed fault-tolerant cooperative control for multi-UAVs under actuator fault and input saturation, IEEE Transactions on Control Systems Technology, 2019, 27(6): 2417–2429." The authors appreciate the permission from the Institute of Electrical and Electronics Engineers to reuse the results published in the relevant Journal.)

4.2 Preliminaries and Problem Formulation

4.2.1 UAV Longitudinal Dynamics

Consider a group of N follower UAVs and one leader UAV, by adopting the similar longitudinal UAV model as (3.1)–(3.5) in Chap. 3, the velocity subsystem of the ith follower UAV can be derived as

$$\dot{V}_i = f_{iv} + g_{iv}T_i + \Delta_{i1} \tag{4.1}$$

where $f_{iv} = -\bar{q}_i s_i(C_{iD0} + C_{iD\alpha}\alpha_i + C_{iD\alpha^2}\alpha_i^2)/m_i - g\sin\gamma_i$, $g_{iv} = \cos\alpha_i/m_i$, $\Delta_{i1} = d_{i1}$.

In the model transformation, the thrust term $T_i \sin\alpha_i$ can be ignored because it is generally much smaller than the lift force L_i [31]. Therefore, the attitude subsystem of the ith follower UAV can be written as

$$\dot{z}_i = V_i\gamma_i \tag{4.2}$$

$$\dot{\gamma}_i = f_{i\gamma} + g_{i\gamma}\theta_i + \Delta_{i2} \tag{4.3}$$

$$\dot{\theta}_i = q_i \tag{4.4}$$

$$\dot{q}_i = f_{iq} + g_{iq}\delta_{ie} + \Delta_{i3} \tag{4.5}$$

where

$$f_{i\gamma} = \frac{-m_i g \cos\gamma_i + \bar{q}_i s_i C_{iL0} - \bar{q}_i s_i C_{iL\alpha}\gamma_i}{m_i V_i}$$

$$g_{i\gamma} = \bar{q}_i s_i C_{iL\alpha}/(m_i V_i),\ \Delta_{i2} = d_{i2}$$

$$f_{iq} = \frac{\bar{q}_i s_i c_i (C_{im0} + C_{im\alpha}\alpha_i + C_{imq}\frac{c_i}{2V_i}q_i)}{I_{iy}}$$

$$g_{iq} = \frac{\bar{q}_i s_i c_i C_{im\delta_e}}{I_{iy}},\ \Delta_{i3} = d_{i3}$$

where the small angle approximation $\sin\gamma_i \approx \gamma_i$ has been applied to obtain (4.2) [8].

In order to reduce the influence caused by the input saturation in the controller design, the saturation function is first defined as

$$\mathrm{sat}(\overline{\ell}) = \begin{cases} \ell_{\max}, & \overline{\ell} \geq \ell_{\max} \\ \overline{\ell}, & \ell_{\min} < \overline{\ell} < \ell_{\max} \\ \ell_{\min}, & \overline{\ell} \leq \ell_{\min} \end{cases} \tag{4.6}$$

where $\ell_{\min}$ and $\ell_{\max}$ are the minimum and maximum allowed values, respectively.

Assumption 4.1 The system (3.1)–(3.5) with input saturation (4.6) is input-to-state stable.

Remark 4.1 Assumption 4.1 is reasonable because a feasible control that can stabilize an unstable plant with saturated actuator is not available [10, 28].

4.2.2 *Actuator Fault Model*

For practical engineering control systems, actuator faults may cause system performance degradation or even system instability. This situation is becoming more worse in formation flight since the information exchange among multi-UAVs can send the instable state information of the faulty UAVs to its neighboring UAVs in the distributed communication network. The collisions in the group of UAVs may be caused if the faulty UAVs cannot be re-stabilized in a timely manner in the post-fault stage. Therefore, actuator faults should be further investigated to enhance the flight safety of multi-UAVs. In this chapter, elevator faults including loss-of-effectiveness and bias faults are further considered, and the corresponding fault model can be described as [3]

$$\delta_{ie} = \rho_{fi}\delta_{ie0} + u_{fi} \tag{4.7}$$

where δ_{ie0} represents the desired control signal commanded by the controller. δ_{ie} is the actual control signal generated by the elevator. ρ_{fi} is the unknown actuator effectiveness factor and $0 < \rho_{fi} \leq 1$. u_{fi} is the bounded bias fault.

Remark 4.2 Consider the loss-of-effectiveness and bias faults, the closed-loop system performance can be kept at an acceptable level with faulty actuators by adjusting the control signals with advanced FTC methodologies. Actually, the leakage of hydraulic fluid can lead to the degradation of the actuator effectiveness. Moreover, the sensor fault in the actuator system can cause the bias fault. Specifically, if the amplitude sensor in the actuator system is subjected to a bias fault, the measured amplitude (actual amplitude plus bias value) will be forced to be equal to the reference signal. In such circumstances, the actual value of the actuator amplitude is different from the reference signal [17].

By considering the actuator fault model (4.7), (4.5) can be written as

$$\begin{aligned}\dot{q}_i &= f_{iq} + g_{iq}(\rho_{fi}\delta_{ie0} + u_{fi}) + \Delta_{i3} \\ &= f_{iq} + g_{iq}\delta_{ie0} + \Delta_i\end{aligned} \tag{4.8}$$

where $\Delta_i = g_{iq}(\rho_{fi} - 1)\delta_{ie0} + g_{iq}u_{fi} + \Delta_{i3}$ is the lumped uncertainty.

4.2.3 Basic Graph Theory

The topology of the information flows among multi-UAVs is represented by an undirected network. Let $\mathcal{G}(\mathcal{V}, \mathcal{E}, \mathcal{A})$ be the graph with a set of N nodes, denoted by $\mathcal{V} = \{r_i\}$, $i = 1, 2, \ldots, N$, a set of edges $\mathcal{E} \subseteq \mathcal{V} \times \mathcal{V}$, and an adjacency matrix $\mathcal{A} = [a_{ij}] \in R^{N\times N}$. r_i represents the ith follower UAV, and the edge in $\mathcal{G}$ is denoted by an unordered pair (r_i, r_j). $(r_i, r_j) \in \mathcal{E}$ if and only if there is an information exchange between the ith UAV and the jth UAV, i.e., $(r_i, r_j) \in \mathcal{E} \Longleftrightarrow (r_j, r_i) \in \mathcal{E} \Longleftrightarrow a_{ij} > 0$. The set of neighbors of the node r_i is denoted as $N_i = \{r_j | (r_i, r_j) \in \mathcal{E}, i \neq j\}$. Self-edges are not allowed, which means $a_{ii} = 0$. $\mathcal{A}$ is a time-invariant, symmetric matrix due to the undirected, fixed topology. A path from node r_{i_1} to node r_{i_l} is a sequence of edges (r_{i_1}, r_{i_2}), (r_{i_2}, r_{i_3}), ..., $(r_{i_{l-1}}, r_{i_l})$, where $(r_{i_{k-1}}, r_{i_k}) \in \mathcal{E}$, $k \in \{2, 3, \ldots, l\}$. An undirected graph is connected if there is a path between every two nodes.

The degree matrix is defined by $\mathcal{D} = \text{diag}\{d_1, d_2, \ldots, d_N\}$, where $d_i = \sum_{j=1}^{N} a_{ij}$, $j \in N_i$. Then, the Laplacian matrix $\mathcal{L}$ of the fixed topology is given by

$$\mathcal{L} = \mathcal{D} - \mathcal{A} \tag{4.9}$$

Next, another graph $\bar{\mathcal{G}}$ is considered, which consists of a leader UAV r_0 and N follower UAVs. Denote the leader adjacency matrix by $\mathcal{B} = \text{diag}\{b_1, b_2, \ldots, b_N\}$, and $b_i > 0$ if the ith follower UAV has access to the leader UAV, otherwise, $b_i = 0$. It should be noted that for the fixed topology, $\mathcal{A}$, $\mathcal{D}$, $\mathcal{L}$, and $\mathcal{B}$ are all constant matrices.

Assumption 4.2 The communication topology $\bar{\mathcal{G}}$ is connected, and there exists at least one path from the leader to the follower.

Theorem 4.1 *Under Assumption 4.2, the matrix $\mathcal{L} + \mathcal{B}$ is symmetric and positive definite [13].*

Assumption 4.3 The velocity and altitude references V_0, z_0 from the leader UAV and their derivatives are smooth bounded functions.

Assumption 4.4 The derivatives of disturbances Δ_{i1}, Δ_{i2}, Δ_i are bounded such that $|\dot{\Delta}_{i1}| \leq \bar{\Delta}_{i1}$, $|\dot{\Delta}_{i2}| \leq \bar{\Delta}_{i2}$, $|\dot{\Delta}_i| \leq \bar{\Delta}_i$.

Assumption 4.5 For the ith follower UAV, the control gain functions g_{iv}, $g_{i\gamma}$, g_{iq} are bounded and have nonzero values, i.e., $0 < |g_{iv}| \leq \bar{g}_{iv}$, $0 < |g_{i\gamma}| \leq \bar{g}_{i\gamma}$, $0 < |g_{iq}| \leq \bar{g}_{iq}$.

Remark 4.3 It should be noted that it is difficult and challenging to solve the distributed FTCC problem for multi-UAVs by using the traditional centralized FTC scheme. This is mainly due to the fact that not all follower UAVs can obtain the exact knowledge of the leader UAV. Moreover, the information exchange via the distributed communication network may send the instable states of the faulty UAVs to other healthy UAVs, causing instabilities of healthy UAVs.

4.2.4 Control Objective

The control objective is to design a set of distributed fault-tolerant tracking control laws T_i, δ_{ie0} for UAVs (4.1)–(4.5) based on the local information of neighboring UAVs. Hence, under the supervision of the proposed distributed FTCC scheme, the outputs of all follower UAVs can asymptotically converge to the references V_0, z_0 from the leader UAV even in the presence of actuator faults, external disturbances, and input saturation. Furthermore, the tracking errors are UUB.

4.3 Main Results

In this section, a distributed FTCC scheme is designed for all follower UAVs in the communication network based on the neighboring UAVs' information. First, the references for all follower UAVs are constructed based on the local information from the neighboring UAVs. Second, based on the knowledge of constructed references, the DO technique is utilized to estimate the lumped uncertainties including internal actuator faults and external disturbances. Then, by using the DSC architecture, a virtual control signal is designed at each step with the estimated lumped uncertainties. Furthermore, to reduce the influence of the prolonged input saturation, auxiliary dynamic signals are employed as compensation terms to pull back the control signals from saturated region to unsaturated region.

Inspired by the work in [29], the following Lemma is first proposed.

Lemma 4.1 *Suppose the graph $\bar{\mathcal{G}}$ involving N follower UAVs and one leader UAV is connected and there exists at least one path from the leader to the follower (Assumption 4.2). Design the reference signals V_{id}, z_{id} for the ith follower UAV as*

$$V_{id} = \frac{\sum\limits_{j \in N_i} a_{ij} V_j + b_i V_0}{\sum\limits_{j \in N_i} a_{ij} + b_i} \tag{4.10}$$

$$z_{id} = \frac{\sum\limits_{j \in N_i} a_{ij} z_j + b_i z_0}{\sum\limits_{j \in N_i} a_{ij} + b_i} \tag{4.11}$$

and then the velocity and altitude of the ith follower UAV V_i, z_i can track the velocity and altitude of the leader UAV V_0, z_0 with bounded errors if velocity and altitude of the ith follower UAV V_i, z_i can track the generated velocity and altitude reference signals V_{id}, z_{id} with bounded errors, i.e., the tracking errors $e_{i1} = V_i - V_0$ and $e_{i2} = z_i - z_0$ with respect to the leader UAV will be UUB if the tracking errors $S_{i1} = V_i - V_{id}$ and $S_{i2} = z_i - z_{id}$ with respect to V_{id}, z_{id} are UUB.

Proof With (4.10), it can be derived that

$$\begin{aligned}
S_{i1} &= V_i - V_{id} \\
&= V_i - \frac{\sum_{j=1}^{N} a_{ij} V_j + b_i V_0}{\sum_{j=1}^{N} a_{ij} + b_i} \\
&= \frac{\sum_{j=1}^{N} a_{ij}(V_i - V_j) + b_i (V_i - V_0)}{\sum_{j=1}^{N} a_{ij} + b_i} \\
&= \frac{\sum_{j=1}^{N} l_{ij} e_{j1} + b_i e_{i1}}{\sum_{j=1}^{N} a_{ij} + b_i}
\end{aligned} \tag{4.12}$$

where l_{ij} represents the element of Laplacian matrix $\mathcal{L}$.

From Theorem 4.1, it can be concluded that $\sum_{j\in N_i} a_{ij} + b_i$ is not zero. Then, multiplying both sides of (4.12) with $\sum_{j=1}^{N} a_{ij} + b_i$ yields

$$\left(\sum_{j=1}^{N} a_{ij} + b_i\right) S_{i1} = \sum_{j=1}^{N} l_{ij} e_{j1} + b_i e_{i1} \tag{4.13}$$

Define $\boldsymbol{V} = [V_1, V_2, \ldots, V_N]^T$, $\boldsymbol{V_d} = [V_{1d}, V_{2d}, \ldots, V_{Nd}]^T$, $\boldsymbol{z} = [z_1, z_2, \ldots, z_N]^T$, $\boldsymbol{z_d} = [z_{1d}, z_{2d}, \ldots, z_{Nd}]^T$, $\boldsymbol{S}_1 = [S_{11}, S_{21}, \ldots, S_{N1}]^T$, $\boldsymbol{S}_2 = [S_{12}, S_{22}, \ldots, S_{N2}]^T$, $\boldsymbol{e}_1 = [e_{11}, e_{21}, \ldots, e_{N1}]^T$, $\boldsymbol{e}_2 = [e_{12}, e_{22}, \ldots, e_{N2}]^T$, where $\boldsymbol{V}$, $\boldsymbol{V_d}$, $\boldsymbol{z}$, $\boldsymbol{z_d}$, $\boldsymbol{S}_1$, $\boldsymbol{S}_2$, $\boldsymbol{e}_1$, and $\boldsymbol{e}_2$ are vectors with N elements, and then (4.13) can be rewritten as

$$(\mathcal{D} + \mathcal{B})\boldsymbol{S}_1 = (\mathcal{L} + \mathcal{B})\boldsymbol{e}_1 \tag{4.14}$$

From (4.14), it can be concluded that $||\boldsymbol{e}_1|| \le \frac{\lambda_{\max}(\mathcal{D}+\mathcal{B})}{\underline{\sigma}(\mathcal{L}+\mathcal{B})}||\boldsymbol{S}_1||$, where $\lambda_{\max}(\cdot)$ represents the maximum eigenvalue of a positive definite matrix and $\underline{\sigma}(\cdot)$ denotes the minimum singular value of a matrix. Therefore, $\boldsymbol{e}_1$ is bounded if $\boldsymbol{S}_1$ is bounded. The proof of $||\boldsymbol{e}_2|| \le \frac{\lambda_{\max}(\mathcal{D}+\mathcal{B})}{\underline{\sigma}(\mathcal{L}+\mathcal{B})}||\boldsymbol{S}_2||$ is similar to the above proof. This concludes the proof. □

Remark 4.4 The signals V_{id}, z_{id} in (4.10), (4.11) are obtained for the ith follower UAV in a distributed manner, since V_{id} and z_{id} only use the neighboring UAVs' information and the reference signals V_0, z_0 if the ith follower UAV has access to the leader UAV.

4.3.1 Distributed Controller Design for Velocity Subsystem

According to (4.1), the time derivative of S_{i1} is calculated as

$$\begin{aligned} \dot{S}_{i1} &= \dot{V}_i - \dot{V}_{id} \\ &= f_{iv} + g_{iv}T_i + \Delta_{i1} - \dot{V}_{id} \end{aligned} \tag{4.15}$$

To compensate for the unknown disturbance Δ_{i1}, the DO is designed as

$$\begin{cases} \hat{\Delta}_{i1} = \hat{x}_{i1} + k_{11}S_{i1} \\ \dot{\hat{x}}_{i1} = -k_{11}\left(f_{iv} + g_{iv}T_i + \hat{x}_{i1} + k_{11}S_{i1}\right) + \dfrac{\sum\limits_{j\in N_i} a_{ij}\dot{V}_j + b_i\dot{V}_0}{\sum\limits_{j\in N_i} a_{ij} + b_i} k_{11} \end{cases} \tag{4.16}$$

where k_{11} is a positive design parameter. $\hat{x}_{i1}$ is the state of the DO dynamic system.

Then, design the desired thrust input as

$$\overline{T}_i = g_{iv}^{-1}\left(-k_{12}S_{i1} - f_{iv} - \hat{\Delta}_{i1} + \frac{k_{12}}{2}\xi_{i1}\right) + \frac{\sum\limits_{j\in N_i} a_{ij}\dot{V}_j + b_i\dot{V}_0}{g_{iv}\left(\sum\limits_{j\in N_i} a_{ij} + b_i\right)} \tag{4.17}$$

where k_{12} is a positive design parameter and ξ_{i1} is an auxiliary signal to be designed later.

By taking the time derivative of (4.16), one has

$$\begin{aligned}\dot{\hat{\Delta}}_{i1} =& -k_{11}\left(f_{iv} + g_{iv}T_i + \hat{x}_{i1} + k_{11}S_{i1}\right) \\ &+ \frac{\sum\limits_{j\in N_i} a_{ij}\dot{V}_j + b_i\dot{V}_0}{\sum\limits_{j\in N_i} a_{ij} + b_i}k_{11} \\ &+ k_{11}\left(f_{iv} + g_{iv}T_i + \Delta_{i1} - \dot{V}_{id}\right) \\ =& -k_{11}\hat{\Delta}_{i1} + k_{11}\Delta_{i1} = k_{11}\tilde{\Delta}_{i1}\end{aligned} \tag{4.18}$$

where $\tilde{\Delta}_{i1} = \Delta_{i1} - \hat{\Delta}_{i1}$ is the estimation error of Δ_{i1} and the estimation vector is defined as $\hat{\mathbf{\Delta}}_1 = [\hat{\Delta}_{11}, \hat{\Delta}_{21}, \ldots, \hat{\Delta}_{N1}]^T$.

To handle the control input saturation in the controller design, an auxiliary dynamic system is employed as

$$\dot{\xi}_{i1} = -\frac{k_{12}}{2}\xi_{i1} + g_{iv}\Delta T_i \tag{4.19}$$

where $\Delta T_i = \text{sat}(\overline{T}_i) - \overline{T}_i$.

Remark 4.5 The control system performance may be severely degraded if the control input saturation is ignored in the controller design. Instability can be caused especially when the control input signal is in the saturation region for a long time [6]. To solve such a problem, the auxiliary dynamic system (4.19) is introduced to reduce the influence caused by the input saturation. The compensation term $\frac{k_{12}}{2}\xi_{i1}$ in (4.17) will be activated to alleviate the input saturation influence once the thrust input is saturated, i.e., $\text{sat}(\overline{T}_i) - \overline{T}_i \neq 0 \rightarrow \Delta T_i \neq 0$. Then, the saturated input signal will be pulled back into the unsaturated region under the action of the compensation term until $\Delta T_i = 0$.

Define $E_{i1} = S_{i1} - \xi_{i1}$ as the compensated tracking error and take the time derivative of E_{i1} along the trajectories of (4.15), (4.17), (4.19), and one has

$$\begin{aligned}\dot{E}_{i1} &= f_{iv} + g_{iv}T_i + \Delta_{i1} - \dot{V}_{id} - \dot{\xi}_{i1} \\ &= f_{iv} + g_{iv}(\bar{T}_i + \Delta T_i) + \Delta_{i1} - \dot{V}_{id} - \dot{\xi}_{i1} \\ &= -k_{12}E_{i1} + \tilde{\Delta}_{i1}\end{aligned} \tag{4.20}$$

Choose the Lyapunov function candidate as

$$L_{i1} = \frac{1}{2}E_{i1}^2 + \frac{1}{2}\tilde{\Delta}_{i1}^2 \tag{4.21}$$

By using (4.18) and (4.20), the time derivative of (4.21) is given by

$$\begin{aligned}\dot{L}_{i1} =& E_{i1}\dot{E}_{i1} + \tilde{\Delta}_{i1}\dot{\tilde{\Delta}}_{i1} \\ =& -k_{12}E_{i1}^2 + E_{i1}\tilde{\Delta}_{i1} + \tilde{\Delta}_{i1}\dot{\Delta}_{i1} - k_{11}\tilde{\Delta}_{i1}^2\end{aligned} \tag{4.22}$$

4.3.2 Distributed FTCC Design for Attitude Subsystem

In this section, the distributed FTCC design will focus on the attitude subsystem. Define the flight path angle error as $S_{i3} = \gamma_i - \gamma_{id}$, the pitch angle error as $S_{i4} = \theta_i - \theta_{id}$, and the pitch rate error as $S_{i5} = q_i - q_{id}$. The pitch angle and pitch rate vectors are denoted by $\boldsymbol{\theta} = [\theta_1, \theta_2, \ldots, \theta_N]^T$ and $\boldsymbol{q} = [q_1, q_2, \ldots, q_N]^T$, respectively. To this end, the DSC architecture can be applied to design the control law. The detailed procedures are listed as follows:

Step 1 The flight path angle intermediate control signal is chosen as

$$\bar{\gamma}_{id} = \frac{-k_{21}\left(\sum\limits_{j\in N_i} a_{ij} + b_i\right)S_{i2} + \sum\limits_{j\in N_i} a_{ij}\dot{z}_j + b_i\dot{z}_0}{V_i\left(\sum\limits_{j\in N_i} a_{ij} + b_i\right)} \tag{4.23}$$

where $k_{21} > 0$ is a positive design parameter.

To avoid taking the repetitive derivatives of $\bar{\gamma}_{id}$, the DSC technique is introduced to filter the intermediate signal. Hence, let $\bar{\gamma}_{id}$ pass the first-order filter with time constant τ_{i1} to obtain γ_{id}.

$$\tau_{i1}\dot{\gamma}_{id} + \gamma_{id} = \bar{\gamma}_{id},\ \gamma_{id}(0) = \bar{\gamma}_{id}(0) \tag{4.24}$$

where $\bar{\gamma}_{id}$ and γ_{id} are the intermediate and virtual flight path angle control signals of the ith follower UAV, respectively. The intermediate and virtual flight path angle control signal vectors of N follower UAVs are denoted by $\bar{\boldsymbol{\gamma}}_d = [\bar{\gamma}_{1d}, \bar{\gamma}_{2d}, \ldots, \bar{\gamma}_{Nd}]^T$ and $\boldsymbol{\gamma_d} = [\gamma_{1d}, \gamma_{2d}, \ldots, \gamma_{Nd}]^T$, respectively.

Define the filter error as $y_{i1} = \gamma_{id} - \bar{\gamma}_{id}$, and one has

$$\dot{y}_{i1} = -\frac{y_{i1}}{\tau_{i1}} + M_{i1} \tag{4.25}$$

where $M_{i1} = -\dot{\bar{\gamma}}_{id}$, denoted by

$$M_{i1} = \frac{k_{21}\left(\sum\limits_{j\in N_i} a_{ij} + b_i\right)\dot{S}_{i2} - \sum\limits_{j\in N_i} a_{ij}\ddot{z}_j - b_i\ddot{z}_0}{V_i\left(\sum\limits_{j\in N_i} a_{ij} + b_i\right)} \tag{4.26}$$

Remark 4.6 In the traditional backstepping control design procedure, the derivative of the virtual control input at each step must be provided. However, the derivative of the virtual control signal is not easily obtained due to the existence of unknown external disturbance and actuator fault in this chapter, and it can induce the "differential explosion" phenomenon. For better solving this problem, a first-order filter given by (4.24) is introduced to obtain the first-order derivative of the virtual control signal as $\dot{\gamma}_{id} = (\bar{\gamma}_{id} - \gamma_{id})/\tau_{i1}$. With this filter, the differential operation is replaced by the algebraic operation, making the distributed FTCC simple to be employed in practice.

Choose the Lyapunov function candidate as

$$L_{i2} = \frac{1}{2\bar{V}_i^2}S_{i2}^2 + \frac{1}{2}y_{i1}^2 \tag{4.27}$$

where $\bar{V}_i$ is the maximum of V_i, which is only used for the stability analyses.

Taking the time derivative of (4.27) gives

$$\begin{aligned}
\dot{L}_{i2} =& \frac{1}{\bar{V}_i^2}S_{i2}(\dot{z}_i - \dot{z}_{id}) + y_{i1}\dot{y}_{i1} \\
=& \frac{1}{\bar{V}_i^2}S_{i2}[V_i(\bar{\gamma}_{id} + y_{i1} + S_{i3}) - \dot{z}_{id}] + y_{i1}\dot{y}_{i1} \\
=& \frac{1}{\bar{V}_i^2}S_{i2}\left[-k_{21}S_{i2} + \frac{\sum\limits_{j\in N_i} a_{ij}\dot{z}_j + b_i\dot{z}_0}{\sum\limits_{j\in N_i} a_{ij} + b_i} - \dot{z}_{id}\right] \\
&+ \frac{V_i}{\bar{V}_i^2}S_{i2}y_{i1} + \frac{V_i}{\bar{V}_i^2}S_{i2}S_{i3} - \frac{y_{i1}^2}{\tau_{i1}} + y_{i1}M_{i1} \\
=& -\frac{k_{21}}{\bar{V}_i^2}S_{i2}^2 + \frac{V_i}{\bar{V}_i^2}S_{i2}y_{i1} + \frac{V_i}{\bar{V}_i^2}S_{i2}S_{i3} - \frac{y_{i1}^2}{\tau_{i1}} + y_{i1}M_{i1}
\end{aligned} \tag{4.28}$$

Remark 4.7 In [27, 33], the flight path angle command for the altitude dynamics of UAV is directly given, and the Lyapunov stability analyses are conducted for the pitch rate, pitch angle, and flight path angle dynamics. In this chapter, stability analyses including pitch rate, pitch angle, flight path angle, and altitude dynamics are carried out by introducing the velocity maximum $\bar{V}_i$ in the Lyapunov function candidate (4.27).

Step 2 Considering (4.3) and differentiating the flight path angle tracking error $S_{i3} = \gamma_i - \gamma_{id}$ with respect to time yield

$$\begin{aligned} \dot{S}_{i3} &= \dot{\gamma}_i - \dot{\gamma}_{id} \\ &= f_{i\gamma} + g_{i\gamma}\theta_i + \Delta_{i2} - \dot{\gamma}_{id} \end{aligned} \tag{4.29}$$

Then, a DO can be designed as follows to estimate Δ_{i2}.

$$\begin{cases} \hat{\Delta}_{i2} = \hat{x}_{i2} + k_{31}S_{i3} \\ \dot{\hat{x}}_{i2} = -k_{31}(f_{i\gamma} + g_{i\gamma}\theta_i + \hat{x}_{i2} + k_{31}S_{i3} - \dot{\gamma}_{id}) \end{cases} \tag{4.30}$$

where k_{31} is a positive design parameter. $\hat{x}_{i2}$ is the state of the DO dynamic system.

Design the intermediate pitch angle control signal as

$$\bar{\theta}_{id} = g_{i\gamma}^{-1}\left(-k_{32}S_{i3} - f_{i\gamma} - \hat{\Delta}_{i2} + \dot{\gamma}_{id}\right) \tag{4.31}$$

where $k_{32} > 0$ is a design parameter.

To avoid taking the repetitive derivatives of $\bar{\theta}_{id}$, let $\bar{\theta}_{id}$ pass the first-order filter with time constant τ_{i2} to obtain θ_{id}.

$$\tau_{i2}\dot{\theta}_{id} + \theta_{id} = \bar{\theta}_{id}, \theta_{id}(0) = \bar{\theta}_{id}(0) \tag{4.32}$$

where $\bar{\theta}_{id}$ and θ_{id} are the intermediate and virtual pitch angle control signals of the ith follower UAV. The intermediate and virtual pitch angle control signal vectors of N follower UAVs are denoted by $\bar{\boldsymbol{\theta}}_d = [\bar{\theta}_{1d}, \bar{\theta}_{2d}, \ldots, \bar{\theta}_{Nd}]^T$ and $\boldsymbol{\theta_d} = [\theta_{1d}, \theta_{2d}, \ldots, \theta_{Nd}]^T$, respectively.

Define the filter error as $y_{i2} = \theta_{id} - \bar{\theta}_{id}$, and then one has

$$\dot{y}_{i2} = -\frac{y_{i2}}{\tau_{i2}} + M_{i2} \tag{4.33}$$

where $M_{i2} = -\dot{\bar{\theta}}_{id}$, denoted by

$$\begin{aligned} M_{i2} =& g_{i\gamma}^{-2}\dot{g}_{i\gamma}\left(-k_{32}S_{i3} - f_{i\gamma} - \hat{\Delta}_{i2} + \frac{\bar{\gamma}_{id} - \gamma_{id}}{\tau_{i1}}\right) \\ &- g_{i\gamma}^{-1}\left(-k_{32}\dot{S}_{i3} - \dot{f}_{i\gamma} - \dot{\hat{\Delta}}_{i2} + \frac{\dot{\bar{\gamma}}_{id} - \dot{\gamma}_{id}}{\tau_{i1}}\right) \end{aligned} \tag{4.34}$$

By taking the time derivative of (4.30), one has

$$\begin{aligned}\dot{\hat{\Delta}}_{i2} =& - k_{31}\left(f_{i\gamma} + g_{i\gamma}\theta_i + \hat{x}_{i2} + k_{31}S_{i3} - \dot{\gamma}_{id}\right) \\ & + k_{31}\left(f_{i\gamma} + g_{i\gamma}\theta_i + \Delta_{i2} - \dot{\gamma}_{id}\right) \\ =& - k_{31}\hat{\Delta}_{i2} + k_{31}\Delta_{i2} = k_{31}\tilde{\Delta}_{i2}\end{aligned} \tag{4.35}$$

where $\tilde{\Delta}_{i2} = \Delta_{i2} - \hat{\Delta}_{i2}$ is the estimation error of Δ_{i2} and the estimation vector is defined as $\hat{\mathbf{\Delta}}_2 = [\hat{\Delta}_{12}, \hat{\Delta}_{22}, \ldots, \hat{\Delta}_{N2}]^T$.

Substituting (4.31) into (4.29) gives

$$\begin{aligned}\dot{S}_{i3} &= f_{i\gamma} + g_{i\gamma}(\bar{\theta}_{id} + y_{i2} + S_{i4}) + \Delta_{i2} - \dot{\gamma}_{id} \\ &= -k_{32}S_{i3} + g_{i\gamma}y_{i2} + g_{i\gamma}S_{i4} + \tilde{\Delta}_{i2}\end{aligned} \tag{4.36}$$

Choose the Lyapunov function candidate as

$$L_{i3} = \frac{1}{2}S_{i3}^2 + \frac{1}{2}\tilde{\Delta}_{i2}^2 + \frac{1}{2}y_{i2}^2 \tag{4.37}$$

By taking the time derivative of (4.37), one has

$$\begin{aligned}\dot{L}_{i3} =& S_{i3}\dot{S}_{i3} + \tilde{\Delta}_{i2}\dot{\tilde{\Delta}}_{i2} + y_{i2}\dot{y}_{i2} \\ =& S_{i3}(-k_{32}S_{i3} + g_{i\gamma}y_{i2} + g_{i\gamma}S_{i4} + \tilde{\Delta}_{i2}) \\ & + \tilde{\Delta}_{i2}\dot{\tilde{\Delta}}_{i2} + y_{i2}\dot{y}_{i2} \\ =& - k_{32}S_{i3}^2 + g_{i\gamma}y_{i2}S_{i3} + g_{i\gamma}S_{i3}S_{i4} + \tilde{\Delta}_{i2}S_{i3} \\ & + \tilde{\Delta}_{i2}\dot{\Delta}_{i2} - k_{31}\tilde{\Delta}_{i2}^2 - \frac{y_{i2}^2}{\tau_{i2}} + y_{i2}M_{i2}\end{aligned} \tag{4.38}$$

Step 3 By using (4.4), the time derivative of the pitch angle tracking error $S_{i4} = \theta_i - \theta_{id}$ is given by

$$\dot{S}_{i4} = \dot{\theta}_i - \dot{\theta}_{id} = q_i - \dot{\theta}_{id} \tag{4.39}$$

One can design the intermediate pitch angle control signal as

$$\bar{q}_{id} = -k_{42}S_{i4} + \dot{\theta}_{id} - \xi_{i2} \tag{4.40}$$

where k_{42} is a positive design parameter. ξ_{i2} is an auxiliary variable to be designed in the next step. $\boldsymbol{q_d} = [q_{1d}, q_{2d}, \ldots, q_{Nd}]^T$.

By using the DSC architecture, q_{id} can be obtained by passing $\bar{q}_{id}$ into the first-order filter with time constant τ_{i3}. The first-order filter is given by

$$\tau_{i3}\dot{q}_{id} + q_{id} = \bar{q}_{id}, q_{id}(0) = \bar{q}_{id}(0) \tag{4.41}$$

where $\bar{q}_{id}$ and q_{id} are the intermediate and virtual pitch rate control signals of the ith follower UAV. The intermediate and virtual pitch rate control signal vectors of N follower UAVs are denoted by $\bar{\boldsymbol{q}}_d = [\bar{q}_{1d}, \bar{q}_{2d}, \ldots, \bar{q}_{Nd}]^T$ and $\boldsymbol{q_d} = [q_{1d}, q_{2d}, \ldots, q_{Nd}]^T$, respectively.

Define the filter error as $y_{i3} = q_{id} - \bar{q}_{id}$, and then one has

$$\dot{y}_{i3} = -\frac{y_{i3}}{\tau_{i3}} + M_{i3} \tag{4.42}$$

where $M_{i3} = -\dot{\bar{q}}_{id}$, denoted by

$$M_{i3} = k_{42}\dot{S}_{i4} - \frac{\dot{\bar{\theta}}_{id} - \dot{\theta}_{id}}{\tau_{i2}} + \dot{\xi}_{i2} \tag{4.43}$$

By differentiating S_{i4} with respect to time and applying (4.40), one has

$$\begin{aligned} \dot{S}_{i4} &= \bar{q}_{id} + y_{i3} + S_{i5} - \dot{\theta}_{id} \\ &= -k_{42}S_{i4} + E_{i5} + y_{i3} \end{aligned} \tag{4.44}$$

where $E_{i5} = S_{i5} - \xi_{i2}$ is the compensated error.

Choose the Lyapunov function candidate as

$$L_{i4} = \frac{1}{2}S_{i4}^2 + \frac{1}{2}y_{i3}^2 \tag{4.45}$$

By recalling (4.39) and (4.40), the time derivative of L_{i4} is given by

$$\begin{aligned} \dot{L}_{i4} =& S_{i4}\dot{S}_{i4} + y_{i3}\dot{y}_{i3} \\ =& -k_{42}S_{i4}^2 + S_{i4}E_{i5} + y_{i3}S_{i4} - \frac{y_{i3}^2}{\tau_{i3}} + y_{i3}M_{i3} \end{aligned} \tag{4.46}$$

Step 4 Differentiating the pitch rate tracking error $S_{i5} = q_i - q_{id}$ with respect to time and applying (4.8), the following expression is obtained:

$$\begin{aligned} \dot{S}_{i5} &= \dot{q}_i - \dot{q}_{id} \\ &= f_{iq} + g_{iq}\delta_{ie0} + \Delta_i - \dot{q}_{id} \end{aligned} \tag{4.47}$$

To estimate the lumped disturbance Δ_i, the DO can be designed as

$$\begin{cases} \hat{\Delta}_i = \hat{x}_{i3} + k_{51} S_{i5} \\ \dot{\hat{x}}_{i3} = -k_{51}(f_{iq} + g_{iq}\delta_{ie0} + \hat{x}_{i3} + k_{51} S_{i5} - \dot{q}_{id}) \end{cases} \tag{4.48}$$

where $k_{51} > 0$ is a design parameter. $\hat{x}_{i3}$ is the state of the DO dynamic system.

Then, design the desired control input as

$$\overline{\delta_{ie0}} = g_{iq}^{-1}\left(-k_{52} S_{i5} - f_{iq} - \hat{\Delta}_i + \dot{q}_{id}\right) \tag{4.49}$$

where k_{52} is a positive design parameter.

Similar to (4.19), to deal with the input saturation of the ith follower UAV in the controller design, an auxiliary dynamic system is constructed as

$$\dot{\xi}_{i2} = -k_{52}\xi_{i2} + g_{iq}\Delta\delta_{ie0} \tag{4.50}$$

where $\Delta\delta_{ie0} = \delta_{ie0} - \overline{\delta_{ie0}}$.

By taking the time derivative of (4.48), one has

$$\begin{aligned} \dot{\hat{\Delta}}_i &= -k_{51}\left(f_{iq} + g_{iq}\delta_{ie0} + \hat{x}_{i3} + k_{51} S_{i5} - \dot{q}_{id}\right) \\ &\quad + k_{51}\left(f_{iq} + g_{iq}\delta_{ie0} + \Delta_i - \dot{q}_{id}\right) \\ &= -k_{51}\hat{\Delta}_i + k_{51}\Delta_i = k_{51}\tilde{\Delta}_i \end{aligned} \tag{4.51}$$

where $\tilde{\Delta}_i = \Delta_i - \hat{\Delta}_i$ is the estimation error of Δ_i, and the estimation vector is defined as $\hat{\mathbf{\Delta}} = [\hat{\Delta}_1, \hat{\Delta}_2, \ldots, \hat{\Delta}_N]^T$.

Define the compensated error as $E_{i5} = S_{i5} - \xi_{i2}$, by recalling (4.49), and the time derivative of E_{i5} along the trajectories of (4.47) and (4.50) gives

$$\begin{aligned} \dot{E}_{i5} &= f_{iq} + g_{iq}\delta_{ie0} + \Delta_i - \dot{q}_{id} - \dot{\xi}_{i2} \\ &= f_{iq} + g_{iq}\Delta\delta_{ie0} + g_{iq}\overline{\delta_{ie0}} + \Delta_i - \dot{q}_{id} - \dot{\xi}_2 \\ &= -k_{52}E_{i5} + \tilde{\Delta}_i \end{aligned} \tag{4.52}$$

Choose the Lyapunov function candidate as

$$L_{i5} = \frac{1}{2}E_{i5}^2 + \frac{1}{2}\tilde{\Delta}_i^2 \tag{4.53}$$

Taking the time derivative of (4.53) gives

$$\begin{aligned} \dot{L}_{i5} &= E_{i5}\dot{E}_{i5} + \tilde{\Delta}_i\dot{\tilde{\Delta}}_i \\ &= -k_{52}E_{i5}^2 + \tilde{\Delta}_i E_{i5} + \tilde{\Delta}_i\dot{\Delta}_i - k_{51}\tilde{\Delta}_i^2 \end{aligned} \tag{4.54}$$

4.3.3 Stability Analysis

Theorem 4.2 *Consider a group of one leader UAV and N follower UAVs described by (4.1)–(4.5) with Assumptions 4.1–4.5 satisfied. In view of any positive constant p, for initial conditions satisfying* $\sum_{i=1}^{N}[E_{i1}^2(0)+S_{i2}^2(0)+S_{i3}^2(0)+S_{i4}^2(0)+E_{i5}^2(0)+\tilde{\Delta}_{i1}^2(0)+\tilde{\Delta}_{i2}^2(0)+\tilde{\Delta}_i^2(0)+y_{i1}^2(0)+y_{i2}^2(0)++y_{i3}^2(0)] \leq 2p$, *if the distributed FTCC laws are designed as (4.10), (4.11), (4.17), (4.23), (4.31), (4.40), (4.49), the DOs are designed as (4.16), (4.30), (4.48), and the auxiliary dynamic systems are designed as (4.19), (4.50), then it is guaranteed that the velocity tracking errors* $\boldsymbol{e}_1 = [e_{11}, e_{21}, \ldots, e_{N1}]^T = [V_1 - V_0, V_2 - V_0, \ldots, V_N - V_0]^T$ *and the altitude tracking errors* $\boldsymbol{e}_2 = [e_{12}, e_{22}, \ldots, e_{N2}]^T = [z_1 - z_0, z_2 - z_0, \ldots, z_N - z_0]^T$ *are UUB in the presence of external disturbances* d_{i1}, d_{i2}, d_{i3}, *and actuator faults (4.7).*

Proof Choose the Lyapunov function candidate as

$$L = \sum_{i=1}^{N} L_i = \sum_{i=1}^{N}(L_{i1} + L_{i2} + L_{i3} + L_{i4} + L_{i5}) \tag{4.55}$$

where L_{i1}, L_{i2}, L_{i3}, L_{i4}, and L_{i5} are given in (4.21), (4.27), (4.37), (4.45), and (4.53), respectively.

By taking the time derivative of L_i and using (4.22), (4.28), (4.38), (4.46), (4.54), one has

$$\begin{aligned}
\dot{L}_i =& -k_{12}E_{i1}^2 + E_{i1}\tilde{\Delta}_{i1} + \tilde{\Delta}_{i1}\dot{\Delta}_{i1} - k_{11}\tilde{\Delta}_{i1}^2 - \frac{k_{21}}{\bar{V}_i^2}S_{i2}^2 + \frac{V_i}{\bar{V}_i^2}S_{i2}y_{i1} \\
&+ \frac{V_i}{\bar{V}_i^2}S_{i2}S_{i3} - \frac{y_{i1}^2}{\tau_{i1}} + y_{i1}M_{i1} - k_{32}S_{i3}^2 + g_{i\gamma}y_{i2}S_{i3} + g_{i\gamma}S_{i3}S_{i4} \\
&+ \tilde{\Delta}_{i2}S_{i3} + \tilde{\Delta}_{i2}\dot{\Delta}_{i2} - k_{31}\tilde{\Delta}_{i2}^2 - \frac{y_{i2}^2}{\tau_{i2}} + y_{i2}M_{i2} - k_{42}S_{i4}^2 + S_{i4}E_{i5} \\
&+ y_{i3}S_{i4} - \frac{y_{i3}^2}{\tau_{i3}} + y_{i3}M_{i3} - k_{52}E_{i5}^2 + \tilde{\Delta}_iE_{i5} + \tilde{\Delta}_i\dot{\Delta}_i - k_{51}\tilde{\Delta}_i^2
\end{aligned} \tag{4.56}$$

Then, by using Young's inequality, one has

$$\begin{aligned}\dot{L}_i \leq &-\left(k_{12}-\frac{1}{2}\right)E_{i1}^2-(k_{11}-1)\tilde{\Delta}_{i1}^2+\frac{\dot{\Delta}_{i1}}{2}-\frac{k_{21}-1}{\bar{V}_i^2}S_{i2}^2+\frac{M_{i1}^2}{2}\\ &-\left(\frac{1}{\tau_{i1}}-\frac{V_i^2}{2\bar{V}_i^2}-\frac{1}{2}\right)y_{i1}^2-\left(k_{32}-|\bar{g}_{i\gamma}|-\frac{1}{2}-\frac{V_i^2}{2\bar{V}_i^2}\right)S_{i3}^2+\frac{\dot{\Delta}_{i2}^2}{2}\\ &-\left(\frac{1}{\tau_{i2}}-\frac{1}{2}-\frac{|\bar{g}_{i\gamma}|}{2}\right)y_{i2}^2-(k_{31}-1)\tilde{\Delta}_{i2}^2-(k_{52}-1)E_{i5}^2+\frac{M_{i2}^2}{2}\\ &-\left(k_{42}-1-\frac{|\bar{g}_{i\gamma}|}{2}\right)S_{i4}^2-\left(\frac{1}{\tau_{i3}}-1\right)y_{i3}^2-(k_{51}-1)\tilde{\Delta}_i^2+\frac{M_{i3}^2}{2}+\frac{\dot{\Delta}_i^2}{2}\end{aligned} \tag{4.57}$$

According to the intermediate control signals (4.23), (4.31), (4.40), (4.49) together with Assumption 4.3, there exist constants $M_{i1m} > 0$, $M_{i2m} > 0$, $M_{i3m} > 0$, such that [23, 25]

$$|M_{i1}| \leq M_{i1m},\ |M_{i2}| \leq M_{i2m},\ |M_{i3}| \leq M_{i3m} \tag{4.58}$$

When λ and σ are chosen as

$$\lambda_i = \min\left\{\begin{array}{l}(2k_{12}-1),(2k_{11}-2),(2k_{21}-2),\\ \left(2k_{32}-2|\bar{g}_{i\gamma}|-1-\frac{V_i^2}{\bar{V}_i^2}\right),\\ (2k_{31}-2),(2k_{42}-2-|\bar{g}_{i\gamma}|),\\ (2k_{52}-2),(2k_{51}-2),\\ \left(\frac{2}{\tau_{i1}}-\frac{V_i^2}{\bar{V}_i^2}-1\right),\left(\frac{2}{\tau_{i2}}-1-|\bar{g}_{i\gamma}|\right)\\ \left(\frac{2}{\tau_{i3}}-2\right)\end{array}\right\} > 0 \tag{4.59}$$

$$\sigma_i = \frac{\bar{\Delta}_{i1}^2}{2}+\frac{\bar{\Delta}_{i2}^2}{2}+\frac{\bar{\Delta}_i^2}{2}+\frac{M_{i1m}^2}{2}+\frac{M_{i2m}^2}{2}+\frac{M_{i3m}^2}{2} \tag{4.60}$$

then one has

$$\dot{L}_i \leq -\lambda_i L_i + \sigma_i \tag{4.61}$$

By recalling (4.55), the following inequality can be obtained as

$$\dot{L} \leq -\lambda L + \sigma \tag{4.62}$$

where $\lambda = \min\{\lambda_i\}$, $\sigma = \sum_{i=1}^{N} \sigma_i$.

From (4.59), it can be concluded that the design parameters k_{12}, k_{11}, k_{21}, k_{32}, k_{31}, k_{42}, k_{52}, k_{51}, τ_{i1}, τ_{i2}, τ_{i3} satisfy

$$\begin{cases} k_{12} > 0.5,\ k_{11} > 1,\ k_{21} > 1,\ k_{32} > |\bar{g}_{i\gamma}| + 1 \\ k_{31} > 1,\ k_{42} > 1 + 0.5|\bar{g}_{i\gamma}|,\ k_{52} > 1, k_{51} > 1 \\ \frac{1}{\tau_{i1}} > 1,\ \frac{1}{\tau_{i2}} > 0.5 + 0.5|\bar{g}_{i\gamma}|,\ \frac{1}{\tau_{i3}} > 1 \end{cases} \tag{4.63}$$

Multiplying both sides of (4.62) by $e^{\lambda t}$ yields

$$\frac{d}{dt}(Le^{\lambda t}) \leq \sigma e^{\lambda t} \tag{4.64}$$

Integrating (4.64) over $[0, t]$ gives

$$L \leq \frac{\sigma}{\lambda} + \left(L(0) - \frac{\sigma}{\lambda}\right) e^{-\lambda t} \tag{4.65}$$

Select proper design parameters k_{12}, k_{11}, k_{21}, k_{32}, k_{31}, k_{42}, k_{52}, k_{51}, τ_{i1}, τ_{i2}, τ_{i3} such that $\lambda > \sigma/p$. Therefore, $\dot{L} \leq 0$ on $L = p$, which implies that $L \leq p$ is an invariant set and guarantees $L(t) \leq p$ for $L(0) \leq p$, $t \geq 0$. From (4.65), it can be seen that L will be ultimately confined in the closed region $0 \leq L \leq \sigma/\lambda$. Therefore, it can be concluded that E_{i1}, S_{i2}, S_{i3}, S_{i4}, E_{i5}, $\tilde{\Delta}_{i1}$, $\tilde{\Delta}_{i2}$, $\tilde{\Delta}_i$, y_{i1}, y_{i2}, y_{i3} are UUB. When the thrust inputs and elevator deflection inputs of all UAVs are not saturated, $\Delta T_i = 0$, $\Delta \delta_{ie0} = 0$. As a result, E_{i1} and E_{i5} are UUB once the ith UAV is free of input saturation. Moreover, V_i and q_i are also bounded because the inputs are bounded based on Assumption 4.1. Therefore, ξ_{i1} and ξ_{i2} are guaranteed bounded, and S_{i1} and S_{i5} can be guaranteed bounded as well. This ends the proof.

□

Remark 4.8 The errors S_{i1} and S_{i2} can be made enough small by setting enough large k_{12}, k_{11}, k_{21}, k_{32}, k_{31}, k_{42}, k_{52}, k_{51} and enough small τ_{i1}, τ_{i2}, τ_{i3}, based on the parameters selection criteria (4.63). In practical applications, such control parameters are chosen by trial and error under (4.63) until a good performance is obtained. Therefore, the ith follower UAV can track its reference commands V_{id}, z_{id} with enough small tracking errors under the proper selection of control parameters. By recalling Lemma 4.1, the tracking errors e_{i1}, e_{i2} with respect to the leader UAV V_0, z_0 are UUB.

4.4 Simulations Results

In this section, a cooperative system of four follower UAVs with the same configuration and one leader UAV is constructed, which is illustrated in Fig. 4.1. The structure parameters and coefficients of the ith follower UAV can be referred to Chap. 3. Input saturation limits of the ith follower UAV are listed in Table 4.1. The leader UAV is used to give the velocity and altitude references V_0, z_0. The communication weights are selected as $a_{ij} = 1$ if $(r_i, r_j) \in \mathcal{E}$, $i = 1, 2, 3, 4$, $j = 1, 2, 3, 4$, and only follower UAV#1 and UAV#4 have accesses to the leader UAV, i.e., $\mathcal{B} = \text{diag}[1, 0, 0, 1]$. The commands from the leader UAV are $V_0 = 30$ m/s, $z_0 = 1000$ m at $t \in [0\ 10]$ s. Then, the velocity command V_0 and altitude command z_0 from the leader UAV change 10 m/s and 100 m every 150 s. The command signals are generated by utilizing $\omega_n^2/s^2 + 2\omega_n\xi_n s + \omega_n^2$ to filter step commands V_c, z_c, where $\omega_n = 0.05$, $\xi_n = 0.95$. When follower UAVs and leader UAV are flying in a formation team, it is assumed that elevator faults and external disturbances are encountered by the follower UAVs. Based on the fault model (4.7), the elevator faults and external disturbances imposed on the follower UAV#2 and UAV#4 are described as Fig. 4.2, which are mainly used to verify the effectiveness of the proposed distributed FTCC scheme. The initial values of the ith follower UAV are listed in Table 4.2.

In the simulation, the control parameters are selected as $k_{11} = 5$, $k_{12} = 20$, $k_{21} = 10.4$, $k_{31} = 6.4$, $k_{32} = 2.2$, $k_{42} = 14.4$, $k_{51} = 46.6$, and $k_{52} = 39.4$. The filter parameters are selected as $\tau_{i1} = 0.05$, $\tau_{i2} = 0.05$, and $\tau_{i3} = 0.05$. The initial values of DOs are selected as $\hat{x}_{i1}(0) = 0$, $\hat{x}_{i2}(0) = 0$, $\hat{x}_{i3}(0) = 0$, $i = 1, 2, 3, 4$.

Figure 4.3 shows the constructed velocity references V_{1d}, V_{2d}, V_{3d}, V_{4d} and altitude references z_{1d}, z_{2d}, z_{3d}, z_{4d} for four follower UAVs based on the nearest neighbor rule. It can be observed that the constructed references can converge to

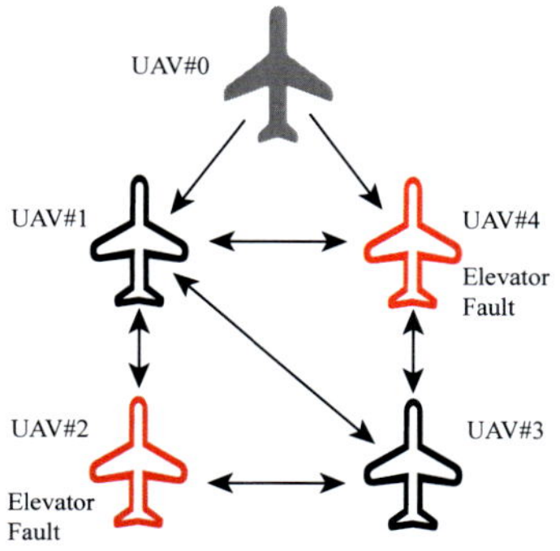

Fig. 4.1 Communication topology

Table 4.1 Input saturation limits of the ith UAV

Lower limit	Upper limit
$T_{i\min} = 0\,\text{N}$	$T_{i\max} = 100\,\text{N}$
$\delta_{ie0\min} = -0.96\,\text{rad}$	$\delta_{ie0\max} = 0.96\,\text{rad}$

Fig. 4.2 External disturbances and actuator faults encountered by UAVs#1–4

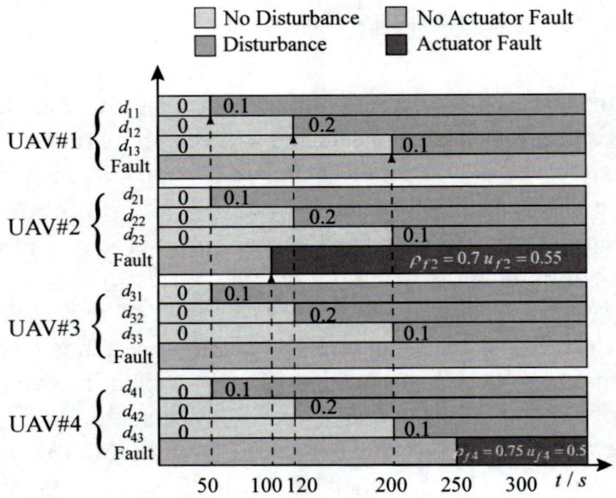

Table 4.2 Initial values of four follower UAVs

	$V_i(0)$	$z_i(0)$	$\gamma_i(0)$	$\theta_i(0)$	$q_i(0)$
UAV#1	31.5 m/s	998 m	0.4 rad	0.03 rad	0 rad/s
UAV#2	27 m/s	1000 m	0 rad	0.03 rad	0 rad/s
UAV#3	30 m/s	1000 m	0 rad	0.03 rad	0 rad/s
UAV#4	33 m/s	1000 m	0 rad	0.1 rad	0.5 rad/s

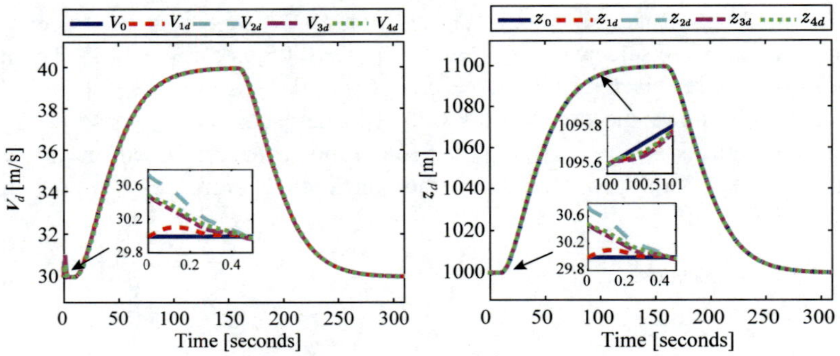

Fig. 4.3 Velocity and altitude references of UAVs#1–4 constructed by using the nearest neighbor rule

the leader UAV's references V_0, z_0. The generated altitude references in Fig. 4.3 are affected when the follower UAV#2 is encountered by the elevator fault at $t = 100$ s. Such a phenomenon can be attributed to the fact that the references are based on the outputs from neighboring UAVs. Figure 4.4 shows the velocity and altitude responses. It is observed that the velocities and altitudes of four follower UAVs are bounded and can track the leader UAV's references V_0, z_0 very well. Figure 4.5 illustrates the velocity and altitude tracking errors of four follower UAVs with respect to the generated references V_{id}, z_{id}, $i = 1, 2, 3, 4$. It can be found that the track errors are bounded even when external disturbances are encountered

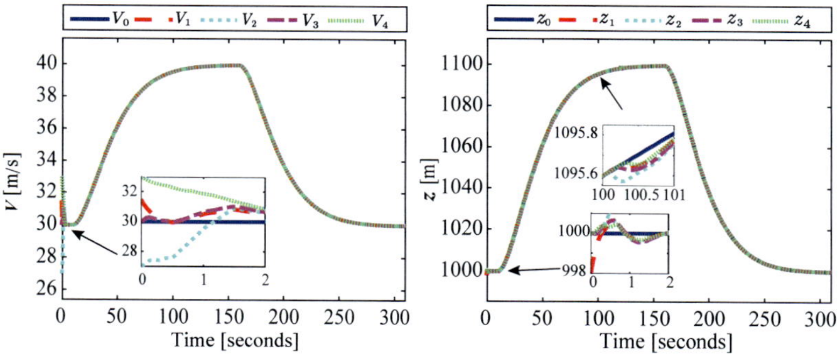

Fig. 4.4 Velocities and altitudes of UAVs#1–4

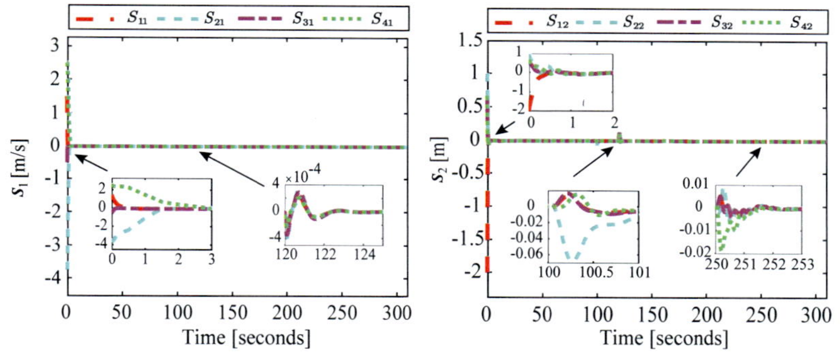

Fig. 4.5 Velocity and altitude tracking errors of UAVs#1–4 with respect to respective references

by UAVs#1–4, and actuator faults are encountered by UAV#2 and UAV#4 in the formation team. This indicates that the designed distributed FTCC scheme can make each UAV track respective reference commands V_{id}, z_{id} in the connected communication network.

Figure 4.6 shows that the velocities and altitudes of all follower UAVs can converge to the references V_0, z_0 from the leader UAV. Under the supervision of the proposed distributed FTCC, the tracking errors of each follower UAV are UUB. Besides, Fig. 4.6 further validates Lemma 4.1, which states that if velocity and altitude tracking errors S_{i1}, S_{i2} with respect to individual references V_{id}, z_{id} are UUB, the tracking errors e_{i1}, e_{i2} with respect to the references V_0, z_0 from the leader UAV are UUB. The neighboring velocity errors and altitude errors are shown in Fig. 4.7, and it can be found that four follower UAVs reach velocity and altitude consensus, which indicates that the proposed distributed FTCC can maintain the relative velocity errors and relative altitude errors of all UAVs in a very small region containing zero.

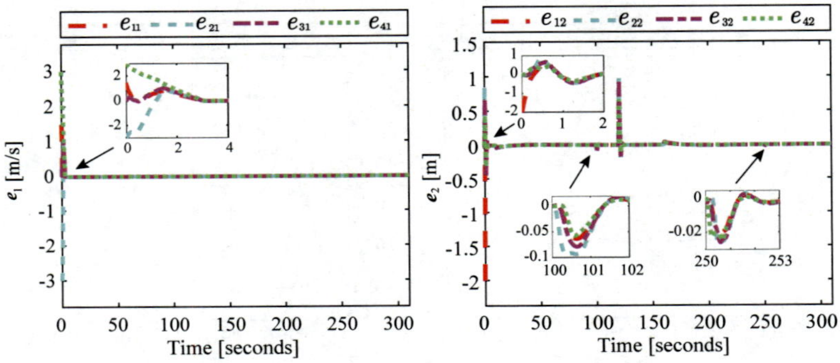

Fig. 4.6 Velocity and altitude tracking errors of UAVs#1–4 with respect to the leader UAV

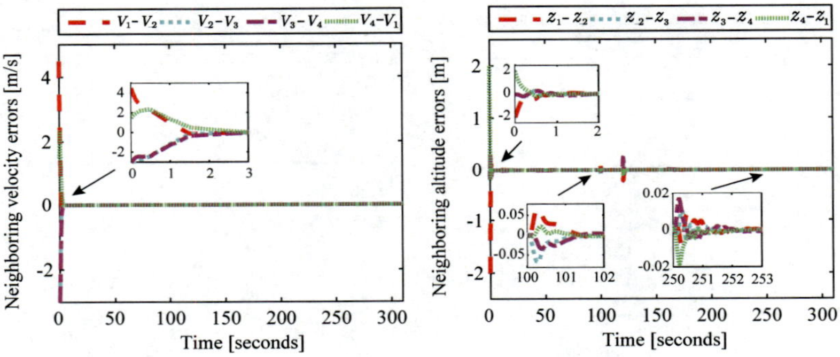

Fig. 4.7 Neighboring velocity errors and altitude errors between UAVs#1–4

Figure 4.8 indicates the states including flight path angles, pitch angles, angles of attack, and pitch rates. It is observed that all states of each UAV are bounded even when the external disturbances are imposed on each UAV and actuators in UAV#2 and UAV#4 become faulty. The control input signals of four follower UAVs are shown in Fig. 4.9. It is illustrated that the elevator deflection angle of UAV#2 reaches the lower limit when UAV#2 becomes faulty at $t = 100$ s and encounters external disturbance $d_{22} = 0.2$ at $t = 120$ s abruptly. When the elevator of UAV#2 stays at the lower limit, i.e., $\Delta\delta_{2e0} \neq 0$, the auxiliary dynamic system (4.50) starts working to pull the elevator from the lower limit back to the unsaturated region. It can also be seen from Fig. 4.9 that when the actuator of UAV#2 becomes faulty at $t = 100$ s and external disturbance d_{22} is encountered at $t = 120$ s, the elevator will compensate for the actuator fault by adjusting the deflection angle in a timely manner until the stability of the follower UAV#2 is regained. By using this response mode, the faulty UAVs will not have adverse effects on the networked UAVs.

The estimated lumped uncertainties are shown in Fig. 4.10. It can be seen that when actuator faults are encountered by UAV#2 and UAV#4, the DOs can approximate the lumped uncertainties including external disturbances and actuator faults in a timely manner.

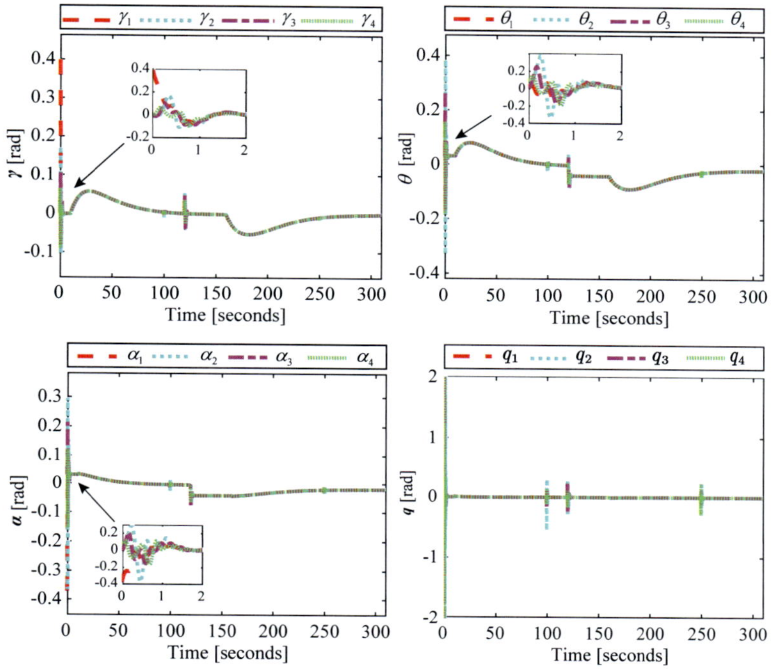

Fig. 4.8 States of UAVs#1–4

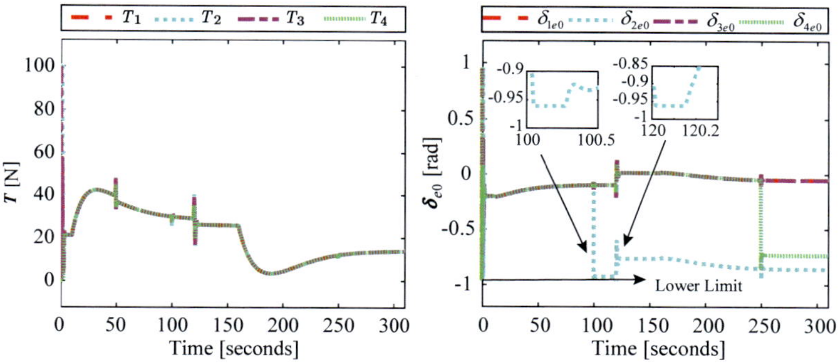

Fig. 4.9 Thrust inputs and elevator deflection angles of UAVs#1–4

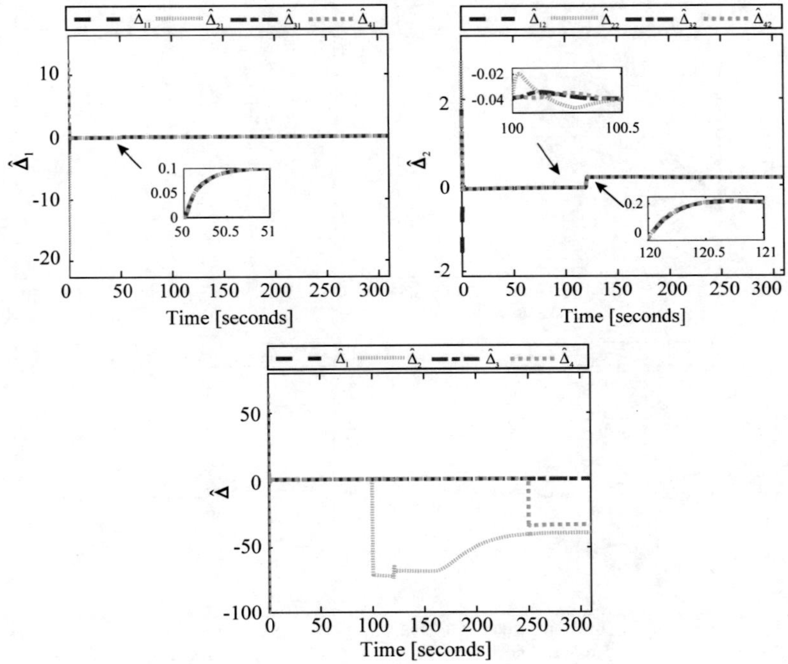

Fig. 4.10 Estimated lumped uncertainties

4.5 Conclusions

Although there have been significant investigations on the cooperative control of multi-UAVs, very few researches have considered the difficult problem of distributed FTCC design for multi-UAVs with external disturbances, actuator faults, and input saturation. By integrating the nearest neighbor rule, DO technique, auxiliary dynamic system, and DSC technique, a distributed FTCC scheme has been proposed for the longitudinal dynamics of multi-UAVs in the presence of actuator faults, external disturbances, and control input saturation. Compared with traditional centralized control, the proposed distributed FTCC scheme has been designed to reduce the communication cost and increase the robustness significantly. Lyapunov stability analysis and simulation results have revealed the effectiveness of the proposed distributed FTCC scheme.

References

1. Q. Ali, N. Gageik, S. Montenegro, A review on distributed control of cooperating mini UAVs. Int. J. Artificial Intell. Appl. **5**(4), 1–13 (2014)

2. J.R. Azinheira, A. Moutinho, Hover control of an UAV with backstepping design including input saturations. IEEE Trans. Control Syst. Technol. **16**(3), 517–526 (2008)
3. J.D. Boskovic, S. Bergstrom, R.K. Mehra, Robust integrated flight control design under failures, damage, and state-dependent disturbances. J. Guid. Control Dynam. **28**(5), 902–917 (2005)
4. M. Chen, G. Tao, Adaptive fault-tolerant control of uncertain nonlinear large-scale systems with unknown dead zone. IEEE Trans. Cybern. **46**(8), 1851–1862 (2016)
5. M. Chen, S.Z. Sam Ge, B.B. Ren, Adaptive tracking control of uncertain MIMO nonlinear systems with input constraints. Automatica **47**(3), 452–465 (2011)
6. R.X. Cui, C.G. Yang, Y. Li, S. Sharma, Adaptive neural network control of AUVs with control input nonlinearities using reinforcement learning. IEEE Trans. Syst. Man Cybern. **47**(6), 1019–1029 (2017)
7. Z.T. Ding, Consensus disturbance rejection with disturbance observers. IEEE Trans. Ind. Electron. **62**(9), 5829–5837 (2015)
8. L. Fiorentini, A. Serrani, M.A. Bolender, D.B. Doman, Nonlinear robust adaptive control of flexible air-breathing hypersonic vehicles. J. Guid. Control Dyn. **32**(2), 402–417 (2009)
9. G. Gao, J.Z. Wang, Observer-based fault-tolerant control for an air-breathing hypersonic vehicle model. Nonlinear Dyn. **76**(1), 409–430 (2014)
10. S.G. Gao, H.R. Dong, B. Ning, L. Chen, Neural adaptive control for uncertain nonlinear system with input saturation: state transformation based output feedback. Neurocomputing **159**, 117–125 (2015)
11. Y. Gu, B. Seanor, G. Campa, M.R. Napolitano, L. Rowe, S. Gururajan, S. Wan, Design and flight testing evaluation of formation control laws. IEEE Trans. Control Syst. Technol. **14**(6), 1105–1112 (2006)
12. L. He, X.X. Sun, Y. Lin, Distributed output-feedback formation tracking control for unmanned aerial vehicles. Int. J. Syst. Sci. **47**(16), 3919–3928 (2016)
13. Y.G. Hong, J.P. Hu, L.X. Gao, Tracking control for multi-agent consensus with an active leader and variable topology. Automatica **42**(7), 1177–1182 (2006)
14. G.P. Kladis, P.P. Menon, C. Edwards, Fuzzy distributed cooperative tracking for a swarm of unmanned aerial vehicles with heterogeneous goals. Int. J. Syst. Sci. **47**(16), 3803–3811 (2016)
15. Y. Kuriki, T. Namerikawa, Consensus-based cooperative formation control with collision avoidance for a multi-UAV system, in *American Control Conference*, Portland (2014)
16. X.J. Li, G.H. Yang, Robust adaptive fault-tolerant control for uncertain linear systems with actuator failures. IET Control Theory Appl. **6**(10), 1544–1551 (2012)
17. P. Li, X. Yu, Y.M. Zhang, X.Y. Peng, Adaptive multivariable integral TSMC of a hypersonic gliding vehicle with actuator faults and model uncertainties. IEEE-ASME Trans. Mechatron. **22**(6), 2723–2735 (2017)
18. F. Liao, R. Teo, J.L. Wang, X.X. Dong, F. Lin, K.M. Peng, Distributed formation and reconfiguration control of VTOL UAVs. IEEE Trans. Control Syst. Technol. **25**(1), 270–277 (2017)
19. W. Lin, Distributed UAV formation control using differential game approach. Aerosp. Sci. Technol. **35**(1), 54–62 (2014)
20. P. Lu, E.J. Van Kampen, C. De Visser, Q.L. Chu, Aircraft fault-tolerant trajectory control using incremental nonlinear dynamic inversion. Control Eng. Pract. **57**, 126–141 (2016)
21. S. Rao, D. Ghose, Sliding mode control-based autopilots for leaderless consensus of unmanned aerial vehicles. IEEE Trans. Control Syst. Technol. **22**(5), 1964–1972 (2014)
22. A. Sutton, B. Fidan, D. Van d. Walle, Hierarchical UAV formation control for cooperative surveillance, in *IFAC World Congress*, Seoul, Korea (2008)
23. D. Swaroop, J.K. Hedrick, P.P. Yip, J.C. Gerdes, Dynamic surface control for a class of nonlinear systems. IEEE Trans. Autom. Control **45**(10), 1893–1899 (2000)
24. D. Van D. Walle, B. Fidan, A. Sutton, C.B. Yu, B.D. Anderson, Non-hierarchical UAV formation control for surveillance tasks, in *American Control Conference*, Seattle (2008)
25. D. Wang, Neural network-based adaptive dynamic surface control of uncertain nonlinear pure-feedback systems. Int. J. Robust Nonlinear Control **21**(5), 527–541 (2011)

26. X.H. Wang, V. Yadav, S.N. Balakrishnan, Cooperative UAV formation flying with obstacle/collision avoidance. IEEE Trans. Control Syst. Technol. **15**(4), 672–679 (2007)
27. F. Wang, Q. Zou, C.C. Hua, Q. Zong, Disturbance observer-based dynamic surface control design for a hypersonic vehicle with input constraints and uncertainty. Proc. Inst. Mech. Eng. Part I-J Syst. Control Eng. **230**(6), 522–536 (2016)
28. C.Y. Wen, J. Zhou, Z.T. Liu, H.Y. Su, Robust adaptive control of uncertain nonlinear systems in the presence of input saturation and external disturbance. IEEE Trans. Autom. Control **56**(7), 1672–1678 (2011)
29. S.X. Weng, D. Yue, Z.G. Sun, L. Xiao, Distributed robust finite-time attitude containment control for multiple rigid bodies with uncertainties. Int. J. Robust Nonlinear Control **25**(15), 2561–2581 (2015)
30. B. Xiao, S. Yin, Velocity-free fault-tolerant and uncertainty attenuation control for a class of nonlinear systems. IEEE Trans. Ind. Electron. **63**(7), 4400–4411 (2016)
31. B. Xu, Robust adaptive neural control of flexible hypersonic flight vehicle with dead-zone input nonlinearity. Nonlinear Dyn. **80**(3), 1509–1520 (2015)
32. B. Xu, Disturbance observer-based dynamic surface control of transport aircraft with continuous heavy cargo airdrop. IEEE Trans. Syst. Man Cybern. Syst. **47**(1), 161–170 (2017)
33. B. Xu, C.G. Yang, Y.P. Pan, Global neural dynamic surface tracking control of strict-feedback systems with application to hypersonic flight vehicle. IEEE Trans. Neural Netw. Learn. Syst. **26**(10), 2563–2575 (2015)
34. R.B. Xue, J.M. Song, G.H. Cai, Distributed formation flight control of multi-UAV system with nonuniform time-delays and jointly connected topologies. Proc. Inst. Mech. Eng. Part G.J. Aerosp. Eng. **230**(10), 1871–1881 (2016)
35. X. Yu, Z.X. Liu, Y.M. Zhang, Fault-tolerant formation control of multiple UAVs in the presence of actuator faults. Int. J. Robust Nonlinear Control **26**(12), 2668–2685 (2016)
36. X. Yu, Z.X. Liu, Y.M. Zhang, Fault-tolerant flight control design with finite-time adaptation under actuator stuck failures. IEEE Trans. Control Syst. Technol. **25**(4), 1431–1440 (2017)
37. C. Yuan, Y.M. Zhang, Z.X. Liu, A survey on technologies for automatic forest fire monitoring, detection, and fighting using unmanned aerial vehicles and remote sensing techniques. Can. J. For. Res. **45**(7), 783–792 (2015)
38. Y.M. Zhang, J. Jiang, Bibliographical review on reconfigurable fault-tolerant control systems. Ann. Rev. Control **32**(2), 229–252 (2008)
39. Y.J. Zhang, Y. Yang, Y. Zhao, G.H. Wen, Distributed finite-time tracking control for nonlinear multi-agent systems subject to external disturbances. Int. J. Control **86**(1), 1–12 (2013)
40. Y.M. Zhang, X. Yu, Y.H. Qu, D. Liu, New developments on key techniques in UAV autonomous control. Sci. Technol. Rev. **35**(7), 39–48 (2017)
41. N. Zhou, Y.Q. Xia, M.Y. Fu, Y. Li, Distributed cooperative control design for finite-time attitude synchronisation of rigid spacecraft. IET Contr. Theory Appl. **9**(10), 1561–1570 (2015)
42. S.Q. Zhu, D.W. Wang, Adversarial ground target tracking using UAVs with input constraints. J. Intell. Robot. Syst. **65**(1), 521–532 (2012)
43. S.Q. Zhu, D.W. Wang, C.B. Low, Cooperative control of multiple UAVs for source seeking. J. Intell. Robot. Syst. **70**(1–4), 293–301 (2013)
44. A.M. Zou, K.D. Kumar, Distributed attitude coordination control for spacecraft formation flying. IEEE Trans. Aerosp. Electron. Syst. **48**(2), 1329–1346 (2012)
45. A.M. Zou, K.D. Kumar, Robust attitude coordination control for spacecraft formation flying under actuator failures. J. Guid. Control Dynam. **35**(4), 1247–1255 (2012)

Chapter 5
Distributed FTCC of Multi-UAVs with Multiple Leader UAVs

5.1 Introduction

In general, cooperative control problem can be classified into three aspects: regulation problem, leader–follower tracking control problem with one leader, and containment control problem with multiple leaders [2, 3, 7, 13, 15, 18]. For the regulation problem (also known as leaderless consensus), each UAV in the communication network is driven to an unprescribed state, and all UAVs eventually achieve consensus. For the tracking control problem with one leader, all follower UAVs try to track the trajectory of leader UAV. Compared with the regulation problem and the tracking control problem with one leader, containment control problem involves multiple leader UAVs.

In the containment control framework, all follower UAVs are driven into the convex hull spanned by the trajectories of all leader UAVs. As illustrated in Fig. 5.1, when multi-UAVs are needed to perform a task cooperatively in an unknown dangerous region, follower UAVs may be safer and easier to avoid obstacles if they converge into the convex hull. In such circumstances, only leader UAVs are equipped with high-precision sensors to sense the obstacles and dangers, resulting in low sensor cost. If each UAV is equipped with high-precision sensor, the task execution will be costly. Despite the numerous results on the cooperative control of multi-UAVs, very few results on the cooperative tracking control problem with multiple leaders have been reported in the literature. Actually, cooperative control involving multiple leaders can reduce the overall cost if only leaders are equipped with high-precision sensors, especially when the team performs a task cooperatively in an unknown dangerous/complex environment. In [8], the containment control problem of multi-UAVs was considered, and a control protocol was proposed by only using relative state information. In [9], the distributed formation containment control of multi-UAVs under both fixed and switching topologies was investigated. By combining the algebraic graph theory and stability theory, it was guaranteed that follower UAVs can converge into the convex hull spanned by leader UAVs. Although

Z. Yu et al., *Fault-Tolerant Cooperative Control of Unmanned Aerial Vehicles*,
https://doi.org/10.1007/978-981-99-7661-4_5

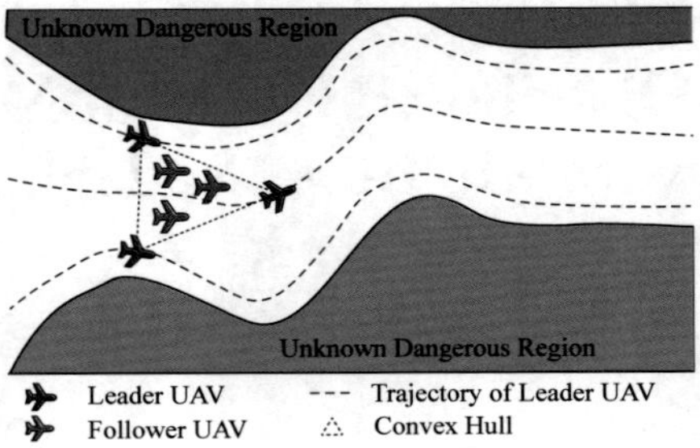

Fig. 5.1 An illustrative example of containment control

several literatures have studied the containment control of multi-UAVs, it is still an open topic and needs further research. Furthermore, to the authors' best knowledge, the containment control schemes for multi-UAVs against actuator faults should also be developed.

As analyzed in Chaps. 2 and 3, in real-world applications, unlimited control signals are not available due to physical actuator characteristics [1, 6, 10, 14, 17]. The input saturation in applications may lead to system performance degradation or even system instability if the input saturation is not rapidly compensated. From the above discussions, one question will be naturally arisen: Can multi-UAVs (including multiple leader UAVs and multiple follower UAVs) achieve the containment flying in the presence of actuator faults, input saturation, and external disturbances by using distributed containment control protocol? To the best of the authors' knowledge, very few solutions have been reported to solve the above question, which in fact motivates this chapter.

Inspired by the above analysis, this chapter presents a distributed FTCC scheme for multi-UAVs consisting of multiple leader UAVs and multiple follower UAVs. Distributed SMO, DO, and high-gain observer (HGO) are integrated to facilitate the control design. Compared with other existing works, the main contributions of this chapter are as follows:

(1) A distributed FTCC scheme is designed to steer multiple follower UAVs into the convex hull spanned by the altitudes of all leader UAVs even when follower UAVs are encountered by actuator faults, external disturbances, and input saturation simultaneously.
(2) A distributed SMO is proposed for each follower UAV to estimate its unknown altitude reference, which is in the convex hull. Based on the estimated altitude reference, each follower UAV will try to track the reference under the supervision of the proposed control laws.
(3) DO technique is utilized to estimate the external disturbances and actuator faults as a lumped uncertainty. To avoid the repetitive derivatives inherent in the traditional backstepping framework, the HGO technique is used to iteratively design the control law. Moreover, an auxiliary dynamic system is constructed to compensate for the input saturation.

The rest of this chapter is organized as follows. Faulty UAV longitudinal dynamics, basic graph theory, HGO, and control objective are given in Sect. 5.2. Section 5.3 describes the distributed SMO design, the distributed FTCC design, and the velocity controller design. Simulation results are presented in Sect. 5.4 to demonstrate the effectiveness of the proposed distributed control scheme, followed by some concluding remarks in Sect. 5.5.

(Remark: The main control schemes and contents of this chapter are from the published journal paper "Z. Q. Yu, Y. H. Qu, and Y. M. Zhang. Fault-tolerant containment control of multiple unmanned aerial vehicles based on distributed sliding-mode observer, Journal of Intelligent & Robotic Systems, 2019, 93(4): 163–177." The authors appreciate the permission from the Springer Nature to reuse the results published in the relevant Journal.)

5.2 Preliminaries and Problem Formulation

5.2.1 *Faulty UAV Longitudinal Dynamics*

Consider a group of UAVs including N follower UAVs and M leader UAVs. The leader UAVs are used to form the convex hull. $F = \{1, \ldots, N\}$ and $L = \{N+1, \ldots, N+M\}$ represent the follower set and the leader set, respectively. By recalling the longitudinal model (2.8)–(2.12) and considering external disturbances, the velocity subsystem of the ith follower UAV can be transformed as

$$\dot{V}_i = f_{iv} + g_{iv}T_i + d_{i1} \tag{5.1}$$

where $f_{iv} = -\bar{q}_i s_i (C_{iD0} + C_{iD\alpha}\alpha_i + C_{iD\alpha^2}\alpha_i^2)/m_i - g\sin\gamma_i$, $g_{iv} = \cos\alpha_i/m_i$.

Similar to Chaps. 3 and 4, the attitude subsystem of the ith UAV can be transformed as

$$\dot{z}_i = g_{iz}\gamma_i \tag{5.2}$$

$$\dot{\gamma}_i = f_{i\gamma} + g_{i\gamma}\theta_i + d_{i2} \tag{5.3}$$

$$\dot{\theta}_i = q_i \tag{5.4}$$

$$\dot{q}_i = f_{iq} + g_{iq}\delta_{ie} + d_{i3} \tag{5.5}$$

where g_{iz}, $f_{i\gamma}$, $g_{i\gamma}$, f_{iq}, g_{iq} are given by

$$g_{iz} = V_i$$

$$f_{i\gamma} = -m_i g\cos\gamma_i + \bar{q}_i s_i C_{iL0} - \bar{q}_i s_i C_{iL\alpha}\gamma_i/m_i V_i$$

$$g_{i\gamma} = \bar{q}_i s_i C_{iL\alpha}/(m_i V_i)$$

$$f_{iq} = \bar{q}_i s_i c_i \left(C_{im0} + C_{im\alpha}\alpha_i + C_{imq}\frac{c_i}{2V_i}q_i \right) / I_{iy}$$

$$g_{iq} = \frac{\bar{q}_i s_i c_i C_{im\delta_e}}{I_{iy}}$$

In this chapter, elevator faults including loss-of-effectiveness and bias faults are considered for the follower UAVs. Due to such faults, the actuator input signal is not identical to the designed control input signal. Similar to the fault models (3.6) in Chap. 3 and (4.7) in Chap. 4, to handle the actuator faults encountered by the follower UAVs, the actuator fault is modeled as

$$\delta_{ie} = \rho_i \delta_{ie0} + u_{if} \tag{5.6}$$

where $i \in F$, $0 < \rho_i \leq 1$ is the unknown actuator effectiveness factor and u_{if} is the bounded bias signal. δ_{ie} and δ_{ie0} are the applied control input signal and designed control input signal of the ith follower UAV, respectively.

Then, substituting (5.6) into (5.5) yields

$$\begin{aligned} \dot{q}_i &= f_{iq} + g_{iq}(\rho_i \delta_{ie0} + u_{if}) + d_{i3} \\ &= f_{iq} + g_{iq}\delta_{ie0} + g_{iq}(\rho_i - 1)\delta_{ie0} + g_{iq}u_{if} + d_{i3} \\ &= f_{iq} + g_{iq}\delta_{ie0} + \Delta_i \end{aligned} \tag{5.7}$$

where $\Delta_i = g_{iq}(\rho_i - 1)\delta_{ie0} + g_{iq}u_{if} + d_{i3}$ is the lumped uncertainty including actuator faults and external disturbances.

Assumption 5.1 The derivatives of external disturbances d_{i1}, d_{i2}, and lumped uncertainty Δ_i are bounded such that $|\dot{d}_{i1}| \leq \bar{d}_{i1}$, $|\dot{d}_{i2}| \leq \bar{d}_{i2}$, $|\dot{\Delta}_i| \leq \bar{\Delta}_i$, where $\bar{d}_{i1}$, $\bar{d}_{i2}$, and $\bar{\Delta}_i$ are unknown positive constants and will be used for stability analysis. Moreover, it is assumed that the control gain functions g_{iv}, $g_{i\gamma}$, g_{iq} are bounded and have non-zero values, i.e., $0 < |g_{iv}| \leq g_{ivm}$, $0 < |g_{i\gamma}| \leq g_{i\gamma m}$, $0 < |g_{iq}| \leq g_{iqm}$.

5.2.2 Basic Graph Theory

The communication network of $N + M$ UAVs in this chapter is an undirected, fixed topology $\mathcal{G} = (\mathcal{V}, \mathcal{E})$, where $\mathcal{V} = \{v_1, \ldots, v_i, \ldots, v_{N+M}\}$ represents N follower UAVs ($i = 1, \ldots, N$) and M leader UAVs ($i = N+1, \ldots, N+M$), and $\mathcal{E} = \mathcal{V} \times \mathcal{V}$ is the edge set representing the information flows among UAVs. $(v_i, v_j) \in \mathcal{E}$ indicates that there is an edge from UAV#j to UAV#i. The set of neighbors of UAV#i is represented by $N_i = \{v_j | (v_i, v_j) \in \mathcal{E}\}$, which is the set of UAVs with edges incoming to UAV#i. It is assumed that the motion of leader UAVs is independent from that of the follower UAVs and the leader UAVs are used to give references to the follower UAVs. The follower UAVs#1-#N have at least one neighbor, and the leader UAVs#$(N+1)$-#$(N+M)$ have no neighbors. $\mathcal{A} = [a_{ij}] \in R^{(N+M)\times(N+M)}$ is the adjacency matrix with nonnegative elements, where $a_{ij} > 0$ if $(v_i, v_j) \in \mathcal{E}$, otherwise $a_{ij} = 0$, $i = 1, \ldots, N+M$, $j = 1, \ldots, N+M$. It should be noted that $a_{ij} = 0$, $i = N+1, \ldots, N+M$, $j = 1, \ldots, N+M$ due to the fact that leader UAVs have no neighbors. The Laplacian matrix $\mathcal{L}$ is defined as $\mathcal{L} = \mathcal{D} - \mathcal{A} \in R^{(N+M)\times(N+M)}$, where $\mathcal{D} = \text{diag}\{d_1, d_2, \ldots, d_N\}$, $d_i = \sum_{j=1, j\neq i}^{N+M} a_{ij}$, $i = 1, \ldots, N+M$. By considering the information flows among follower UAVs and the information flows between follower UAVs and leader UAVs, the Laplacian matrix $\mathcal{L}$ is given by

$$\mathcal{L} = \begin{bmatrix} \mathcal{L}_1 & \mathcal{L}_2 \\ 0_{M\times N} & 0_{M\times M} \end{bmatrix} \tag{5.8}$$

where $\mathcal{L}_1 \in R^{N\times N}$ is the matrix related to the information flows among N follower UAVs. $\mathcal{L}_2 \in R^{N\times M}$ is the matrix associated with the information flows from leader UAVs to follower UAVs.

Assumption 5.2 For each follower UAV in the communication network, there exists at least one leader UAV that has a path to the follower UAV.

Lemma 5.1 *With Assumption 5.2, the matrix $\mathcal{L}_1$ is positive definite, and each entry of $-\mathcal{L}_1^{-1}\mathcal{L}_2$ is nonnegative. Moreover, each row sum of $-\mathcal{L}^{-1}\mathcal{L}_2$ is equal to one [12].*

Definition 5.1 Let X be a set in a real vector space $V \subseteq R^P$, where P is a positive integer. The convex hull $Co(X)$ is the minimal convex set containing all points in X, which is defined as [11]

$$Co(X) = \left\{ \sum_{i=1}^{P} c_i x_i | x_i \in X, c_i \geq 0, \sum_{i=1}^{P} c_i = 1 \right\} \tag{5.9}$$

5.2.3 High-Gain Observer

To eliminate the "differential explosion" problem in the traditional backstepping architecture, the HGO technique is used to estimate the virtual control signals and their derivatives.

Lemma 5.2 *Consider the following linear system:*

$$\begin{aligned} \epsilon\dot{\pi}_i &= \pi_{i+1}, \quad i = 1, 2, \ldots, n-1 \\ \epsilon\dot{\pi}_n &= -l_1\pi_n - l_2\pi_{n-1} - \cdots - l_{n-1}\pi_2 - \pi_1 + y(t) \end{aligned} \tag{5.10}$$

where ϵ is a small positive constant. $\pi_1, \pi_2, \ldots, \pi_n$ are the states of the HGO. $l_1, \ldots,$ l_{n-1} are chosen such that $s^n + l_1 s^{n-1} + \cdots + l_{n-1}s + 1$ is Hurwitz. If the input signal

$y(t)$ and its n derivatives are bounded, then the following expression holds [4, 5]:

$$\zeta_k := \frac{\pi_k}{\epsilon^{k-1}} - y^{(k-1)} = -\epsilon \varsigma^{(k)}, \quad k = 1, 2, \ldots, n \tag{5.11}$$

where $\varsigma = \pi_n + l_1 \pi_{i+1n-1} + \ldots + l_{n-1}\pi_1$. $\varsigma^{(k)}$ represents the kth derivative of ς.

Remark 5.1 If the HGO states are bounded, the estimation error ζ_k is bounded. Lemma 5.2 shows that π_{k+1}/ϵ^k can converge to $y^{(k)}$ with a bounded error.

5.2.4 Control Objective

Consider the fact that only a subset of follower UAVs has access to the leader UAVs, and the control objective here is to design a set of distributed FTCC laws for follower UAVs (5.1)–(5.5) based on the neighboring UAVs' information. Therefore, the altitudes of follower UAVs can converge into the convex hull spanned by the altitudes of leader UAVs in the presence of actuator faults, external disturbances, and input saturation. Moreover, the velocities of follower UAVs can track the velocities of leader UAVs with bounded errors.

5.3 Main Results

In this section, the distributed SMO is first designed to estimate the unknown individual altitude reference for each follower UAV. Then, the FTC law is designed for each UAV to track the estimated reference. Finally, velocity controller is proposed for each UAV to track the velocities of leader UAVs.

5.3.1 Distributed Sliding-Mode Observer

To estimate the individual altitude reference for each follower UAV, which is in the convex hull, the distributed SMO is designed as

$$\begin{aligned} \dot{\hat{z}}_{id} =& -\beta_1 \left[\sum_{j \in F \cup L} a_{ij} \left(\hat{z}_{id} - \hat{z}_{jd} \right) \right] \\ & - \beta_2 \text{sign} \left[\sum_{j \in F \cup L} a_{ij} \left(\hat{z}_{id} - \hat{z}_{jd} \right) \right], i \in F \end{aligned} \tag{5.12}$$

where $\hat{z}_{id}$ denotes the estimation of the ith follower UAV's unknown altitude reference z_{id}. $\hat{z}_{jd} = z_{j0}$, $j \in L$. $\beta_1 > 0$, $\beta_2 > \bar{z}_{dl}$ with $\bar{z}_{dl}$ being the upper bound of $\dot{z}_{j0}$, $j \in L$.

It should be noted that the SMO is designed in a distributed manner such that the observer is embedded in each follower UAV. Denote $z_L = [z_{(N+1)0}, \ldots, z_{(N+M)0}]^T$, $z_d = [z_{1d}, \ldots, z_{Nd}]^T = -\mathcal{L}_1^{-1}\mathcal{L}_2 z_L$, $\hat{z}_d = [\hat{z}_{1d}, \ldots, \hat{z}_{Nd}]^T$, $z = [z_1, \ldots, z_N]^T$. From Lemma 5.1 and Definition 5.1, $z \to z_d$ means that z_i, $i \in F$, converges into the convex hull $Co\{z_{i0}, i \in L\}$. Actually, $z \to z_d$ can be divided into two parts: $z \to \hat{z}_d$ and $\hat{z}_d \to z_d$. For the convergence of z to $\hat{z}_d$, a distributed controller will be constructed to achieve it. For the convergence of $\hat{z}_d$ to z_d, one has the following lemma.

Lemma 5.3 *If the communication network is connected and the distributed SMO is designed as (5.12) for each UAV to estimate its individual unknown altitude reference z_{id}, then the estimated altitude reference $\hat{z}_{id}$ can converge to z_{id}, $i \in F$.*

Proof Let $\tilde{z}_{id} = \hat{z}_{id} - z_{id}$, $i \in F$, and $\tilde{z}_d = [\tilde{z}_{1d}, \ldots, \tilde{z}_{Nd}]^T$ denote the observation error vector. Then, taking the time derivative of $\tilde{z}_d$ gives

$$\dot{\tilde{z}}_d = -\beta_1\mathcal{L}_1\tilde{z}_d - \beta_2 \text{sign}\left(\mathcal{L}_1\tilde{z}_d\right) + \left(\mathcal{L}_1^{-1}\mathcal{L}_2\right)\dot{z}_L \tag{5.13}$$

Choose the Lyapunov function as

$$L_0 = \frac{1}{2}\tilde{z}_d^T\mathcal{L}_1\tilde{z}_d \tag{5.14}$$

By taking the time derivative of L_0, one has

$$\begin{aligned}\dot{L}_0 =& \tilde{z}_d^T\mathcal{L}_1\dot{\tilde{z}}_d \\ =& \tilde{z}_d^T\mathcal{L}_1\left[-\beta_1\mathcal{L}_1\tilde{z}_d - \beta_2 \text{sign}\left(\mathcal{L}_1\tilde{z}_d\right) + \left(\mathcal{L}_1^{-1}\mathcal{L}_2\right)\dot{z}_L\right] \\ =& -\beta_1\tilde{z}_d^T\mathcal{L}_1\mathcal{L}_1\tilde{z}_d - \beta_2\tilde{z}_d^T\mathcal{L}_1 \text{sign}\left(\mathcal{L}_1\tilde{z}_d\right) + \tilde{z}_d^T\mathcal{L}_2\dot{z}_L\end{aligned} \tag{5.15}$$

To this end, the following inequality can be derived:

$$\begin{aligned}\tilde{z}_d^T\mathcal{L}_2\dot{z}_L =& \tilde{z}_d^T\mathcal{L}_1\mathcal{L}_1^{-1}\mathcal{L}_2\dot{z}_L \\ \leq& ||\mathcal{L}_1\tilde{z}_d||_1 \cdot ||\mathcal{L}_1^{-1}\mathcal{L}_2\dot{z}_L||_\infty \leq \bar{z}_{dl}||\mathcal{L}_1\tilde{z}_d||_1\end{aligned} \tag{5.16}$$

Substituting (5.16) into (5.15) yields

$$\begin{aligned}\dot{L}_0 \leq &- \beta_1 \tilde{z}_d^T \mathcal{L}_1 \mathcal{L}_1 \tilde{z}_d - \beta_2 \mathcal{L}_1 \tilde{z}_d \text{sign}(\mathcal{L}_1 \tilde{z}_d) \\ &+ \bar{z}_{dl} ||\mathcal{L}_1 \tilde{z}_d||_1 \\ \leq &- \beta_1 \tilde{z}_d^T \mathcal{L}_1 \mathcal{L}_1 \tilde{z}_d - (\beta_2 - \bar{z}_{dl}) ||\mathcal{L}_1 \tilde{z}_d||_1 \\ \leq &- \frac{\beta_1 \lambda_{\min}(\mathcal{L}_1 \mathcal{L}_1)}{\lambda_{\max}(\mathcal{L}_1)} L_0 - \frac{2^{\frac{1}{2}} (\beta_2 - \bar{z}_{dl}) \lambda_{\min}(\mathcal{L}_1)}{\lambda_{\max}^{\frac{1}{2}}} L_0^{\frac{1}{2}}\end{aligned} \tag{5.17}$$

According to (5.17) and finite-time stability theory, it can be seen that if $\beta_1 > 0$, $\beta_2 > \bar{z}_{dl}$, then $\tilde{z}_d \to 0$ in finite time. This completes the proof. □

Remark 5.2 Since $\bar{z}_{dl}$ is unknown to a subset of follower UAVs, to guarantee the convergence of the distributed SMO, β_2 can be set relatively large.

Remark 5.3 Lemma 5.3 states that the estimated altitude reference for each follower UAV can converge into the convex hull spanned by all leader UAVs. In the next section, a distributed controller will be designed to achieve $z - \hat{z}_d \to \mathbf{0}$. After that, $z - z_d \to \mathbf{0}$ will be achieved by the combination of $z - \hat{z}_d \to \mathbf{0}$ and $\hat{z}_d - z_d \to \mathbf{0}$.

5.3.2 Distributed FTCC for Attitude Subsystem

In this section, the distributed FTCC scheme is designed for attitude subsystem including altitude, flight path angle, pitch angle, and pitch rate dynamics. The DO technique is utilized to estimate the actuator faults and external disturbances. In the control scheme, backstepping design architecture with the integration of HGO is used to recursively design the distributed control law at each step.

Define the altitude tracking error of the ith follower UAV with respect to the estimated altitude reference $\hat{z}_{id}$ as $S_{i1} = z_i - \hat{z}_{id}$, $i \in F$. Define $\boldsymbol{S}_1 = [S_{11}, \ldots, S_{N1}]^T$ as the error vector. The flight path angle command γ_{id} of the ith follower UAV is chosen as

$$\gamma_{id} = \frac{-k_{11} S_{i1} - k_{12} \int S_{i1} dt + \dot{\hat{z}}_{id}}{V_i} \tag{5.18}$$

where k_{11} and k_{12} are positive design parameters.

Remark 5.4 If $k_{11} > 0$ and $k_{12} > 0$ are chosen and γ_i is controlled to follow γ_{id}, then the error S_{i1} will be regulated to zero [16].

Then, the control signals for γ_i, θ_i, and q_i will be the focus, and three steps are involved to obtain the overall control signal δ_{ie0}.

Step 1 Define $\boldsymbol{\gamma} = [\gamma_1, \ldots, \gamma_N]^T$ and the flight path angle tracking error of the ith follower UAV as

$$S_{i2} = \gamma_i - \gamma_{id} \tag{5.19}$$

Taking the time derivative of (5.19) along the trajectory of (5.3) yields

$$\begin{aligned} \dot{S}_{i2} &= \dot{\gamma}_i - \dot{\gamma}_{id} \\ &= f_{i\gamma} + g_{i\gamma}\theta_i + d_{i2} - \dot{\gamma}_{id} \end{aligned} \tag{5.20}$$

Then, based on the flight path angle tracking error S_{i2}, one designs the virtual control signal $\bar{\theta}_{id}$ as

$$\bar{\theta}_{id} = g_{i\gamma}^{-1}\left(-k_{21}S_{i2} - f_{i\gamma} + \dot{\gamma}_{id} - \hat{d}_{i2}\right) \tag{5.21}$$

where $k_{21} > 0$ is the design parameter and $\hat{d}_{i2}$ is the estimated value of d_{i2}.

To estimate the external disturbance, DO is designed as

$$\begin{cases} \hat{d}_{i2} = \hat{\bar{z}}_{i1} + k_{22}S_{i2} \\ \dot{\hat{\bar{z}}}_{i1} = -k_{22}(f_{i\gamma} + g_{i\gamma}\theta_i + \hat{\bar{z}}_{i1} + k_{22}S_{i2} - \dot{\gamma}_{id}) \end{cases} \tag{5.22}$$

where k_{22} is the positive design parameter and $\hat{\bar{z}}_{i1}$ is the state variable of the DO. The estimation error is defined as $\tilde{d}_{i2} = d_{i2} - \hat{d}_{i2}$ and $\hat{\boldsymbol{d}}_2 = [\hat{d}_{12}, \ldots, \hat{d}_{N2}]^T$.

Then, taking the time derivative of $\hat{d}_{i2}$ gives

$$\begin{aligned} \dot{\hat{d}}_{i2} =& \dot{\hat{\bar{z}}}_{i1} + k_{22}\dot{S}_2 \\ =& -k_{22}(f_{i\gamma} + g_{i\gamma}\theta_i + \hat{\bar{z}}_{i1} + k_{22}S_{i2} - \dot{\gamma}_{id}) \\ & + k_{22}(f_{i\gamma} + g_{i\gamma}\theta_i + d_{i2} - \dot{\gamma}_{id}) \\ =& k_{22}\tilde{d}_{i2} \end{aligned} \tag{5.23}$$

To solve the repetitive time derivatives of the virtual control signal $\bar{\theta}_{id}$ in traditional backstepping architecture, the HGO is employed to estimate the first derivative of the virtual control signal, which is designed as

$$\begin{cases} \epsilon_1\dot{\theta}_{id} = \pi_1 \\ \epsilon_1\dot{\pi}_1 = -l_1\pi_1 - \theta_{id} + \bar{\theta}_{id} \end{cases} \tag{5.24}$$

where ϵ_1 and l_1 are positive constants to be designed. θ_{id} and $\frac{\pi_1}{\epsilon_1}$ are the estimated values of $\bar{\theta}_{id}$ and $\dot{\bar{\theta}}_{id}$, respectively. The estimation error with respect to $\bar{\theta}_{id}$ is defined as $\Delta\theta_{id} = \theta_{id} - \bar{\theta}_{id}$.

Step 2 Define the pitch angle tracking error as $S_{i3} = \theta_i - \theta_{id}$, and taking the time derivative of S_{i3} along the trajectory of (5.4) yields

$$\dot{S}_{i3} = q_i - \dot{\theta}_{id} \tag{5.25}$$

Then, the virtual control signal is designed as

$$\bar{q}_{id} = -k_{31}S_{i3} + \dot{\theta}_{id} - \zeta_{i1} \tag{5.26}$$

where $k_{31} > 0$ is the design parameter. ζ_{i1} is an auxiliary dynamic signal to be designed at the next step. $\boldsymbol{\theta} = [\theta_1, \ldots, \theta_N]^T$ and $\boldsymbol{q} = [q_1, \ldots, q_N]^T$ are the pitch angle and pitch rate vectors, respectively.

Similar to (5.24), $\bar{q}_{id}$ is passed through the HGO to obtain q_{id} and estimate $\dot{\bar{q}}_{id}$, and the HGO is given by

$$\begin{cases} \epsilon_2 \dot{q}_{id} = \pi_2 \\ \epsilon_2 \dot{\pi}_2 = -l_2 \pi_2 - q_{id} + \bar{q}_{id} \end{cases} \tag{5.27}$$

where ϵ_2 and l_2 are positive design parameters. q_{id} is the estimation of $\bar{q}_{id}$, and the estimation error is defined as $\Delta_{qid} = q_{id} - \bar{q}_{id}$. π_2 is the state variable of the HGO.

Step 3 Define the tracking error between q_i and q_{id} as $S_{i4} = q_i - q_{id}$. Taking the time derivatives of S_{i4} along (5.7) gives

$$\dot{S}_{i4} = f_{iq} + g_{iq}\delta_{ie0} + \Delta_i - \dot{q}_{id} \tag{5.28}$$

Then, design the elevator input signal as

$$\bar{\delta}_{ie0} = g_{iq}^{-1}(-k_{41}S_{i4} - f_{iq0} + \dot{q}_{id} - \hat{\Delta}_i + \frac{k_{41}}{2}\zeta_{i1}) \tag{5.29}$$

where $k_{41} > 0$ is the design parameter. $\hat{\Delta}_i$ is the estimated value of Δ_i. ζ_i is an auxiliary dynamic signal to be designed to compensate for the actuator input saturation.

To estimate the lumped uncertainty Δ_i, the DO is designed as

$$\begin{cases} \hat{\Delta}_i = \hat{\bar{z}}_{i2} + k_{42}S_{i4} \\ \dot{\hat{\bar{z}}}_{i2} = -k_{42}(f_{iq} + g_{iq}\delta_{ie0} + \hat{\bar{z}}_{i2} + k_{42}S_{i4} - \dot{q}_{id}) \end{cases} \tag{5.30}$$

where k_{42} is the positive parameter to be designed and $\hat{\bar{z}}_{i2}$ is the state variable of the DO. Define the estimation error as $\tilde{\Delta}_i = \Delta_i - \hat{\Delta}_i$ and $\hat{\Delta} = [\hat{\Delta}_1, \ldots, \hat{\Delta}_N]^T$.

Similar to (5.23), the time derivative of $\hat{\Delta}_i$ is given by

$$\dot{\hat{\Delta}}_i = \dot{\hat{\bar{z}}}_{i2} + k_{42}\dot{S}_{i4} = k_{42}\tilde{\Delta}_i \tag{5.31}$$

In engineering applications, the actuator input is usually encountered by saturation nonlinearities, and the saturated control signal is described as

$$\delta_{ie0} = \text{sat}(\bar{\delta}_{ie0}) = \begin{cases} \delta_{i\max}, & \bar{\delta}_{ie0} \geq \delta_{i\max} \\ \bar{\delta}_{ie0}, & \delta_{i\min} < \bar{\delta}_{ie0} < \delta_{i\max} \\ \delta_{i\min}, & \bar{\delta}_{ie0} \leq \delta_{i\min} \end{cases} \tag{5.32}$$

where $\delta_{i\min}$ and $\delta_{i\max}$ are the minimum and maximum elevator deflections, respectively. $\boldsymbol{\delta}_{e0} = [\delta_{1e0}, \ldots, \delta_{Ne0}]^T$ is the elevator deflection vector.

To deal with the elevator saturation in the controller design, an auxiliary dynamic system is designed as

$$\dot{\zeta}_{i1} = -\frac{k_{41}}{2}\zeta_{i1} + g_{iq}\Delta\delta_{ie0} \tag{5.33}$$

where $\Delta\delta_{ie0} = \delta_{ie0} - \bar{\delta}_{ie0}$ is the difference between the designed input signal and applied signal.

5.3.3 *Velocity Controller Design*

In this section, all follower UAVs attempt to track the leader UAVs' velocities when follower UAVs converge into the convex hull. Define the velocity tracking error as $S_{i5} = V_i - V_{id}$, $i \in F$. Then, taking the time derivative of S_{i5} along (5.1) yields

$$\begin{aligned} \dot{S}_{i5} &= \dot{V}_i - \dot{V}_{id} \\ &= f_{iv} + g_{iv}T_i + d_{i1} - \dot{V}_{id} \end{aligned} \tag{5.34}$$

where V_{id} is the velocity command for the ith follower UAV and $V_{id} = V_{j0}$, $i \in F$, $j \in L$. $\boldsymbol{V} = [V_1, \ldots, V_N]^T$ is the velocity vector.

Design the thrust input as

$$T_{id} = g_{iv}^{-1}\left(-k_{51}S_{i5} - f_{iv} + \dot{V}_{id} - \hat{d}_{i1} + \frac{k_{51}}{2}\zeta_{i2}\right) \tag{5.35}$$

where $k_{51} > 0$ is the parameter to be designed. $\hat{d}_{i1}$ is the estimated value of d_{i1}. ζ_{i2} is the auxiliary dynamic signal designed to compensate for the input saturation.

Design the DO as

$$\begin{cases} \hat{d}_{i1} = \hat{z}_{i3} + k_{52}S_{i5} \\ \dot{\hat{z}}_{i3} = -k_{52}(f_{iv} + g_{iv}T_i + \hat{z}_{i3} + k_{52}S_{i5} - \dot{V}_{id}) \end{cases} \tag{5.36}$$

where k_{52} is the positive parameter and $\hat{\hat{z}}_{i3}$ is the state variable. Define the estimation error as $\tilde{d}_{i1} = d_{i1} - \hat{d}_{i1}$ and $\hat{\boldsymbol{d}}_1 = [\hat{d}_{11}, \ldots, \hat{d}_{N1}]^T$.

Similar to (5.23) and (5.31), the time derivative of $\hat{d}_{i1}$ is calculated as

$$\dot{\hat{d}}_{i1} = \dot{\hat{\hat{z}}}_{i3} + k_{52}\dot{S}_{i5} = k_{52}\tilde{d}_{i1} \tag{5.37}$$

The applied thrust signal subject to input saturation is given by

$$T_i = \text{sat}(T_{id}) = \begin{cases} T_{i\max}, & T_{id} \geq T_{i\max} \\ T_{id}, & T_{i\min} < T_{id} < T_{i\max} \\ T_{i\min}, & T_{id} \leq T_{i\min} \end{cases} \tag{5.38}$$

where $T_{i\min}$ and $T_{i\max}$ are the minimum thrust and maximum thrust inputs of the ith follower UAV, respectively, and $\boldsymbol{T} = [T_1, \ldots, T_N]^T$ is the thrust input vector.

To deal with the thrust input saturation, auxiliary dynamic system is constructed as

$$\dot{\zeta}_{i2} = -\frac{k_{51}}{2}\zeta_{i2} + g_{iv}\Delta T_i \tag{5.39}$$

where $\Delta T_i = T_i - T_{id}$.

The distributed FTCC scheme is constructed by (5.12), (5.18), (5.21), (5.22), (5.24), (5.26), (5.27), (5.29), (5.30), (5.33), (5.35), (5.36), and (5.39), which is illustrated as Fig. 5.2. Furthermore, based on the estimated altitude reference $\hat{z}_{id}$ and the velocity reference V_{id}, $i \in F$, the control scheme of the ith follower UAV is illustrated as Fig. 5.3.

Remark 5.5 The cooperative control variables in this chapter are the altitudes of follower UAVs. The leader UAVs are used to generate references (altitude reference

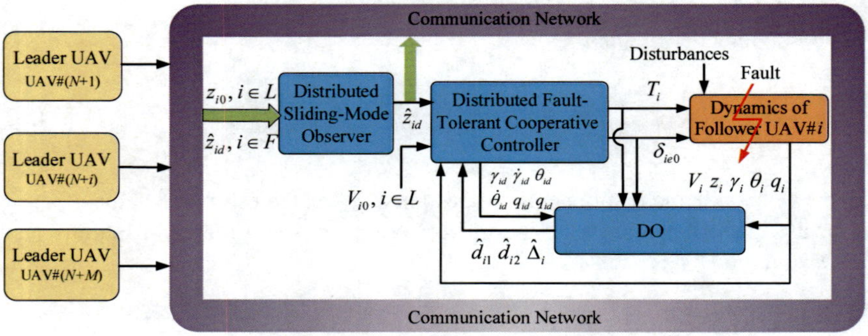

Fig. 5.2 Distributed FTCC scheme

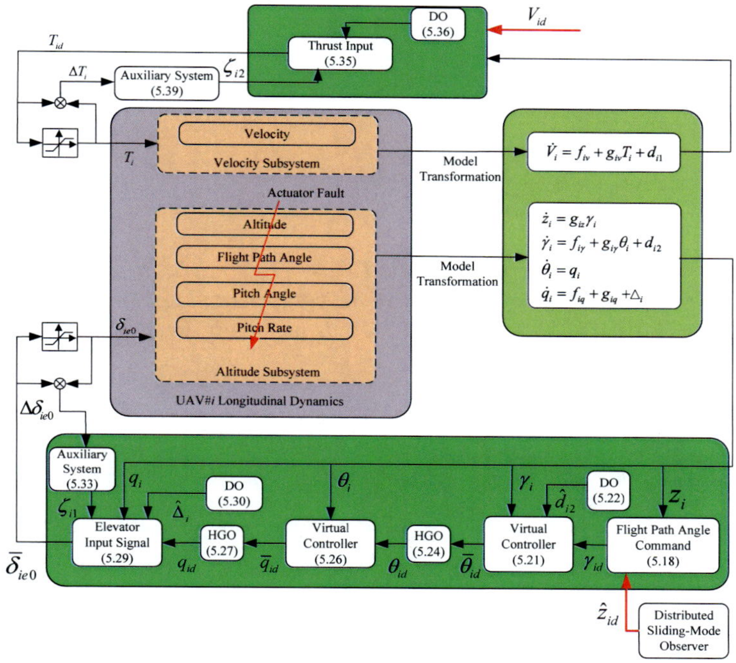

Fig. 5.3 The developed control scheme for the ith follower UAV based on the velocity reference V_{id} and the estimated altitude reference $\hat{z}_{id}$

z_{i0} and velocity reference V_{i0}, $i \in L$) to follower UAVs. To achieve the altitude cooperative control, the altitude controller (5.18), flight path angle controller (5.21), pitch angle controller (5.26), and the pitch rate controller (5.29) are constructed to regulate the tracking errors. Furthermore, to regulate the velocity tracking errors of all follower UAVs with respect to the leader UAVs' velocities, i.e., $V_i \to V_{id}$, where $V_{id} = V_{j0}$, $i \in F$, $j \in L$, the velocity controller is constructed as (5.35).

5.3.4 *Stability Analysis*

Theorem 5.1 *Consider a group of N follower UAVs and M leader UAVs. The longitudinal dynamics of follower UAVs is described as (5.1)–(5.5), and the actuator fault is given by (5.6), and then the altitudes of all follower UAVs converge into the convex hull spanned by the altitudes of all leader UAVs if the distributed SMO is designed as (5.12), the control laws are designed as (5.18), (5.21), (5.26), (5.29), (5.35), the DOs are designed as (5.22), (5.30), (5.36), the HGOs are designed as (5.24), (5.27), and auxiliary dynamic signals are constructed as (5.33), (5.39). Moreover, the* *velocities of all follower UAVs can track the leader UAVs with bounded tracking errors.*

Proof Choose the Lyapunov function candidate as

$$L = \sum_{i=1}^{N} L_i, \ L_i = L_{i1} + L_{i2} + L_{i3} + L_{i4} \tag{5.40}$$

where $L_{i1} = \frac{1}{2}S_{i2}^2 + \frac{1}{2}\tilde{d}_{i2}^2$, $L_{i2} = \frac{1}{2}S_{i3}^2$, $L_{i3} = \frac{1}{2}E_{i4}^2 + \frac{1}{2}\tilde{\Delta}_i^2$, $L_{i4} = \frac{1}{2}E_{i5}^2 + \frac{1}{2}\tilde{d}_{i1}^2$. $E_{i4} = S_{i4} - \zeta_{i1}$, and $E_{i5} = S_{i5} - \zeta_{i2}$ are the compensated tracking errors.

By using (5.20), (5.21), (5.23), and Young's inequality, the time derivative of L_{i1} is given by

$$\begin{aligned}
\dot{L}_{i1} =& S_{i2}(f_{i\gamma} + g_{i\gamma}\theta_i + d_{i2} - \dot{\gamma}_{id}) + \tilde{d}_{i2}(\dot{d}_{i2} - k_{22}\tilde{d}_{i2}) \\
=& S_{i2}[f_{i\gamma} + g_{i\gamma}(\bar{\theta}_{id} + \Delta\theta_{id} + S_{i3}) + d_{i2} - \dot{\gamma}_{id}] + \tilde{d}_{i2}\dot{d}_{i2} - k_{22}\tilde{d}_{i2}^2 \\
=& S_{i2}[-k_{21}S_{i2} + g_{i\gamma}\Delta\theta_{id} + g_{i\gamma}S_{i3}] + \tilde{d}_{i2}\dot{d}_{i2} - k_{22}\tilde{d}_{i2}^2 \\
\leq& -(k_{21} - |g_{i\gamma m}|)S_{i2}^2 - \left(k22 - \frac{1}{2}\right)\tilde{d}_{i2}^2 + \frac{|g_{i\gamma m}|}{2}S_{i3}^2 \\
&+ \frac{|g_{i\gamma m}|}{2}\Delta\theta_{id}^2 + \frac{\bar{d}_{i2}^2}{2}
\end{aligned} \tag{5.41}$$

Differentiating L_{i2} with respect to time and applying (5.25), (5.26), the following expression can be obtained:

$$\begin{aligned}
\dot{L}_{i2} =& S_{i3}\dot{S}_{i3} \\
=& S_{i3}(\bar{q}_{id} + \Delta q_{id} + S_{i4} - \dot{\theta}_{id}) \\
=& -k_{31}S_{i3}^2 + S_{i3}\Delta q_{id} + S_{i3}E_{i4} \\
\leq& -(k_{31} - 1)S_{i3}^2 + \frac{1}{2}E_{i4}^2 + \frac{1}{2}\Delta q_{id}^2
\end{aligned} \tag{5.42}$$

Considering (5.28), (5.29), (5.31), (5.33) and differentiating L_{i3} with respect to time yield

$$\begin{aligned}
\dot{L}_{i3} =& E_{i4}[f_{iq} + g_{iq}(\bar{\delta}_{ie0} + \Delta\delta_{ie0}) + \Delta_i - \dot{q}_{id} - \dot{\zeta}_{i1}] + \tilde{\Delta}_i(\dot{\Delta}_i - k_{42}\tilde{\Delta}_i) \\
\leq& -\left(k_{41} - \frac{1}{2}\right)E_{i4}^2 - (k_{42} - 1)\tilde{\Delta}_i^2 + \frac{\bar{\Delta}_i^2}{2}
\end{aligned} \tag{5.43}$$

In view of (5.34), (5.35), (5.36), (5.39), and Young's inequality, one has

$$\begin{aligned}
\dot{L}_{i4} =& E_{i5}\dot{E}_{i5} + \tilde{d}_{i1}(\dot{d}_{i1} - \dot{\hat{d}}_{i1}) \\
=& E_{i5}(f_{iv} + g_{iv}T_i + d_{i1} - \dot{V}_{id} - \dot{\zeta}_{i2}) + \tilde{d}_{i1}(\dot{d}_{i1} - k_{52}\tilde{d}_{i1}) \\
=& E_{i5}(f_{iv} + g_{iv}T_{id} + g_{iv}\Delta T_i + d_{i1} - \dot{V}_{id} - \dot{\zeta}_{i2}) + \tilde{d}_{i1}\dot{d}_{i1} - k_{52}\tilde{d}_{i1}^2 \\
=& E_{i5}\left(-k_{51}S_{i5} + \tilde{d}_{i1} + k_{51}\zeta_{i2}\right) + \tilde{d}_{i1}\dot{d}_{i1} - k_{52}\tilde{d}_{i1}^2 \\
\leq & -\left(k_{51} - \frac{1}{2}\right)E_{i5}^2 - (k_{52} - 1)\tilde{d}_{i1}^2 + \frac{\bar{d}_{i1}^2}{2}
\end{aligned} \tag{5.44}$$

By combining (5.41), (5.42), (5.43), and (5.44), the following inequality can be obtained:

$$\begin{aligned}
\dot{L}_i =& \dot{L}_{i1} + \dot{L}_{i2} + \dot{L}_{i3} + \dot{L}_{i4} \\
\leq & -(k_{21} - |g_{i\gamma m}|)S_{i2}^2 - \left(k_{22} - \frac{1}{2}\right)\tilde{d}_{i2}^2 + \frac{|g_{i\gamma m}|}{2}S_{i3}^2 \\
& + \frac{|g_{i\gamma m}|}{2}\Delta\theta_{id}^2 + \frac{\bar{d}_{i2}^2}{2} - (k_{31} - 1)S_{i3}^2 + \frac{1}{2}E_{i4}^2 + \frac{1}{2}\Delta q_{id}^2 \\
& - \left(k_{41} - \frac{1}{2}\right)E_{i4}^2 - (k_{42} - 1)\tilde{\Delta}_i^2 + \frac{\bar{\Delta}_i^2}{2} \\
& - \left(k_{51} - \frac{1}{2}\right)E_{i5}^2 - (k_{52} - 1)\tilde{d}_{i1}^2 + \frac{\bar{d}_{i1}^2}{2} \\
\leq & -(k_{21} - |g_{i\gamma m}|)S_{i2}^2 - \left(k_{31} - 1 - \frac{|g_{i\gamma m}|}{2}\right)S_{i3}^2 \\
& - (k_{41} - 1)E_{i4}^2 - \left(k_{51} - \frac{1}{2}\right)E_{i5}^2 \\
& - (k_{52} - 1)\tilde{d}_{i1}^2 - \left(k_{22} - \frac{1}{2}\right)\tilde{d}_{i2}^2 - (k_{42} - 1)\tilde{\Delta}_i^2 + \frac{|g_{i\gamma m}|}{2}\Delta\theta_{id}^2 + \frac{\bar{d}_{i2}^2}{2} \\
& + \frac{\Delta q_{id}^2}{2} \\
& + \frac{\bar{\Delta}_i^2}{2} + \frac{\bar{d}_{i1}^2}{2}
\end{aligned} \tag{5.45}$$

Choose λ_i and σ_i as

$$\lambda_i = \min \left\{ \begin{array}{l} 2(k_{21} - |g_{i\gamma m}|), 2\left(k_{31} - 1 - \frac{|g_{i\gamma m}|}{2}\right), \\ 2(k_{41} - 1), 2\left(k_{51} - \frac{1}{2}\right), \\ 2(k_{52} - 1), 2(k_{22} - 1), 2(k_{42} - 1) \end{array} \right\} > 0 \tag{5.46}$$

$$\sigma_i = \frac{|g_{i\gamma m}|}{2} \Delta\theta_{id}^2 + \frac{\bar{d}_{i2}^2}{2} + \frac{\Delta q_{id}^2}{2} + \frac{\bar{\Delta}_i^2}{2} + \frac{\bar{d}_{i1}^2}{2} \tag{5.47}$$

Then, one has

$$\dot{L}_i \leq -\lambda_i L_i + \sigma_i \tag{5.48}$$

Therefore, the time derivative of L is given by

$$\dot{L} \leq -\lambda L + \sigma \tag{5.49}$$

where $\lambda = \min\{\lambda_i\}$, $\sigma = \sum_{i=1}^{N} \sigma_i$.

From (5.49), Remark 5.4, and Lyapunov stability theory, it can be concluded that S_{i1}, S_{i2}, S_{i3}, E_{i4}, E_{i5}, $\tilde{d}_{i1}$, $\tilde{d}_{i2}$, and $\tilde{\Delta}_i$ are UUB. Therefore, the altitudes of all follower UAVs can track the estimated altitude references $\hat{z}_{id}$, $i \in F$. When the thrust input is not saturated, $\Delta T_i = 0$. As a result, S_{i5} is UUB once the ith follower UAV is free of input saturation and velocities of all follower UAVs can track the velocity references V_{id}, $i \in F$. The velocity references are the same as the leader UAVs' velocities. Furthermore, by recalling Lemma 5.3, it can be seen that the altitudes of all follower UAVs can converge into the convex hull $Co(z_L)$. This ends the proof. □

Remark 5.6 Since sign(·) function is used in (5.12), the control chattering may be introduced due to the discontinuous design. In order to overcome the chattering problem, $\text{sign}\left[\sum_{j\in F\cup L} a_{ij}\left(\hat{z}_{id} - \hat{z}_{jd}\right)\right]$ is replaced by $\tanh\left[\sum_{j\in F\cup L} a_{ij}\left(\hat{z}_{id} - \hat{z}_{jd}\right)/\phi_i\right]$, $i \in F$.

5.4 Simulation Results

In this section, numerical simulations are performed to demonstrate the effectiveness of the proposed distributed FTCC scheme for multi-UAVs against actuator faults and external disturbances. In the communication network, six follower UAVs (UAV#1–UAV#6) and two leader UAVs (UAV#7–UAV#8) are involved. The leader

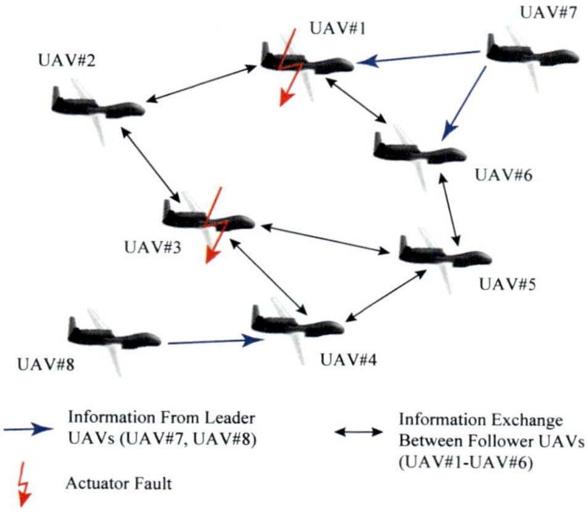

Fig. 5.4 Communication network consisting of eight UAVs (including six follower UAVs and two leader UAVs)

Table 5.1 Input saturation limits of the ith follower UAV, $i \in F$

Lower limit	Upper limit
$T_{i\min} = 0$ N	$T_{i\max} = 40$ N
$\delta_{i\min} = -0.7$ rad	$\delta_{i\max} = 0.7$ rad

UAVs are used to generate altitude references to form the convex hull $Co(z_L)$ and velocity references for all follower UAVs to track. The topology is illustrated in Fig. 5.4, and the adjacency matrix $\mathcal{A}$ is given by (5.50). In the communication network, UAV#1 and UAV#6 have access to the leader UAV#7, and UAV#4 has access to the leader UAV#8. In the simulation, all follower UAVs are identical. The parameters and coefficients of the ith follower UAV can be referred to Chap. 3, and the input saturation limits are listed in Table 5.1.

$$\mathcal{A} = \begin{bmatrix} 0 & 0.4 & 0 & 0 & 0 & 0.5 & 0.9 & 0 \\ 0.4 & 0 & 0.2 & 0 & 0 & 0 & 0 & 0 \\ 0 & 0.2 & 0 & 0.3 & 0.2 & 0 & 0 & 0 \\ 0 & 0 & 0.3 & 0 & 0.2 & 0 & 0 & 0.8 \\ 0 & 0 & 0.2 & 0.2 & 0 & 0.3 & 0 & 0 \\ 0.5 & 0 & 0 & 0 & 0.3 & 0 & 0.2 & 0 \\ 0 & 0 & 0 & 0 & 0 & 0 & 0 & 0 \\ 0 & 0 & 0 & 0 & 0 & 0 & 0 & 0 \end{bmatrix} \tag{5.50}$$

To demonstrate the robustness and fault-tolerant capability, external disturbances encountered by all follower UAVs and actuator faults subjected by follower UAV#1 and UAV#3 are assumed as

$$\begin{cases} d_{i1} = d_{i2} = d_{i3} = 0 & t < 50 \text{ s} \\ d_{i1} = 0.2, d_{i2} = 0.2, d_{i3} = 0.1 & t \geq 50 \text{ s} \end{cases}, \quad i \in F \tag{5.51}$$

$$\text{UAV\#1 Fault}: \begin{cases} \rho_1 = 1, u_{1f} = 0 & t < 80\text{ s} \\ \rho_1 = 0.7, u_{1f} = 0.35 & t \geq 80\text{ s} \end{cases} \tag{5.52}$$

$$\text{UAV\#3 Fault}: \begin{cases} \rho_3 = 1, u_{3f} = 0 & t < 230\text{ s} \\ \rho_3 = 0.7, u_{3f} = 0.2 & t \geq 230\text{ s} \end{cases} \tag{5.53}$$

The initial values of the ith follower UAV are set as $V_i(0) = 30\,\text{m/s}$, $z_i(0) = 1000\,\text{m}$, $\gamma_i(0) = 0\,\text{rad}$, $\theta_i(0) = 0.0322\,\text{rad}$, $q_i(0) = 0\,\text{rad/s}$, $i \in F$. The controller design parameters are chosen as $k_{11} = 7.3$, $k_{12} = 24.1$, $k_{21} = 2.2$, $k_{31} = 14$, $k_{41} = 39$, $k_{51} = 5.4$. The DO design parameters are chosen as $k_{22} = 6.4$, $k_{42} = 46$, $k_{52} = 12.1$, and the initial values are $\hat{\bar{z}}_{i1}(0) = 0$, $\hat{\bar{z}}_{i2}(0) = 0$, $\hat{\bar{z}}_{i3}(0) = 0$. The HGO design parameters are chosen as $\epsilon_1 = 0.01$, $\epsilon_2 = 0.01$, $l_1 = 2$, $l_2 = 2$, and initial values are selected as $\theta_{id}(0) = 0\,\text{rad}$, $\pi_1(0) = 0$, $q_{id}(0) = 0\,\text{rad/s}$, $\pi_2(0) = 0$. The parameters of the distributed SMO are selected as $\beta_1 = 8$, $\beta_2 = 3$, and initial values are set as $\hat{z}_d(0) = [1000, 1000, 1000, 1000, 1000, 1000]^T\,\text{m}$.

The velocity references from leader UAV#7 and UAV#8 are 30 m/s, and the altitude references are generated as

$$z_{70} = \frac{\omega^2}{s^2 + 2\xi\omega s + \omega^2} z_{70c} + 60\cos(0.02(t - 150)) - 60 \tag{5.54}$$

$$z_{80} = \frac{\omega^2}{s^2 + 2\xi\omega s + \omega^2} z_{80c} + 60\cos(0.02(t - 150)) - 60 \tag{5.55}$$

where z_{70c} is a step command from 1000 to 1100 m at $t = 15$ s. z_{80c} is 1000 m in the simulation. $\omega = 0.05\,\text{rad/s}$, $\xi = 0.9$.

Figure 5.5 illustrates the velocities of six follower UAVs. It is demonstrated that the velocities of all follower UAVs track the leader UAVs' velocities ($V_{70} = V_{80} = 30\,\text{m/s}$) with small tracking errors. The altitudes of six follower UAVs are shown in

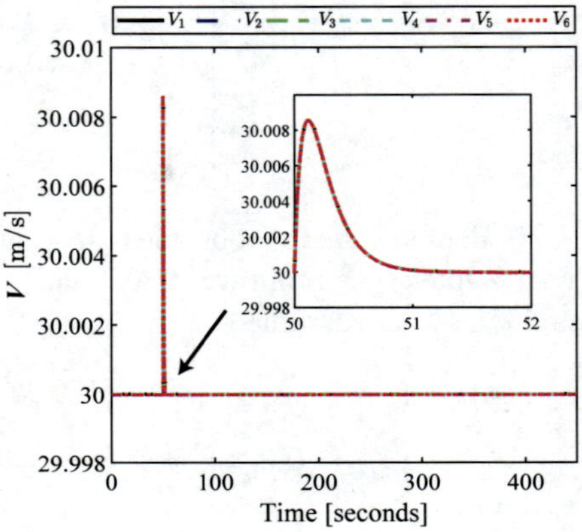

Fig. 5.5 Velocities of six follower UAVs

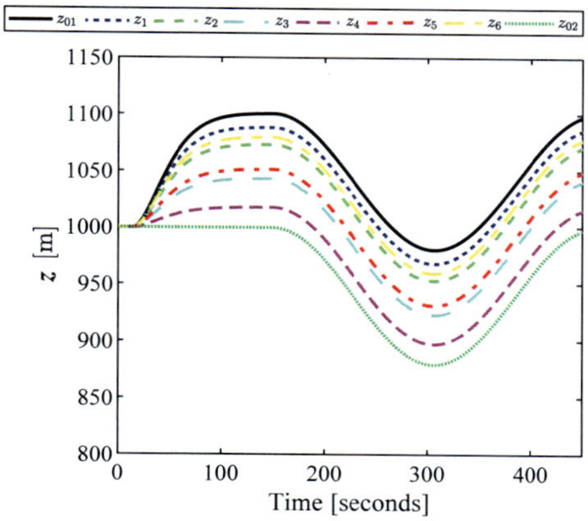

Fig. 5.6 Altitudes of six follower UAVs

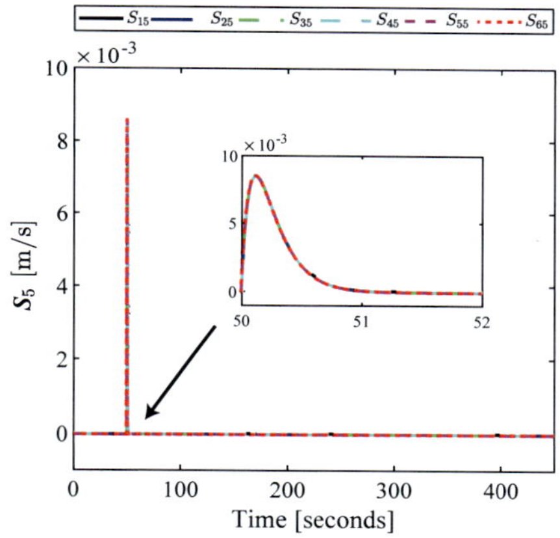

Fig. 5.7 Velocity tracking errors $S_5 = [S_{15}, S_{25}, \ldots, S_{65}]^T$

Fig. 5.6. It can be observed that the altitudes of UAVs#1–#6 are in the convex hull spanned by the altitudes of leader UAVs#7–#8. When the altitude of leader UAV#7 increases to 1100 m at $t = 125$ s and the altitude of leader UAV#8 is still at $z_{02} = 1000$ m, the altitudes of UAVs#1–#6 increase to 1088 m, 1073 m, 1043 m, 1018 m, 1052 m, 1079 m, respectively, which indicates that the altitudes of all follower UAVs are in [1000 1100] m. Then, when the leader UAV#7 and UAV#8 move according to (5.54) and (5.55), respectively, the follower UAVs (UAV#1–UAV#6) make the movements accordingly in the convex hull.

The velocity tracking errors with respect to the velocity references $V_{70} = V_{80} = 30$ m/s and the altitude tracking errors with respect to the estimated altitude references $[\hat{z}_{1d}, \ldots, \hat{z}_{6d}]^T$ are indicated in Figs. 5.7 and 5.8, respectively. It can be seen from Figs. 5.7 to 5.8 that the tracking errors are bounded under the proposed distributed control scheme even when the follower UAVs encounter external disturbances and actuator faults. The altitude displacements between follower UAVs are illustrated in Fig. 5.9, and it is observed that the altitude displacements eventually reach to constants even when the leader UAVs make the movements based on (5.54)

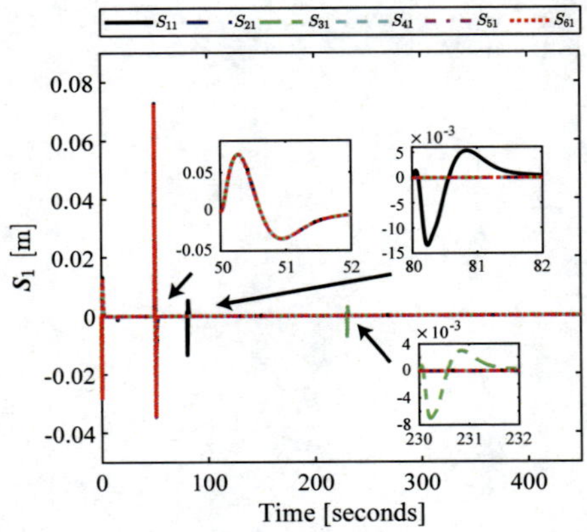

Fig. 5.8 Altitude tracking errors $\boldsymbol{S}_1 = [S_{11}, S_{21}, \ldots, S_{61}]^T$

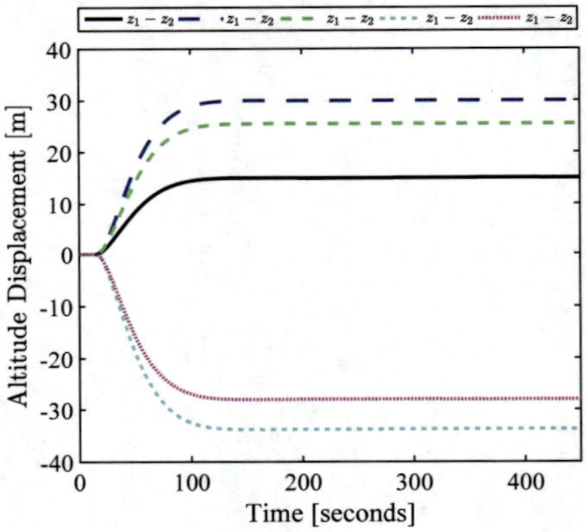

Fig. 5.9 Altitude displacements between six follower UAVs

and (5.55). Figures 5.10, 5.11, 5.12 and 5.13 are the states of six follower UAVs, which are UUB.

The time responses of elevator deflection angles are indicated in Fig. 5.14. It is observed that to deal with the external disturbances and actuator faults, the elevators react to compensate for the disturbances and faults. It should be emphasized that when the actuator of follower UAV#1 becomes faulty ($\rho_1 = 0.7, u_{1f} = 0.35$) at $t = 80$ s, the rapid response of actuator to compensate for the fault leads to the input saturation. Then, the auxiliary dynamic system (5.33) starts working to pull the elevator deflection from the lower limit back to the unsaturated region. The time responses of thrust inputs are shown in Fig. 5.15. It is noted that the thrust inputs react rapidly to make the velocity tracking errors as small as possible once the follower UAVs are encountered by the external disturbances. The estimated external disturbances $\hat{d}_{i1}$, $\hat{d}_{i2}$, and $\hat{\Delta}_i$, $i = 1, 2, \ldots, 6$ are presented in Fig. 5.16, which are bounded by using the proposed DOs.

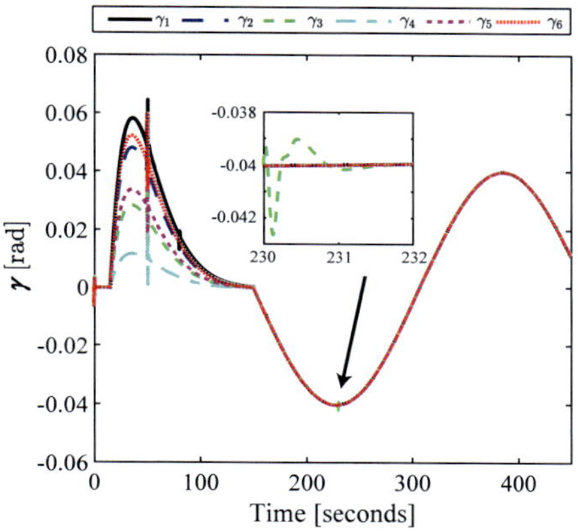

Fig. 5.10 Flight path angles of six follower UAVs

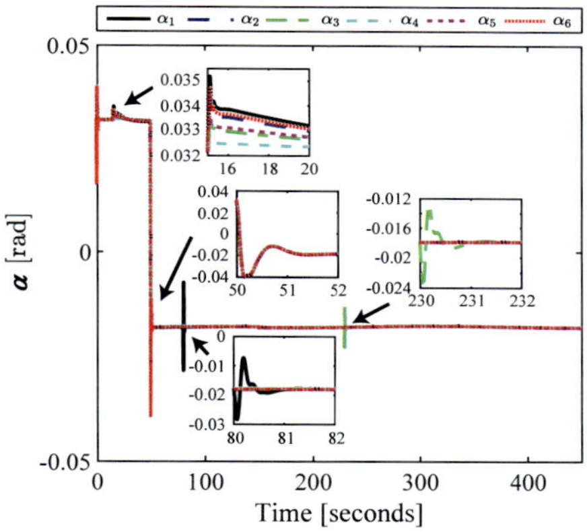

Fig. 5.11 Angles of attack of six follower UAVs

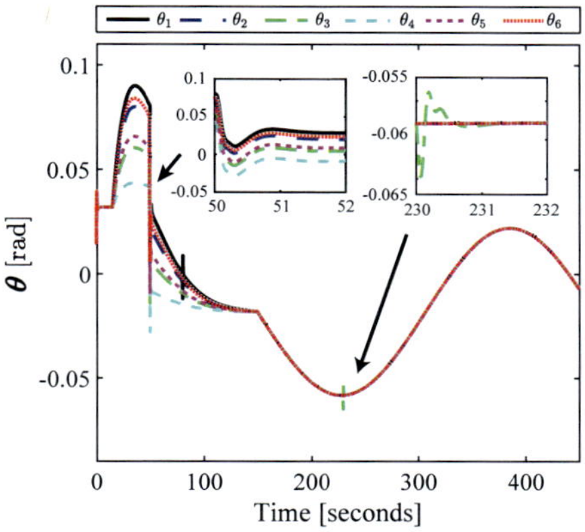

Fig. 5.12 Pitch angles of six follower UAVs

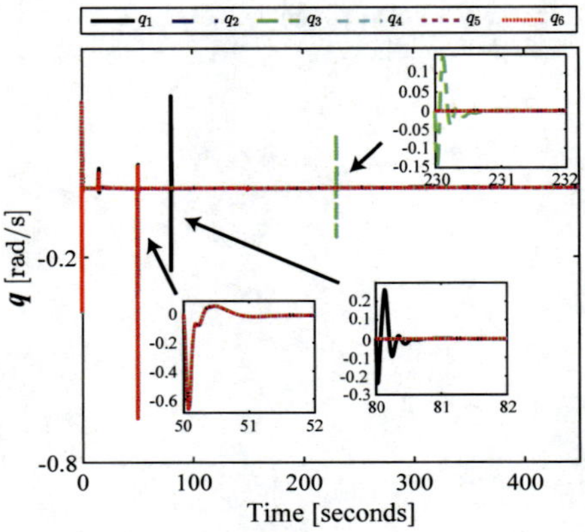

Fig. 5.13 Pitch rates of six follower UAVs

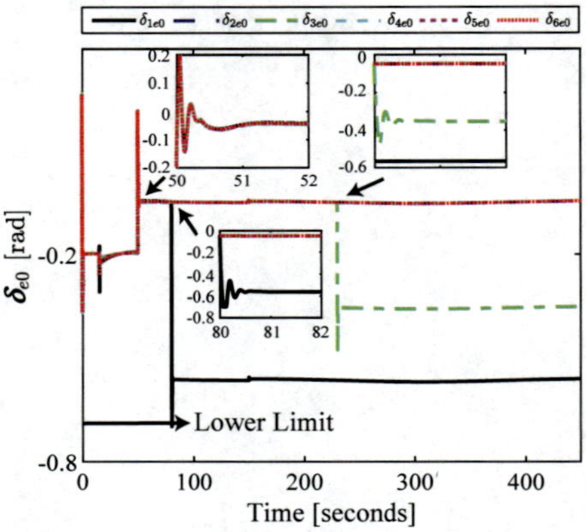

Fig. 5.14 Elevator deflection angles of six follower UAVs

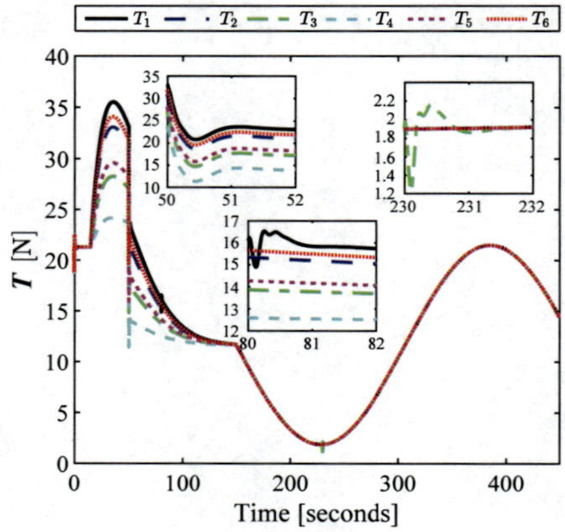

Fig. 5.15 Thrust inputs of six follower UAVs

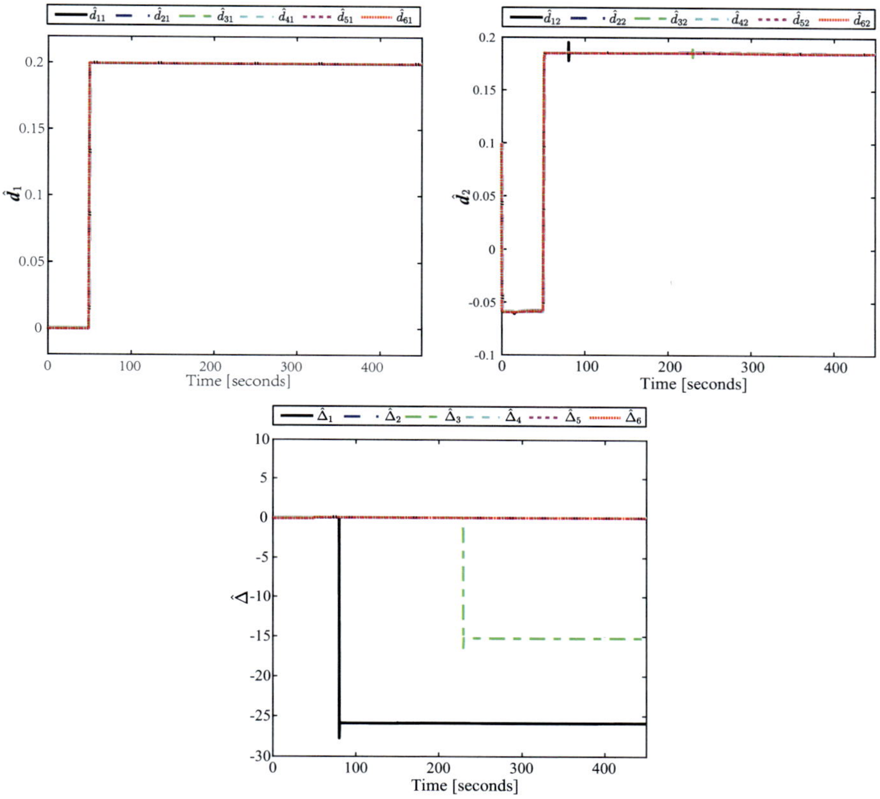

Fig. 5.16 Estimated disturbances $\hat{\boldsymbol{d}}_1$, $\hat{\boldsymbol{d}}_2$, $\hat{\Delta}$

5.5 Conclusions

In this chapter, a distributed FTCC scheme has been proposed for multi-UAVs against actuator faults and external disturbances. In the communication network, only a subset of follower UAVs has access to the leader UAVs. By using distributed SMO technique, unknown altitude references of all follower UAVs are estimated in a distributed manner. On the basis of the estimated altitude references, backstepping architecture, and HGO, the distributed FTCC has been proposed. Moreover, by using Lyapunov method and graph theory, it has shown that all follower UAVs can converge into the convex hull spanned by all leader UAVs. Furthermore, velocity tracking errors and DO estimation errors are UUB. Simulation results have demonstrated the effectiveness of the proposed control scheme.

References

1. J.D. Boskovic, S.M. Li, R.K. Mehra, Robust tracking control design for spacecraft under control input saturation. J Guidance Control Dyn. **27**(4), 627–633 (2004)
2. Y.C. Cao, D. Stuart, W. Ren, Z.Y. Meng, Distributed containment control for multiple autonomous vehicles with double-integrator dynamics: algorithms and experiments. IEEE Trans. Control Syst. Technol. **19**(4), 929–938 (2010)
3. Y.C. Cao, W. Ren, M. Egerstedt, Distributed containment control with multiple stationary or dynamic leaders in fixed and switching directed networks. Automatica **48**(8), 1586–1597 (2012)
4. M. Chen, S.Z. Ge, Direct adaptive neural control for a class of uncertain nonaffine nonlinear systems based on disturbance observer. IEEE Trans. Cybern. **43**(4), 1213–1225 (2013)
5. J.L. Du, X. Hu, H.B. Liu, C.P. Chen, Adaptive robust output feedback control for a marine dynamic positioning system based on a high-gain observer. IEEE Trans. Neural Netw. Learn. Syst. **26**(11), 2775–2786 (2015)
6. J.L. Du, X. Hu, M. Krstić, Y.Q. Sun, Robust dynamic positioning of ships with disturbances under input saturation. Automatica **73**, 207–214 (2016)
7. H. Haghshenas, M.A. Badamchizadeh, M. Baradarannia, Containment control of heterogeneous linear multi-agent systems. Automatica **54**, 210–216 (2015)
8. T. Han, Z.H. Guan, Y.H. Wu, D.F. Zheng, X.H. Zhang, J.W. Xiao, Three-dimensional containment control for multiple unmanned aerial vehicles. J. Frankl. Inst. **353**(13), 2929–2942 (2016)
9. T. Han, M. Chi, Z.H. Guan, B. Hu, J.W. Xiao, Y.H. Huang, Distributed three-dimensional formation containment control of multiple unmanned aerial vehicle systems. Asian J. Control **19**(3), 1103–1113 (2017)
10. T.S. Hu, A.R. Teel, L. Zaccarian, Anti-windup synthesis for linear control systems with input saturation: Achieving regional, nonlinear performance. Automatica **44**(2), 512–519 (2008)
11. Z.Y. Meng, W. Ren, Z. You, Distributed finite-time attitude containment control for multiple rigid bodies. Automatica **46**(12), 2092–2099 (2010)
12. W. Ren, Y.C. Cao, *Distributed Coordination of Multi-agent Networks* (Springer, Berlin, 2011)
13. D. Wang, W. Wang, Necessary and sufficient conditions for containment control of multi-agent systems with time delay. Automatica **103**, 418–423 (2019)
14. C.Y. Wen, J. Zhou, Z.T. Liu, H.Y. Su, Robust adaptive control of uncertain nonlinear systems in the presence of input saturation and external disturbance. IEEE Trans. Autom. Control **56**(7), 1672–1678 (2011)
15. G.H. Wen, Z.S. Duan, G.R. Chen, W.W. Yu, Consensus tracking of multi-agent systems with Lipschitz-type node dynamics and switching topologies. IEEE Trans. Circuits Syst. I-Regul. Pap. **61**(2), 499–511 (2013)
16. B. Xu, Robust adaptive neural control of flexible hypersonic flight vehicle with dead-zone input nonlinearity. Nonlinear Dyn. **80**(3), 1509–1520 (2015)
17. C.G. Yang, D.Y. Huang, W. He, L. Cheng, Neural control of robot manipulators with trajectory tracking constraints and input saturation. IEEE Trans. Neural Netw. Learn. Syst. **32**(9), 4231–4242 (2020)
18. Z.Y. Zuo, B.L. Tian, M. Defoort, Z.T. Ding, Fixed-time consensus tracking for multiagent systems with high-order integrator dynamics. IEEE Trans. Autom. Control **63**(2), 563–570 (2017)

Chapter 6
Distributed Finite-Time FTCC of Multi-UAVs with Multiple Leader UAVs

6.1 Introduction

To incorporate the distributed cooperative concept into the formation flight of UAVs, many results have been obtained [1]. As stated in Chap. 5, if multiple leader UAVs are selected in the formation flight, the problem of containment control of multi-UAVs arises. Actually, it may be safer and easier for follower UAVs to avoid obstacles in the complex and hazardous environment only if the follower UAVs converge into the convex hull spanned by the leader UAVs. Recently, several results have been obtained to achieve the containment control of multi-UAVs by using a point-mass model [9, 10]. To prompt the research progress of containment control for multi-UAVs, the inner dynamics should be further considered in the controller design, such that the actuator constraints and faults can be explicitly handled to make the containment control scheme practical for multi-UAVs.

When numerous UAVs are involved in the formation team, the fault probability significantly increases due to the massive actuators/sensors. Degraded performance or instability may be induced if prompt actions are not activated to compensate for the faults [32]. In practical applications, actuator faults are often encountered by the plant system [16, 34, 35]. Numerous works have been devoted to the investigation of FTC strategies for single UAV [31, 33]. Following that, an increasing number of FTCC approaches for multi-UAVs have also been developed [25, 30]. However, the FTCC schemes presented in numerous existing literatures can solely make the states of the follower UAVs asymptotically/exponentially converge to the convex hull spanned by the leader UAVs' states in infinite time, other than in finite time. Regarding the FTCC design, finite-time convergence is more desirable for the overall stability of the networked systems, especially when a subset of UAVs encounters actuator faults, since the faulty UAVs may first lose their individual stabilities and then cause the instability of networked systems if they are not regulated in a timely manner.

Z. Yu et al., *Fault-Tolerant Cooperative Control of Unmanned Aerial Vehicles*,
https://doi.org/10.1007/978-981-99-7661-4_6

Compared with infinite-time control methods, finite-time control strategy can provide faster convergence rate, higher accuracy, better disturbance rejection ability, and robustness against uncertainties [6, 12]. Due to these advantages, numerous finite-time containment control schemes have been proposed for MASs. In [8], by defining a set of virtual error variables, a finite-time DSC approach was proposed for nonlinear large-scale systems to enhance the convergence rate and robustness. Recently, in [23], a distributed adaptive finite-time containment control protocol was investigated for networked nonlinear systems with uncertain nonlinear dynamics. Almost at the same time, this result was extended to the MASs with the nonaffine pure-feedback form, in which a distributed neural adaptive control scheme was proposed for MASs based on the fraction DSC technique [22]. However, the results on the finite-time FTCC of multi-UAVs within a distributed communication network are very scarce due to the fact that a fixed-wing UAV is a highly complex nonlinear system, which increases the challenge for the control design.

Motivated by the aforementioned observations, this chapter considers the distributed finite-time FTCC problem for multi-UAVs with multiple leader UAVs. To compensate for the adverse effects of input saturation, auxiliary systems are constructed to pull back the saturated input signals into the unsaturated regions. In order to drive the follower UAVs into the convex hull formed by the leader UAVs' trajectories in finite time, a distributed finite-time SMO is first designed, and then a neural adaptive distributed FTCC scheme is devised based on a new set of error variables. In the control scheme, the RBFNN technique is used to deal with the unknown nonlinear functions [36]. Furthermore, to reduce the computational burden, MLPNN and FOSMD are also employed to facilitate the controller design. The main contributions of this chapter are as follows:

(1) The distributed control scheme proposed in this chapter is a containment control strategy, which can make the follower UAVs avoid obstacles safely and efficiently if the leader UAVs are equipped with high-precision sensors. Therefore, different from the case with single leader UAV, the containment control problem researched in this chapter is practical when multi-UAVs are working in the dangerous/complicated environment.
(2) Different from the result presented in Chap. 5, in which the closed-loop systems are asymptotically convergent and the values of aerodynamic parameters are known, this chapter further presents a finite-time FTCC scheme to handle the actuator faults in a timely manner such that the stabilities can be guaranteed during the faults and the post-fault stages. Moreover, the values of aerodynamic parameters in this chapter are assumed to be unknown to reduce the cost of obtaining these values.
(3) Instead of estimating the NN weighting vectors, MLPNN and FOSMD techniques are employed in this chapter to significantly reduce the computational burden caused by estimating the optimal weighting vector of NN and differentiating the virtual control signals. Furthermore, auxiliary systems are constructed to prevent the control inputs from violating their bounds so as to ensure the system stability.

The rest of this chapter is organized as follows. Section 6.2 introduces the longitudinal dynamics of faulty follower UAVs, model transformation, basic graph theory, NN, and control objective. Section 6.3 presents the distributed finite-time SMO design, the distributed finite-time controller design for the forward subsystem, the distributed finite-time FTCC design for the vertical subsystem, and their relevant stability analyses. Section 6.4 addresses the comparative simulation studies of the proposed control methodology. Section 6.5 outlines the concluding remarks.

(Remark: The main control schemes and contents of this chapter are from the published journal paper "Z. Q. Yu, Z. X. Liu, Y. M. Zhang, Y. H. Qu, and C.-Y. Su. Distributed finite-time fault-tolerant containment control for multiple unmanned aerial vehicles, IEEE Transactions on Neural Networks and Learning Systems, 2020, 31(6): 2077–2091." The authors appreciate the permission from the Institute of Electrical and Electronics Engineers to reuse the results published in the relevant Journal.)

6.2 Preliminaries and Problem Statement

6.2.1 Faulty UAV Longitudinal Dynamics

Consider a group of UAVs including N follower UAVs and M leader UAVs. By recalling the longitudinal dynamics (2.8)–(2.12) in Chap. 2 and adopting the similar model transformation as (5.1)–(5.5) in Chap. 5, the longitudinal dynamics of the ith follower UAV is given by Liu and Chen [15]

$$\dot{x}_i = V_i \cos \gamma_i \tag{6.1}$$

$$\dot{V}_i = -\frac{D_i}{m_i} + \frac{T_i \cos \alpha_i}{m_i} - g \sin \gamma_i \tag{6.2}$$

$$\dot{z}_i = V_i \sin \gamma_i \tag{6.3}$$

$$\dot{\gamma}_i = \frac{L_i - m_i g \cos \gamma_i + T_i \sin \alpha_i}{m_i V_i} \tag{6.4}$$

$$\dot{\alpha}_i = q_i - \frac{L_i - m_i g \cos \gamma_i + T_i \sin \alpha_i}{m_i V_i} \tag{6.5}$$

$$\dot{q}_i = \frac{\mathcal{M}_i}{I_{iy}} \tag{6.6}$$

where $i \in F$, $F = \{1, 2, \ldots, N\}$ and $L = \{N + 1, \ldots, N + M\}$ represent the follower set and leader set, respectively. x_i and z_i are the forward position and the vertical position (altitude) with respect to x and z directions of the ground coordinate frame, respectively. The position of the ith follower UAV in the longitudinal plane can be expressed as (x_i, z_i). The leader UAVs, whose dynamics is not considered, are only used to provide their trajectories to the follower UAVs.

By considering the similar actuator fault model as (5.6) in Chap. 5, the following fault model is involved within the expression of $\mathcal{M}_i$ [3]:

$$\delta_{ie} = \rho_i \delta_{ie0} + h_{if}, \quad i \in F \tag{6.7}$$

where δ_{ie} is the actual control input signal and δ_{ie0} is the control signal to be designed. $0 \leq \rho_i \leq 1$ is the actuator fault indicator, and h_{if} represents the bounded bias fault.

6.2.2 Model Transformation

Based on the functional decomposition [27], the dynamics (6.1)–(6.6) can be divided into the forward subsystem (x_i, V_i) and the vertical subsystem $(z_i, \gamma_i, \alpha_i, q_i)$.

For ease of analysis, the notations $x_{i1} = x_i$, $x_{i2} = V_i$, $x_{i3} = z_i$, $x_{i4} = \gamma_i$, $x_{i5} = \alpha_i$, $x_{i6} = q_i$ are used, $i \in F$. For the forward subsystem, one has

$$\dot{x}_{i1} = g_{ix} x_{i2} \tag{6.8}$$

$$\dot{x}_{i2} = f_{i1}(x_{i2}, u_{i1}) + u_{i1} \tag{6.9}$$

where $g_{ix} = \cos\gamma_i$, $f_{i1}(x_{i2}, u_{i1}) = -\frac{D_i}{m_i} + \left(\frac{\cos\alpha_i}{m_i} - 1\right) T_i - g\sin\gamma_i$. $u_{i1} = T_i$ is the input of forward subsystem.

Assumption 6.1 Since γ_i is quite small during the cruise phase, $\sin\gamma_i \approx \gamma_i$ is used in the controller design and $g_{ix} = \cos\gamma_i > 0$.

Remark 6.1 Assumption 6.1 has been used in (3.8) in Chap. 3. Actually, Assumption 6.1 is not strict due to the fact that most missions do not require UAVs to make very large vertical maneuvers in a very short time. Such an assumption can also be found in [26]. Therefore, it is reasonable to assume that $\dot{z}_i = V_i \gamma_i$ and $g_{ix} = \cos\gamma_i > 0$ in the controller design.

For the vertical subsystem, one has

$$\dot{x}_{i3} = V_i x_{i4} \tag{6.10}$$

$$\dot{x}_{i4} = f_{i2}(x_{i4}, x_{i5}) + x_{i5} \tag{6.11}$$

$$\dot{x}_{i5} = f_{i3}(x_{i4}, x_{i5}) + x_{i6} \tag{6.12}$$

$$\dot{x}_{i6} = f_{i4}(x_{i4}, x_{i5}, x_{i6}, u_{i2}) + u_{i2} \tag{6.13}$$

where $i \in F$, $u_{i2} = -\delta_{ie0}$ is the input of the vertical subsystem. $f_{i2}(x_{i4}, x_{i5}) = (L_i - m_i g \cos\gamma_i + T_i \sin\alpha_i)/(m_i V_i) - \alpha_i$, $f_{i3}(x_{i4}, x_{i5}) = -(L_i - m_i g \cos\gamma_i + T_i \sin\alpha_i)/(m_i V_i)$, $f_{i4}(x_{i4}, x_{i5}, x_{i6}, u_{i2}) = \mathcal{M}_{if}/I_{iy} + \delta_{ie0}$. For the vertical subsystem controller design, velocity is considered as constant with the timescale conclusion in [24]. Considering the elevator actuator fault model described as (6.7), $\mathcal{M}_{if}$ is given by $\mathcal{M}_{if} = \bar{q}_i s_i c_i C_{imf}$, where $C_{imf} = C_{im0} + C_{im\alpha}\alpha_i + C_{imq}\frac{c_i}{2V_i}q_i + C_{im\delta_e}(\rho_i\delta_{ie0} + h_{if})$.

From (6.9), (6.11), and (6.13), one can see that algebraic loops will be introduced into the control scheme if x_{i5} is used as the virtual control signal and u_{i1}, u_{i2} are used as the control signals, which make the control scheme not applicable in engineering applications. To remove the algebraic loops, the BLPFs are used as [37]

$$
\begin{cases}
u_{i1f} = B_{iu1}(s)u_{i1} \approx u_{i1} \\
x_{i5f} = B_{i5}(s)x_{i5} \approx x_{i5} \\
u_{i2f} = B_{iu2}(s)u_{i2} \approx u_{i2}
\end{cases}
\tag{6.14}
$$

where B_{iu1}, $B_{i5}(s)$, and $B_{iu2}(s)$ are BLPFs. u_{i1f}, x_{i5f}, and u_{i2f} are filtered signals. The filter errors are defined as $\varepsilon_{i1} = f_{i1}(x_{i2}, u_{i1}) - f_{i1}(x_{i2}, u_{i1f})$, $\varepsilon_{i2} = f_{i2}(x_{i4}, x_{i5}) - f_{i2}(x_{i4}, x_{i5f})$, and $\varepsilon_{i3} = f_{i4}(x_{i4}, x_{i5}, x_{i6}, u_{i2}) - f_{i4}(x_{i4}, x_{i5}, x_{i6}, u_{i2f})$, where $|\varepsilon_{i1}| \le \bar{\varepsilon}_{i1}$, $|\varepsilon_{i2}| \le \bar{\varepsilon}_{i2}$, $|\varepsilon_{i3}| \le \bar{\varepsilon}_{i3}$.

To facilitate the controller design, the following assumptions, definitions, and lemmas are introduced.

Assumption 6.2 It is assumed that $\frac{\partial f_{i1}(x_{i2}, u_{i1})}{\partial u_{i1}} + 1 \neq 0$, $\frac{\partial f_{i2}(x_{i4}, x_{i5})}{\partial x_{i5}} + 1 \neq 0$, $\frac{\partial f_{i4}(x_{i4}, x_{i5}, x_{i6}, u_{i2})}{\partial u_{i2}} + 1 \neq 0$.

Remark 6.2 Assumption 6.2 imposes the controllability condition of forward subsystem and vertical subsystem, which is a normal assumption for the UAV system and guarantees the existences of controllers u_{i1}, u_{i2} and virtual controller in (6.11). A similar assumption can also be found in [20].

Assumption 6.3 For each follower UAV in the communication network, there exists at least one leader UAV that has a path to the follower UAV.

Lemma 6.1 *Under Assumption 6.3, the matrix $\mathcal{L}_1$ is positive definite. In addition, each entry of $-\mathcal{L}_1^{-1}\mathcal{L}_2$ is nonnegative and each row sum of $-\mathcal{L}_1^{-1}\mathcal{L}_2$ is equal to one, where $\mathcal{L}_1$ and $\mathcal{L}_2$ have the same definitions as Chap. 5 [17].*

For brevity, denote the vectors $\boldsymbol{x}_1 = [x_{11}, x_{21}, \ldots, x_{N1}]^T$, $\boldsymbol{x}_2 = [x_{12}, x_{22}, \ldots, x_{N2}]^T$, $\boldsymbol{x}_3 = [x_{13}, x_{23}, \ldots, x_{N3}]^T$, $\boldsymbol{x}_4 = [x_{14}, x_{24}, \ldots, x_{N4}]^T$, $\boldsymbol{x}_5 = [x_{15}, x_{25}, \ldots, x_{N5}]^T$, $\boldsymbol{x}_6 = [x_{16}, x_{26}, \ldots, x_{N6}]^T$ for the follower UAVs, denote the vectors $\boldsymbol{x}_{10} = [x_{(N+1)10}, x_{(N+2)10}, \ldots, x_{(N+M)10}]^T$, and $\boldsymbol{x}_{30} = [x_{(N+1)30}, x_{(N+2)30}, \ldots, x_{(N+M)30}]^T$ as the forward and vertical positions of leader UAVs in the ground coordinate frame, respectively. The position of the ith leader UAV can be expressed as (x_{i10}, x_{i30}), $i \in L$. The weighted averages of leader UAVs' forward and vertical positions can be represented by $\boldsymbol{x}_{1d} = [x_{11d}, x_{21d}, \ldots, x_{N1d}]^T = -\mathcal{L}_1^{-1}\mathcal{L}_2\boldsymbol{x}_{10}$ and $\boldsymbol{x}_{3d} = [x_{13d}, x_{23d}, \ldots, x_{N3d}]^T = -\mathcal{L}_1^{-1}\mathcal{L}_2\boldsymbol{x}_{30}$, respectively. The weighted averages of leader UAVs' forward and vertical velocities can be denoted by $\boldsymbol{V}_{1d} = [V_{11d}, V_{21d}, \ldots, V_{N1d}]^T = \dot{\boldsymbol{x}}_{1d} = -\mathcal{L}_1^{-1}\mathcal{L}_2\dot{\boldsymbol{x}}_{10}$, and $\boldsymbol{V}_{3d} = [V_{13d}, V_{23d}, \ldots, V_{N3d}]^T = \dot{\boldsymbol{x}}_{3d} = -\mathcal{L}_1^{-1}\mathcal{L}_2\dot{\boldsymbol{x}}_{30}$, respectively. From Lemma 6.1 and Definition 5.1 in Chap. 5, $\boldsymbol{x}_1 \to \boldsymbol{x}_{1d}$ and $\boldsymbol{x}_3 \to \boldsymbol{x}_{3d}$ mean that the forward and vertical positions of follower UAVs converge to the convex hull formed by the forward and vertical positions of leader UAVs in the ground coordinate frame. However, the desired signals $\boldsymbol{x}_{1d}$, $\boldsymbol{x}_{3d}$, $\boldsymbol{V}_{1d}$, and $\boldsymbol{V}_{3d}$ contain the global information of communication network and cannot be directly acquired by the follower UAVs in practical applications since each follower UAV only has access to its neighboring UAVs in the distributed communication network. Therefore, the distributed SMOs will be constructed later to estimate x_{i1d}, V_{i1d} x_{i3d}, and V_{i3d} for the ith follower UAV.

Assumption 6.4 The second-order derivatives of the leader UAVs' forward and vertical positions are bounded such that $|\ddot{x}_{i10}| \le \bar{x}_{dd10} < \infty$ and $|\ddot{x}_{i30}| \le \bar{x}_{dd30} < \infty$, $i \in L$.

Remark 6.3 Assumption 6.4 is an acceleration boundedness assumption, which is very reasonable since only limited forces and moments can be provided for the UAVs.

Lemma 6.2 *Consider the system* $\dot{x} = f(x)$, $f(0) = 0$, $x \in U \subset R^n$, *where* $f(x)$ *is continuous on an open neighborhood* U *of the origin, and there exist a positive definite continuous function* $V(x)$ *defined on a neighborhood* $U_1 \in U$ *of the origin, some real numbers* $a \in (0, 1)$, *and* $c > 0$, *such that*

$$\dot{V}(x) + cV^a(x) \le 0, \; x \in U_1/\{0\} \tag{6.15}$$

Then, the origin of the system is finite-time stable, and the settling time satisfies $T \le \frac{V^{1-a}(x(0))}{c(1-a)}$ *[2].*

Lemma 6.3 *Suppose that there exists a continuous function* $V(x)$, *which satisfies the following inequality:*

$$\dot{V}(x) \le -\lambda_1 V(x) - \lambda_2 V^a(x) \tag{6.16}$$

where $\lambda_1 > 0$, $\lambda_2 > 0$, $0 < a < 1$.

Then, $V(x)$ *converges to the equilibrium point in finite time* T, *which is given by Yu et al. [29]*

$$T \le \frac{1}{\lambda_1(1-a)} \ln \frac{\lambda_1 V^{1-a}(x(0)) + \lambda_2}{\lambda_2} \tag{6.17}$$

Lemma 6.4 *Consider the system* $\dot{x} = f(x)$. *Suppose that there exist a continuous function* $V(x)$, *scalars* $\lambda_1 > 0$, $\lambda_2 > 0$, $0 < a < 1$, *and* $0 < \eta < \infty$ *such that*

$$\dot{V}(x) \leq -\lambda_1 V(x) - \lambda_2 V^a(x) + \eta \tag{6.18}$$

Then, the trajectory of system $\dot{x} = f(x)$ *is bounded in finite time.*

Proof Reorganize (6.18) as

$$\dot{V}(x) \leq -\vartheta\lambda_1 V(x) - (1-\vartheta)\lambda_1 V(x) - \lambda_2 V^a(x) + \eta \tag{6.19}$$

or

$$\dot{V}(x) \leq -\lambda_1 V(x) - \vartheta\lambda_2 V^a(x) - (1-\vartheta)\lambda_2 V^a(x) + \eta \tag{6.20}$$

where $0 < \vartheta < 1$.

Define the convergence regions for (6.19) and (6.20) as $\Omega_1 = \{x : V(x) \leq \frac{\eta}{(1-\vartheta)\lambda_1}\}$ and $\Omega_2 = \{x : V^a(x) \leq \frac{\eta}{(1-\vartheta)\lambda_2}\}$, respectively.

Regarding (6.19), if x is outside or escapes from the convergence region Ω_1, i.e., $V(x) > \frac{\eta}{(1-\vartheta)\lambda_1}$, according to Lemma 6.3, x will be pulled back into the region Ω_1 in finite time $T_{f1} \leq \frac{1}{\vartheta\lambda_1(1-a)}\ln\frac{\vartheta\lambda_1 V^{1-a}(x(0))+\lambda_2}{\lambda_2}$, where $V(x(0))$ is the initial value of $V(x)$.

With respect to (6.20), if x is outside or escapes from the convergence region Ω_2, i.e., $V^a(x) > \frac{\eta}{(1-\vartheta)\lambda_2}$, according to Lemma 6.3, x will converge into the region Ω_2 in finite time $T_{f2} \leq \frac{1}{\lambda_1(1-a)}\ln\frac{\lambda_1 V^{1-a}(x(0))+\vartheta\lambda_2}{\vartheta\lambda_2}$.

Based on the analysis mentioned above, one can conclude that the system state x converges into the bounded region $\Omega = \min\{\Omega_1, \Omega_2\}$ in finite time $T = \max\{T_{f1}, T_{f2}\}$. □

Lemma 6.5 *For any real numbers* x_i, $i = 1, 2, \ldots, n$, *and* $0 < r \leq 1$, *the following inequality holds [23]*

$$\left(\sum_{i=1}^{n} |x_i|\right)^r \leq \sum_{i=1}^{n} |x_i|^r \tag{6.21}$$

6.2.3 Control Objective

The control objective of this chapter is to design a set of distributed finite-time FTCC laws for multi-UAVs with multiple leaders, such that the forward and vertical positions of each follower UAV can be steered into the convex hull spanned by the forward and vertical positions of all leader UAVs in finite time.

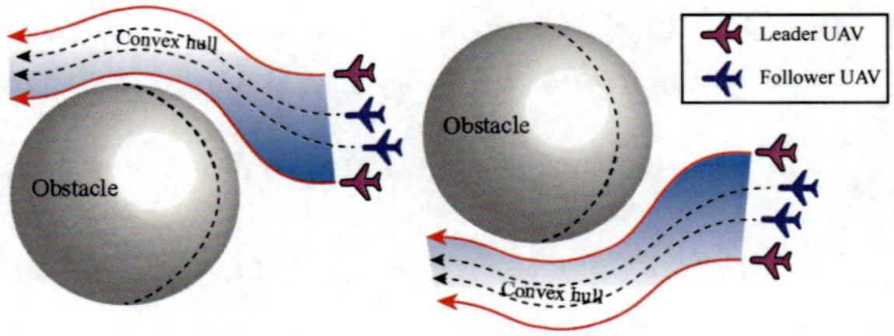

Fig. 6.1 Two successful obstacle avoidance situations of multi-UAVs by using longitudinal cooperative formation flying

Remark 6.4 When a swarm of UAVs is launched from the mothership or the launch vehicle, the longitudinal formation of numerous UAVs in the initial phase is very important and necessary to avoid collisions, since these UAVs are usually continuously launched in the longitudinal plane. Moreover, according to the traffic alert and collision avoidance system II released by the Federal Aviation Administration, which assists in preventing collisions between aircrafts, the motion in the longitudinal dimension can be used to avoid the obstacles [7, 18]. However, the task will be very costly if each UAV in the formation team is equipped with high-precision sensors to detect the obstacles, since the high-precision sensors in the obstacle detection system of an aircraft are usually very expensive. Fortunately, with the cooperative control strategy, all UAVs (leader UAVs and follower UAVs) will be safe, and the overall sensor cost is significantly reduced by placing the high-precision sensors only on the leader UAVs to detect the obstacles. Figure 6.1 illustrates two situations that leader UAVs and follower UAVs can successfully avoid the obstacle if the leader UAVs equipped with high-precision sensors change their relative distances with respect to the obstacle in the longitudinal plane, and the follower UAVs converge into the convex hull spanned by the trajectories of leader UAVs.

6.3 Main Results

In this section, the distributed finite-time control scheme is designed by constructing a distributed SMO to estimate the forward and vertical position references x_{id}, z_{id} for each follower UAV and a set of distributed finite-time control laws based on the estimated references.

6.3.1 Distributed Finite-Time Sliding-Mode Observer Design

Inspired by the works in [4, 14, 21, 38], the distributed SMO embedded in each follower UAV is designed as

$$\begin{aligned}\dot{\hat{x}}_{i\iota d} =& \hat{V}_{i\iota d} - \beta_{1\iota}\mathrm{sig}^{a}\left(\sum_{j=F\cup L} a_{ij}(\hat{x}_{i\iota d} - \hat{x}_{j\iota d})\right) \\ & - \beta_{2\iota}\mathrm{sign}\left(\sum_{j=F\cup L} a_{ij}(\hat{x}_{i\iota d} - \hat{x}_{j\iota d})\right)\end{aligned} \tag{6.22}$$

$$\begin{aligned}\dot{\hat{V}}_{i\iota d} =& - \beta_{3\iota}\mathrm{sig}^{a}\left(\sum_{j=F\cup L} a_{ij}(\hat{V}_{i\iota d} - \hat{V}_{j\iota d})\right) \\ & - \beta_{4\iota}\mathrm{sign}\left(\sum_{j=F\cup L} a_{ij}(\hat{V}_{i\iota d} - \hat{V}_{j\iota d})\right)\end{aligned} \tag{6.23}$$

where $i \in F$, $\iota = 1, 3$. $\hat{x}_{i1d}$ and $\hat{x}_{i3d}$ are the estimations of the weighted averages of leader UAVs' forward and vertical positions, respectively. $\hat{V}_{i1d}$ and $\hat{V}_{i3d}$ are the estimations of V_{i1d} and V_{i3d}, respectively. $\hat{x}_{j\iota d} = x_{j\iota 0}$, $\hat{V}_{j\iota d} = \dot{x}_{j\iota 0}$, $j \in L$. $\beta_{1\iota}$, $\beta_{2\iota}$, $\beta_{3\iota}$, $\beta_{4\iota}$, and $a \in (0, 1)$ are positive design parameters. $\mathrm{sig}^{a}(\cdot) = |\cdot|^{a}\mathrm{sign}(\cdot)$.

Lemma 6.6 *Let $\beta_{1\iota} > 0$, $\beta_{2\iota} > 0$, $\beta_{3\iota} > 0$, $\beta_{4\iota} > \bar{x}_{dd\iota 0}$. If Assumptions 6.3, 6.4 hold, under the distributed SMO (6.22)–(6.23), the follower UAVs can acquire the precise estimations of the weighted averages of leader UAVs' forward and vertical positions, as well as the precise estimations of the weighted averages of corresponding velocities, i.e., $\hat{x}_{i\iota d} \to x_{i\iota d}$, $\hat{V}_{i\iota d} \to V_{i\iota d}$, $i \in F$, $\iota = 1, 3$.*

Proof Define the position and velocity estimation errors of follower UAVs as $\tilde{\boldsymbol{x}}_{\iota d} = [\tilde{x}_{1\iota d}, \tilde{x}_{2\iota d}, \ldots, \tilde{x}_{N\iota d}]^{T} = [\hat{x}_{1\iota d} - x_{1\iota d}, \hat{x}_{2\iota d} - x_{2\iota d}, \ldots, \hat{x}_{N\iota d} - x_{N\iota d}]^{T}$ and $\tilde{\boldsymbol{V}}_{\iota d} = [\tilde{V}_{1\iota d}, \tilde{V}_{2\iota d}, \ldots, \tilde{V}_{N\iota d}]^{T} = [\hat{V}_{1\iota d} - V_{1\iota d}, \hat{V}_{2\iota d} - V_{2\iota d}, \ldots, \hat{V}_{N\iota d} - V_{N\iota d}]^{T}$, respectively. The synchronized position and velocity estimation errors with respect to the ith follower UAV are defined as $e_{i\iota sp} = \sum_{j=F\cup L} a_{ij}(\hat{x}_{i\iota d} - \hat{x}_{j\iota d})$ and $e_{i\iota sv} = \sum_{j=F\cup L} a_{ij}(\hat{V}_{i\iota d} - \hat{V}_{j\iota d})$, respectively. $\boldsymbol{e}_{\iota sp} = [e_{1\iota sp}, e_{2\iota sp}, \ldots, e_{N\iota sp}]^{T}$ and $\boldsymbol{e}_{\iota sv} = [e_{1\iota sv}, e_{2\iota sv}, \ldots, e_{N\iota sv}]^{T}$ are the synchronized estimation error vectors. Then, with Assumption 6.3 and Lemma 6.1, one has $\boldsymbol{e}_{\iota sp} = \mathcal{L}_{1}\hat{\boldsymbol{x}}_{\iota d} + \mathcal{L}_{2}\boldsymbol{x}_{\iota 0} = \mathcal{L}_{1}[\hat{\boldsymbol{x}}_{\iota d} - (-\mathcal{L}_{1}^{-1}\mathcal{L}_{2}\boldsymbol{x}_{\iota 0})] \to 0 \Leftrightarrow \hat{\boldsymbol{x}}_{\iota d} \to -\mathcal{L}_{1}^{-1}\mathcal{L}_{2}\boldsymbol{x}_{\iota 0}$ and $\boldsymbol{e}_{\iota sv} = \mathcal{L}_{1}\hat{\boldsymbol{V}}_{\iota d} + \mathcal{L}_{2}\dot{\boldsymbol{x}}_{\iota 0} = \mathcal{L}_{1}[\hat{\boldsymbol{V}}_{\iota d} - (-\mathcal{L}_{1}^{-1}\mathcal{L}_{2}\dot{\boldsymbol{x}}_{\iota 0})] \to 0 \Leftrightarrow \hat{\boldsymbol{V}}_{\iota d} \to -\mathcal{L}_{1}^{-1}\mathcal{L}_{2}\dot{\boldsymbol{x}}_{\iota 0}$. Therefore, the synchronized estimation errors $\boldsymbol{e}_{\iota sp}$ and $\boldsymbol{e}_{\iota sv}$ are convergent once the estimation errors

$\tilde{\boldsymbol{x}}_{\iota d}$ and $\tilde{\boldsymbol{V}}_{\iota d}$ are convergent. Define the following Lyapunov function according to (6.23):

$$L_1 = \frac{1}{2}\tilde{\boldsymbol{V}}_{\iota d}^T \mathcal{L}_1^T \tilde{\boldsymbol{V}}_{\iota d} \tag{6.24}$$

From Lemma 6.1, one knows that $\mathcal{L}_1$ is symmetric positive definite, and then the time derivative of (6.24) has

$$\begin{aligned} \dot{L}_1 =& \tilde{\boldsymbol{V}}_{\iota d}^T \mathcal{L}_1^T \left[-\beta_{3\iota}\mathrm{sig}^a(\mathcal{L}_1 \tilde{\boldsymbol{V}}_{\iota d}) - \beta_{4\iota}\mathrm{sign}(\mathcal{L}_1 \tilde{\boldsymbol{V}}_{\iota d}) + \mathcal{L}_1^{-1}\mathcal{L}_2 \ddot{\boldsymbol{x}}_{\iota 0} \right] \\ =& -\beta_{3\iota}\sum_{i=1}^{N} |\mathcal{Z}_1[i]|^{a+1} - \beta_{4\iota}\|\mathcal{Z}_1\|_1 + \tilde{\boldsymbol{V}}_{\iota d}^T \mathcal{L}_2 \ddot{\boldsymbol{x}}_{\iota 0} \\ \leq& -\beta_{3\iota}\|\mathcal{Z}_1\|_2^{a+1} - (\beta_{4\iota} - \bar{x}_{dd\iota 0})\|\mathcal{Z}_1\|_2 \end{aligned} \tag{6.25}$$

where $\mathcal{Z}_1 = \mathcal{L}_1 \tilde{\boldsymbol{V}}_{\iota d}$. $\mathcal{Z}_1[i]$ is the ith element of the vector $\mathcal{Z}_1$, $i = 1, 2, \ldots, N$. The inequality $\tilde{\boldsymbol{V}}_{\iota d}^T \mathcal{L}_2 \ddot{\boldsymbol{x}}_{\iota 0} = \tilde{\boldsymbol{V}}_{\iota d}^T \mathcal{L}_1 \mathcal{L}_1^{-1} \mathcal{L}_2 \ddot{\boldsymbol{x}}_{\iota 0} \leq ||\mathcal{L}_1 \tilde{\boldsymbol{V}}_{\iota d}||_1 ||\mathcal{L}_1^{-1}\mathcal{L}_2 \ddot{\boldsymbol{x}}_{\iota 0}||_\infty \leq \bar{x}_{dd\iota 0}||\mathcal{L}_1 \boldsymbol{V}_{\iota d}||_1$ is used in (6.25). $||\boldsymbol{y}||_1 = \sum_{i=1}^N |y_i|$, $||\boldsymbol{y}||_2 = \sqrt{\boldsymbol{y}^T \boldsymbol{y}}$, and $||\boldsymbol{y}||_\infty = \max_{i=1,2,\ldots,N} |y_i|$ denote the 1-norm, Euclidean norm, and infinity norm of vector $\boldsymbol{y} = [y_1, y_2, \ldots, y_N]^T$, respectively.

Due to the fact that $\|\mathcal{Z}_1\|_2 = ||\mathcal{L}_1 \tilde{\boldsymbol{V}}_{\iota d}||_2 = \left(\tilde{\boldsymbol{V}}_{\iota d}^T \mathcal{L}_1^2 \tilde{\boldsymbol{V}}_{\iota d}\right)^{1/2} \geq \lambda_{\min}(\mathcal{L}_1)||\tilde{\boldsymbol{V}}_{\iota d}||_2 \geq \frac{\sqrt{2}\lambda_{\min}(\mathcal{L}_1)L_1^{1/2}}{\sqrt{\lambda_{\max}(\mathcal{L}_1)}}$, (6.25) can be further obtained as

$$\dot{L}_1 \leq -\frac{2^{\frac{a+1}{2}}\beta_{3\iota}\lambda_{\min}^{a+1}(\mathcal{L}_1)}{\lambda_{\max}^{\frac{a+1}{2}}(\mathcal{L}_1)} L_1^{\frac{a+1}{2}} - \frac{2^{\frac{a+1}{2}}(\beta_{4\iota} - \bar{x}_{dd\iota 0})\lambda_{\min}(\mathcal{L}_1)}{\lambda_{\max}^{\frac{1}{2}}(\mathcal{L}_1)} L_1^{\frac{1}{2}} \tag{6.26}$$

According to Lemma 6.2, it can be concluded that $\tilde{\boldsymbol{V}}_{\iota d} \rightarrow \boldsymbol{0}$ in finite time T_1, which is given by

$$T_1 \leq \min\left\{ \frac{2^{\frac{1-a}{2}}\lambda_{\max}^{\frac{a+1}{2}}(\mathcal{L}_1)L_1^{\frac{1-a}{2}}(0)}{(1-a)\beta_{3\iota}\lambda_{\min}^{a+1}(\mathcal{L}_1)}, \frac{\sqrt{2}\lambda_{\max}^{\frac{1}{2}}(\mathcal{L}_1)L_1^{\frac{1}{2}}(0)}{(\beta_{4\iota} - \bar{x}_{dd\iota 0})\lambda_{\min}(\mathcal{L}_1)} \right\} \tag{6.27}$$

where $L_1(0)$ is the initial value of L_1. Then, $\boldsymbol{V}_{\iota d}$ can be utilized to replace $\hat{\boldsymbol{V}}_{\iota d}$ as $t > T_1$.

When $t > T_1$, the following Lyapunov function is constructed.

$$L_2 = \frac{1}{2}\tilde{\boldsymbol{x}}_{\iota d}^T \mathcal{L}_1^T \tilde{\boldsymbol{x}}_{\iota d} \tag{6.28}$$

By taking the time derivative of (6.28), one has

$$
\begin{aligned}
\dot{L}_2 =&\tilde{\boldsymbol{x}}_{\iota d}^T \mathcal{L}_1^T \left[-\beta_{1\iota}\mathrm{sig}^a(\mathcal{L}_1\tilde{\boldsymbol{x}}_{\iota d}) - \beta_{2\iota}\mathrm{sign}(\mathcal{L}_1\tilde{\boldsymbol{x}}_{\iota d})\right] \\
=&-\beta_{1\iota}\sum_{i=1}^{N}|\mathcal{Z}_2[i]|^{a+1} - \beta_{2\iota}\left\|\mathcal{Z}_2\right\|_1 \\
\leq&-\beta_{1\iota}\left\|\mathcal{Z}_2\right\|_2^{a+1} - \beta_{2\iota}\left\|\mathcal{Z}_2\right\|_1 \\
\leq&-\frac{2^{\frac{a+1}{2}}\beta_{1\iota}\lambda_{\min}^{a+1}(\mathcal{L}_1)}{\lambda_{\max}^{\frac{a+1}{2}}(\mathcal{L}_1)}L_2^{\frac{a+1}{2}} - \frac{2^{\frac{a+1}{2}}\beta_{2\iota}\lambda_{\min}(\mathcal{L}_1)}{\lambda_{\max}^{\frac{1}{2}}(\mathcal{L}_1)}L_2^{\frac{1}{2}}
\end{aligned}
\tag{6.29}
$$

where $\mathcal{Z}_2 = \mathcal{L}_1\tilde{\boldsymbol{x}}_{\iota d}$. $\mathcal{Z}_2[i]$ is the ith element of the vector $\mathcal{Z}_2$, $i = 1, 2, \ldots, N$. $\|\mathcal{Z}_2\|_2 = ||\mathcal{L}_1\tilde{\boldsymbol{x}}_{\iota d}||_2 = \left(\tilde{\boldsymbol{x}}_{\iota d}^T\mathcal{L}_1^2\tilde{\boldsymbol{x}}_{\iota d}\right)^{1/2} \geq \lambda_{\min}(\mathcal{L}_1)||\tilde{\boldsymbol{x}}_{\iota d}||_2 \geq \frac{\sqrt{2}\lambda_{\min}(\mathcal{L}_1)L_2^{1/2}}{\sqrt{\lambda_{\max}(\mathcal{L}_1)}}$ is used in (6.29).

By recalling Lemma 6.2, $\tilde{\boldsymbol{x}}_{\iota d} \to \mathbf{0}$ in finite time T_2, given by

$$
T_2 \leq T_1 + \min\left\{\frac{2^{\frac{1-a}{2}}\lambda_{\max}^{\frac{a+1}{2}}(\mathcal{L}_1)L_2^{\frac{1-a}{2}}(T_1)}{(1-a)\beta_{1\iota}\lambda_{\min}^{a+1}(\mathcal{L}_1)}, \frac{\sqrt{2}\lambda_{\max}^{\frac{1}{2}}(\mathcal{L}_1)L_2^{\frac{1}{2}}(T_1)}{\beta_{2\iota}\lambda_{\min}(\mathcal{L}_1)}\right\}
\tag{6.30}
$$

where $L_2(T_1)$ is the value of L_2 at the time T_1. Then, $\boldsymbol{x}_{\iota d}$ can be utilized to replace $\hat{\boldsymbol{x}}_{\iota d}$ as $t > T_2$.

Therefore, $\hat{\boldsymbol{x}}_{\iota d}$ and $\hat{\boldsymbol{V}}_{\iota d}$ can be used to replace the weighted averages of leader UAVs' positions (including forward and vertical positions) and velocities (including forward and vertical velocities) when $t > T_2$. This ends the proof. □

Remark 6.5 It should be stressed that the distributed finite-time SMOs (6.22)–(6.23) reduce to the distributed SMOs in [4, 14] by choosing $\beta_{1\iota} = 0$ and $\beta_{3\iota} = 0$. From (6.26) and (6.29), the SMOs (6.22)–(6.23) can provide a faster convergence speed than the SMOs proposed in [4, 14] due to the additions of $-\beta_{1\iota}\mathrm{sig}^a\left(\sum_{j=F\cup L}a_{ij}(\hat{x}_{i\iota d} - \hat{x}_{j\iota d})\right)$ and $-\beta_{3\iota}\mathrm{sig}^a\left(\sum_{j=F\cup L}a_{ij}(\hat{V}_{i\iota d} - \hat{V}_{j\iota d})\right)$. To guarantee the superiority, it is required that the parameters $\beta_{1\iota}$ and $\beta_{3\iota}$ are positive constants. Moreover, different from the distributed first-order SMOs presented in [21, 38], the distributed second-order SMO (6.22)–(6.23) can be used to simultaneously estimate the weighted averages of the leader UAVs' positions and velocities. Furthermore, only one leader is involved in the SMO presented in [38], and in this chapter, multiple leaders are involved in (6.22)–(6.23). Compared with the SMOs in [13, 19], which are mainly developed to estimate the unmeasured system states or unknown plant parameters, the distributed SMO is constructed to estimate the weighted averages of leader UAVs' positions and velocities for all follower UAVs, and then these estimated signals can be used as references for the subsequent controller.

6.3.2 Finite-Time Controller Design for Forward Subsystem

In this section, by combining the NN technique and adaptive mechanism, the finite-time forward subsystem controller is constructed for the forward subsystem to achieve $x_{i1} \to \hat{x}_{i1d}$ in finite time.

Define a new set of errors as

$$S_{i1} = e_{i1} \tag{6.31}$$

$$S_{i2} = e_{i2} + b_{i1}|S_{i1}|^{r_i}\,\text{sign}(S_{i1}) \tag{6.32}$$

where $i \in F$, $e_{i1} = x_{i1} - \hat{x}_{i1d}$ is the forward position tracking error with respect to the estimation of x_{i1d}, $e_{i2} = x_{i2} - \bar{x}_{i2d}$ is the velocity tracking error, $\bar{x}_{i2d}$ is the virtual control signal constructed by using the backstepping architecture, $b_{i1} > 0$, $r_i = q_{i0}/p_{i0}$, and q_{i0}, p_{i0} are positive odd numbers, $q_{i0} < p_{i0} < 2q_{i0}$.

Consider the case $t > T_2$, and the controller design procedure is presented as follows.

Step 1 Taking the time derivative of S_{i1} gives

$$\dot{S}_{i1} = \dot{x}_{i1} - \dot{\hat{x}}_{i1d} = g_{ix}x_{i2} - \dot{\hat{x}}_{i1d} \tag{6.33}$$

Design the virtual control signal $\bar{x}_{i2d}$ as

$$\bar{x}_{i2d} = g_{ix}^{-1}(-k_{11}S_{i1} + \hat{V}_{i1d}) \tag{6.34}$$

where k_{11} is a positive design parameter and $\hat{V}_{i1d}$ is estimated by the SMO (6.23).

Choose the Lyapunov function candidate as

$$L_{i1} = \frac{1}{2}S_{i1}^2 \tag{6.35}$$

By recalling Assumption 6.1 and using Young's inequality, the time derivative of (6.35) has

$$\dot{L}_{i1} \leq -\left(k_{11} - \frac{1}{2\zeta_{01}^2}\right)S_{i1}^2 + \frac{\zeta_{01}^2 S_{i2}^2}{2} - g_{ix}b_{i1}S_{i1}|S_{i1}|^{r_i}\,\text{sign}(S_{i1}) \tag{6.36}$$

where ζ_{01} is a positive constant.

Step 2 Consider the input saturation, the control input signal, and the adaptive laws are designed as

$$\bar{u}_{i1} = -k_{21}S_{i2} - \frac{S_{i2}\hat{\chi}_{i11}\varphi_{i1}^T\varphi_{i1}}{2\zeta_{11}^2} - \frac{S_{i2}\hat{\chi}_{i12}}{2\zeta_{12}^2}$$

$$- b_{i2}|S_{i2}|^{r_i}\,\text{sign}(S_{i2}) + k_{22}\eta_{i1} \tag{6.37}$$

$$\dot{\hat{\chi}}_{i11} = g_{11}\left(\frac{S_{i2}^2\boldsymbol{\varphi}_{i1}^T\boldsymbol{\varphi}_{i1}}{2\zeta_{11}^2} - h_{11}\hat{\chi}_{i11}\right) \tag{6.38}$$

$$\dot{\hat{\chi}}_{i12} = g_{12}\left(\frac{S_{i2}^2}{2\zeta_{12}^2} - h_{12}\hat{\chi}_{i12}\right) \tag{6.39}$$

where k_{21}, k_{22}, ζ_{11}, ζ_{12}, g_{11}, g_{12}, h_{11}, h_{12}, b_{i2} are positive design parameters and $\boldsymbol{\varphi}_{i1}$ is the basis function vector of RBFNN. η_{i1} is an auxiliary signal constructed to compensate for the input saturation. $\hat{\chi}_{i11}$ and $\hat{\chi}_{i12}$ are the estimations of χ_{i11} and χ_{i12}, respectively, which will be defined later.

Choose the Lyapunov function candidate at this step as

$$L_{i2} = \frac{1}{2}S_{i2}^2 + \frac{1}{2g_{11}}\tilde{\chi}_{i11}^2 + \frac{1}{2g_{12}}\tilde{\chi}_{i12}^2 + \frac{1}{2}\eta_{i1}^2 \tag{6.40}$$

where $\tilde{\chi}_{i11} = \chi_{i11} - \hat{\chi}_{i11}$ and $\tilde{\chi}_{i12} = \chi_{i12} - \hat{\chi}_{i12}$ are the estimation errors.

Taking the time derivative of (6.40) yields

$$\begin{aligned}\dot{L}_{i2} =&S_{i2}\Big[f_{i1}(x_{i2}, u_{i1f}) + \varepsilon_{i1} + u_{i1} - \dot{\bar{x}}_{i2d} + b_{i1}r_i|S_{i1}|^{r_i-1}\dot{S}_{i1}\Big]\\&+\frac{1}{g_{11}}\tilde{\chi}_{i11}\dot{\tilde{\chi}}_{i11} + \frac{1}{g_{12}}\tilde{\chi}_{i12}\dot{\tilde{\chi}}_{i12} + \eta_{i1}\dot{\eta}_{i1}\end{aligned} \tag{6.41}$$

Define $F_{i1}(x_{i2}, u_{i1f}, \dot{\bar{x}}_{i2d}, x_{i4}, x_{i5}) = f_{i1}(x_{i2}, u_{i1f}) - \dot{\bar{x}}_{i2d} + b_{i1}r_i|S_{i1}|^{r_i-1}\dot{S}_{i1}$. By using RBFNN, $F_{i1}(x_{i2}, u_{i1f}, \dot{\bar{x}}_{i2d}, x_{i4}, x_{i5}) = \boldsymbol{\theta}_{i1}^{*T}\boldsymbol{\varphi}_{i1}(x_{i2}, u_{i1f}, \dot{\bar{x}}_{i2d}, x_{i4}, x_{i5}) + w_{i1}$, where $\boldsymbol{\theta}_{i1}^*$ is weighting vector and $\boldsymbol{\varphi}_{i1}(x_{i2}, u_{i1f}, \dot{\bar{x}}_{i2d}, x_{i4}, x_{i5})$ is the Gaussian basis function vector. From the Gaussian function vector $\boldsymbol{\varphi}_{i1}(x_{i2}, u_{i1f}, \dot{\bar{x}}_{i2d}, x_{i4}, x_{i5})$, it can be seen that the time derivative of $\bar{x}_{i2d}$ is needed. However, the time derivative of $\bar{x}_{i2d}$ will increase the computational burden when numerous UAVs are adopted in the communication network. To obtain $\dot{\bar{x}}_{i2d}$, the FOSMD is employed, which is given by Levant [11]

$$\begin{aligned}\dot{\pi}_d &= -\mu_1|\pi_d - l(t)|^{0.5}\text{sign}(\pi_d - l(t)) + \epsilon\\ \dot{\epsilon} &= -\mu_2\text{sign}(\epsilon - \pi_d)\end{aligned} \tag{6.42}$$

where $\mu_1 > 0$ and $\mu_2 > 0$ are the design parameters of FOSMD. $l(t)$ is the input signal, and ϵ and π_d are the states. $\dot{\pi}_d$ can estimate $\dot{l}(t)$ to any arbitrary precision if $\pi_d(0) - l(0)$ and $\dot{\pi}_d(0) - \dot{l}(0)$ are bounded.

Then, by using the FOSMD, one has

$$F_{i1}(x_{i2}, u_{i1f}, \dot{\bar{x}}_{i2d}, x_{i4}, x_{i5}) = \boldsymbol{\theta}_{i1}^{*T}\boldsymbol{\varphi}_{i1}(x_{i2}, u_{i1f}, \dot{\bar{x}}_{i2ds}, x_{i4}, x_{i5}) + w_{i1} + \xi_{i1} \tag{6.43}$$

where ξ_{i1} is bounded by $|\xi_{i1}| \leq \bar{\xi}_{i1}$. $\dot{\bar{x}}_{i2ds}$ can be obtained by passing $\bar{x}_{i2d}$ into the FOSMD. For brevity, $(\cdot)$ in $\boldsymbol{\varphi}_{i1}(\cdot)$ is omitted in the subsequent analysis.

By utilizing Young's inequality, $S_{i2}\boldsymbol{\theta}_{i1}^{*T}\boldsymbol{\varphi}_{i1} \leq \frac{S_{i2}^2\chi_{i11}\boldsymbol{\varphi}_{i1}^T\boldsymbol{\varphi}_{i1}}{2\zeta_{11}^2} + \frac{\zeta_{11}^2}{2}$ and $S_{i2}(\varepsilon_{i1} + \xi_{i1} + w_{i1}) \leq \frac{S_{i2}^2\chi_{i12}}{2\zeta_{12}^2} + \frac{\zeta_{12}^2}{2}$ are obtained, where $\chi_{i11} = \boldsymbol{\theta}_{i1}^{*T}\boldsymbol{\theta}_{i1}^*$, $\chi_{i12} = (\bar{\varepsilon}_{i1} + \bar{\xi}_{i1} + \bar{w}_{i1})^2$.

Then, by using (6.37), (6.38), and (6.39), one has

$$\begin{aligned} \dot{L}_{i2} \leq & -k_{21}S_{i2}^2 + S_{i2}\Delta u_{i1} - b_{i2}S_{i2}|S_{i2}|^{r_i}\operatorname{sign}(S_{i2}) \\ & + k_{22}\eta_{i1}S_{i2} + \eta_{i1}\dot{\eta}_{i1} - \frac{h_{11}}{2}\tilde{\chi}_{i11}^2 - \frac{h_{12}}{2}\tilde{\chi}_{i12}^2 \\ & + \frac{h_{11}}{2}\chi_{i11}^2 + \frac{h_{12}}{2}\chi_{i12}^2 + \frac{\zeta_{11}^2}{2} + \frac{\zeta_{12}^2}{2} \end{aligned} \tag{6.44}$$

where $\Delta u_{i1} = u_{i1} - \bar{u}_{i1}$ is the error between the designed control input signal and the actual control signal.

Actually, in engineering systems, the input signal is often encountered by input saturation. Therefore, the actual signal can be described as

$$u_{i1} = \operatorname{sat}(\bar{u}_{i1}) = \begin{cases} u_{i1\max}, & \bar{u}_{i1} \geq u_{i1\max} \\ \bar{u}_{i1}, & u_{i1\min} < \bar{u}_{i1} < u_{i1\max} \\ u_{i1\min}, & \bar{u}_{i1} \leq u_{i1\min} \end{cases} \tag{6.45}$$

where $i \in F$. $u_{i1\min} = T_{i\min}$, $u_{i1\max} = T_{i\max}$. $T_{i\max}$ and $T_{i\min}$ are the maximum and minimum thrust forces, respectively.

To handle the input saturation, an auxiliary system is constructed as [5]

$$\dot{\eta}_{i1} = \begin{cases} -k_{22}\eta_{i1} - \frac{|S_{i2}\Delta u_{i1}| + 0.5\Delta u_{i1}^2}{|\eta_{i1}|^2}\eta_{i1} + \Delta u_{i1} & |\eta_{i1}| \geq \nu_{i1} \\ 0 & |\eta_{i1}| < \nu_{i1} \end{cases} \tag{6.46}$$

Remark 6.6 Similar to the analysis in Sect. 3.3 of Chap. 3, $\dot{\eta}_{i1} = -k_{22}\eta_{i1}$ if Δu_{i1} becomes 0 in (6.46), indicating that η_{i1} converges into the region $|\eta_{i1}| < \nu_{i1}$. If the input saturation occurs, i.e., $\Delta u_{i1} \neq 0$, when $|\eta_{i1}| < \nu_{i1}$, one can reset η_{i1} to $|\eta_{i1}| \geq \nu_{i1}$. Then, the auxiliary system (6.46) can be reactivated to adjust the control input signal (6.37) to compensate the input saturation, until $\Delta u_{i1} = 0$ and $|\eta_{i1}| < \nu_{i1}$ are regained.

Therefore, (6.44) can be derived as

$$\dot{L}_{i2} \leq -\left(k_{21}-\frac{k_{22}}{2}\right)S_{i2}^2-\left(\frac{k_{22}}{2}-\frac{1}{2}\right)\eta_{i1}^2-\frac{h_{11}}{2}\tilde{\chi}_{i11}^2 \\ -b_{i2}S_{i2}|S_{i2}|^{r_i}\operatorname{sign}(S_{i2})-\frac{h_{12}}{2}\tilde{\chi}_{i12}^2+\varsigma_{i1} \tag{6.47}$$

where $\varsigma_{i1}=\frac{h_{11}}{2}\chi_{i11}^2+\frac{h_{12}}{2}\chi_{i12}^2+\frac{\zeta_{11}^2}{2}+\frac{\zeta_{12}^2}{2}$.

6.3.3 *Finite-Time Fault-Tolerant Controller Design for Vertical Subsystem*

In this section, the finite-time FTC law is constructed for the vertical subsystem to achieve $x_{i3} \rightarrow \hat{x}_{i3d}$ in finite time.

Define a new set of errors as

$$S_{i3} = e_{i3} \tag{6.48}$$

$$S_{i4} = e_{i4}+b_{i3}|S_{i3}|^{r_i}\operatorname{sign}(S_{i3}) \tag{6.49}$$

$$S_{i5} = e_{i5}+b_{i4}|S_{i4}|^{r_i}\operatorname{sign}(S_{i4}) \tag{6.50}$$

$$S_{i6} = e_{i6}+b_{i5}|S_{i5}|^{r_i}\operatorname{sign}(S_{i5}) \tag{6.51}$$

where $i \in F$, $e_{i3}=x_{i3}-\hat{x}_{i3d}$ is the vertical position (altitude) tracking error with respect to the estimation of x_{i3d}, $e_{i4}=x_{i4}-\bar{x}_{i4d}$ is the flight path angle tracking error, $e_{i5}=x_{i5}-\bar{x}_{i5d}$ is the angle of attack tracking error, and $e_{i6}=x_{i6}-\bar{x}_{i6d}$ is the pitch rate tracking error. $\bar{x}_{i4d}$, $\bar{x}_{i5d}$, and $\bar{x}_{i6d}$ are the virtual control signals constructed at each step by using the backstepping design architecture. $b_{ij}>0$, $j=3,4,5$.

Consider the case $t > T_2$, and the finite-time FTC design for the vertical subsystem contains four steps:

Step 1 Taking the time derivative of S_{i3} along with (6.10) gives

$$\dot{S}_{i3}=\dot{x}_{i3}-\dot{\hat{x}}_{i3d}=V_i x_{i4}-\dot{\hat{x}}_{i3d} \tag{6.52}$$

Then, design the virtual control signal $\bar{x}_{i4d}$ as

$$\bar{x}_{i4d}=V_i^{-1}(-k_{31}S_{i3}+\hat{V}_{i3d}) \tag{6.53}$$

where $k_{31}>0$ is a design parameter and $\hat{V}_{i3d}$ is generated by the SMO (6.23).

Choose the Lyapunov function candidate as

$$L_{i3} = \frac{1}{2\bar{V}_i^2} S_{i3}^2 \tag{6.54}$$

where $\bar{V}_i$ is the maximum of V_i and is only used for the stability analysis.

According to Young's inequality, the time derivative of (6.54) can be given by

$$\dot{L}_{i3} \leq -\frac{1}{\bar{V}_i^2}\left(k_{31} - \frac{1}{2\zeta_{02}^2}\right) S_{i3}^2 + \frac{\zeta_{02}^2 S_{i4}^2}{2} - \frac{V_i}{\bar{V}_i^2} b_{i3} S_{i3} |S_{i3}|^{r_i} \operatorname{sign}(S_{i3}) \tag{6.55}$$

where ζ_{02} is a positive constant.

Step 2 By using the backstepping architecture, the virtual control signal $\bar{x}_{i3d}$ and adaptive laws are designed as

$$\bar{x}_{i5d} = -k_{41} S_{i4} - \frac{S_{i4}\hat{\chi}_{i21}\boldsymbol{\varphi}_{i2}^T\boldsymbol{\varphi}_{i2}}{2\zeta_{21}^2} - \frac{S_{i2}\hat{\chi}_{i12}}{2\zeta_{22}^2} \tag{6.56}$$

$$\dot{\hat{\chi}}_{i21} = g_{21}\left(\frac{S_{i4}^2\boldsymbol{\varphi}_{i2}^T\boldsymbol{\varphi}_{i2}}{2\zeta_{21}^2} - h_{21}\hat{\chi}_{i21}\right) \tag{6.57}$$

$$\dot{\hat{\chi}}_{i22} = g_{22}\left(\frac{S_{i4}^2}{2\zeta_{22}^2} - h_{22}\hat{\chi}_{i21}\right) \tag{6.58}$$

where k_{41}, ζ_{21}, ζ_{22}, g_{21}, g_{22}, h_{21}, and h_{22} are positive design parameters. $\hat{\chi}_{i21}$ and $\hat{\chi}_{i22}$ are the estimations of χ_{i21} and χ_{i22}, respectively.

Choose the Lyapunov function candidate as

$$L_{i4} = \frac{1}{2} S_{i4}^2 + \frac{1}{2g_{21}}\tilde{\chi}_{i21}^2 + \frac{1}{2g_{22}}\tilde{\chi}_{i22}^2 \tag{6.59}$$

where $\tilde{\chi}_{i21} = \chi_{i21} - \hat{\chi}_{i21}$ and $\tilde{\chi}_{i22} = \chi_{i22} - \hat{\chi}_{i22}$ are the estimation errors.

Taking the time derivative of (6.59) gives

$$\begin{aligned}\dot{L}_{i4} =& S_{i4}\left[f_{i2}(x_{i4}, x_{i5f}) + \varepsilon_{i2} + \bar{x}_{i5d} + e_{i5} - \dot{\bar{x}}_{i4d} + b_{i3} r_i |S_{i3}|^{r_i - 1} \dot{S}_{i3} \right] \\ &+ \frac{1}{g_{21}}\tilde{\chi}_{i21}\dot{\hat{\chi}}_{i21} + \frac{1}{g_{22}}\tilde{\chi}_{i22}\dot{\hat{\chi}}_{i22}\end{aligned} \tag{6.60}$$

Define $F_{i2}(x_{i2}, x_{i4}, \dot{\bar{x}}_{i4d}, x_{i5f}) = f_{i2}(x_{i4}, x_{i5f}) - \dot{\bar{x}}_{i4d} + b_{i3} r_i |S_{i3}|^{r_i-1}\dot{S}_{i3}$. By using the RBFNN technique, $F_{i2}(x_{i2}, x_{i4}, \dot{\bar{x}}_{i4d}, x_{i5f}) = \boldsymbol{\theta}_{i2}^{*T}\boldsymbol{\varphi}_{i2}(x_{i2}, x_{i4}, \dot{\bar{x}}_{i4d}, x_{i5f})$

$+ w_{i2}$. To obtain $\dot{\bar{x}}_{i4d}$ in the Gaussian function vector $\boldsymbol{\varphi}_{i2}(x_{i2}, x_{i4}, \dot{\bar{x}}_{i4d}, x_{i5f})$, the FOSMD (6.42) is employed.

Then, one has

$$F_{i2}(x_{i2}, x_{i4}, \dot{\bar{x}}_{i4d}, x_{i5f}) = \boldsymbol{\theta}_{i2}^{*T} \boldsymbol{\varphi}_{i2}(x_{i2}, x_{i4}, \dot{\bar{x}}_{i4ds}, x_{i5f}) + w_{i2} + \xi_{i2} \tag{6.61}$$

where ξ_{i2} is bounded by $|\xi_{i2}| \leq \bar{\xi}_{i2}$ and $\dot{\bar{x}}_{i4ds}$ can be obtained by passing $\bar{x}_{i4d}$ to the FOSMD. For brevity, $(\cdot)$ in $\boldsymbol{\varphi}_{i2}(\cdot)$ is omitted in the subsequent analysis.

Then, substituting (6.56), (6.57), (6.58), and (6.61) into (6.60) yields

$$\begin{aligned} \dot{L}_{i4} \leq & -\left(k_{41} - \frac{1}{2}\right) S_{i4}^2 + \frac{1}{2} S_{i5}^2 - \frac{h_{21}}{2} \tilde{\chi}_{i21}^2 \\ & - b_{i4} S_{i4} |S_{i4}|^{r_i} \operatorname{sign}(S_{i4}) - \frac{h_{22}}{2} \tilde{\chi}_{i22}^2 + \varsigma_{i2} \end{aligned} \tag{6.62}$$

where $\varsigma_{i2} = \frac{h_{21}}{2} \chi_{i21}^2 + \frac{h_{22}}{2} \chi_{i22}^2 + \frac{\zeta_{21}^2}{2} + \frac{\zeta_{22}^2}{2}$, $\chi_{i21} = \boldsymbol{\theta}_{i2}^{*T} \boldsymbol{\theta}_{i2}^*$, $\chi_{i22} = (\bar{w}_{i2} + \bar{\varepsilon}_{i2} + \bar{\xi}_{i2})^2$, and the inequalities $S_{i4} \boldsymbol{\theta}_{i2}^{*T} \boldsymbol{\varphi}_{i2} \leq \frac{S_{i4}^2 \chi_{i21} \boldsymbol{\varphi}_{i2}^T \boldsymbol{\varphi}_{i2}}{2\zeta_{21}^2} + \frac{\zeta_{21}^2}{2}$, $S_{i4}(w_{i2} + \varepsilon_{i2} + \xi_{i2}) \leq \frac{S_{i4}^2 \chi_{i22}}{2\zeta_{22}^2} + \frac{\zeta_{22}^2}{2}$ are used in (6.62).

Step 3 Design the virtual control signal $\bar{x}_{i6d}$ and adaptive laws as

$$\bar{x}_{i6d} = -k_{51} S_{i5} - \frac{S_{i5} \hat{\chi}_{i31} \boldsymbol{\varphi}_{i3}^T \boldsymbol{\varphi}_{i3}}{2\zeta_{31}^2} - \frac{S_{i5} \hat{\chi}_{i32}}{2\zeta_{32}^2} \tag{6.63}$$

$$\dot{\hat{\chi}}_{i31} = g_{31} \left(\frac{S_{i5}^2 \boldsymbol{\varphi}_{i3}^T \boldsymbol{\varphi}_{i3}}{2\zeta_{31}^2} - h_{31} \hat{\chi}_{i31} \right) \tag{6.64}$$

$$\dot{\hat{\chi}}_{i32} = g_{32} \left(\frac{S_{i5}^2}{2\zeta_{32}^2} - h_{32} \hat{\chi}_{i32} \right) \tag{6.65}$$

where k_{51}, ζ_{31}, ζ_{32}, g_{31}, g_{32}, h_{31}, and h_{32} are positive design parameters. $\hat{\chi}_{i31}$ and $\hat{\chi}_{i32}$ are the estimations of χ_{i31} and χ_{i32}, respectively.

Choose the Lyapunov function candidate as

$$L_{i5} = \frac{1}{2} S_{i5}^2 + \frac{1}{2g_{31}} \tilde{\chi}_{i31}^2 + \frac{1}{2g_{32}} \tilde{\chi}_{i32}^2 \tag{6.66}$$

where $\tilde{\chi}_{i31} = \chi_{i31} - \hat{\chi}_{i31}$ and $\tilde{\chi}_{i32} = \chi_{i32} - \hat{\chi}_{i32}$ are the estimation errors.

Taking the time derivative of (6.66) gives

$$\dot{L}_{i5} = S_{i5}\left[f_{i3}(x_{i4}, x_{i5}) + \bar{x}_{i6d} + e_{i6} - \dot{\bar{x}}_{i5d} + b_{i4} r_i |S_{i4}|^{r_i - 1} \dot{S}_{i4} \right] + \frac{1}{g_{31}} \tilde{\chi}_{i31} \dot{\hat{\chi}}_{i31} + \frac{1}{g_{32}} \tilde{\chi}_{i32} \dot{\hat{\chi}}_{i32} \tag{6.67}$$

Define $F_{i3}(x_{i2}, x_{i4}, x_{i5}, \dot{\bar{x}}_{i5d}) = f_{i3}(x_{i4}, x_{i5}) - \dot{\bar{x}}_{i5d} + b_{i4} r_i |S_{i4}|^{r_i - 1} \dot{S}_{i4}$. By using RBFNN, $F_{i3}(x_{i2}, x_{i4}, x_{i5}, \dot{\bar{x}}_{i5d}) = \boldsymbol{\theta}_{i3}^{*T} \boldsymbol{\varphi}_{i3}(x_{i2}, x_{i4}, x_{i5}, \dot{\bar{x}}_{i5d}) + w_{i3}$. Similar to (6.42), the FOSMD is used to calculate $\dot{\bar{x}}_{i5d}$ in $\boldsymbol{\varphi}_{i3}(\cdot)$. Therefore, one has

$$F_{i3}(x_{i2}, x_{i4}, x_{i5}, \dot{\bar{x}}_{i5d}) = \boldsymbol{\theta}_{i3}^{*T} \boldsymbol{\varphi}_{i3}(x_{i2}, x_{i4}, x_{i5}, \dot{\bar{x}}_{i5ds}) + \xi_{i3} + w_{i3} \tag{6.68}$$

where ξ_{i3} is bounded by $|\xi_{i3}| \leq \bar{\xi}_{i3}$ and $\dot{\bar{x}}_{i5ds}$ is obtained by passing $\bar{x}_{i5d}$ to the FOSMD.

Then, substituting (6.63), (6.64), (6.65), and (6.68) into (6.67) yields

$$\dot{L}_{i5} \leq -\left(k_{51} - \frac{1}{2}\right) S_{i5}^2 + \frac{1}{2} S_{i6}^2 - \frac{h_{31}}{2} \tilde{\chi}_{i31}^2 - b_{i5} S_{i5} |S_{i5}|^{r_i} \operatorname{sign}(S_{i5}) - \frac{h_{32}}{2} \tilde{\chi}_{i32}^2 + \varsigma_{i3} \tag{6.69}$$

where $\varsigma_{i3} = \frac{h_{31}}{2} \chi_{i31}^2 + \frac{h_{32}}{2} \chi_{i32}^2 + \frac{\zeta_{31}^2}{2} + \frac{\zeta_{32}^2}{2}$, $\chi_{i31} = \boldsymbol{\theta}_{i3}^{*T} \boldsymbol{\theta}_{i3}^*$, $\chi_{i32} = (\bar{w}_{i3} + \bar{\xi}_{i3})^2$. The Young's inequalities $S_{i5} \boldsymbol{\theta}_{i3}^{*T} \boldsymbol{\varphi}_{i3} \leq \frac{S_{i5}^2 \chi_{i31} \boldsymbol{\varphi}_{i3}^T \boldsymbol{\varphi}_{i3}}{2\zeta_{31}^2} + \frac{\zeta_{31}^2}{2}$ and $S_{i5}(w_{i3} + \xi_{i3}) \leq \frac{S_{i5}^2 \chi_{i32}}{2\zeta_{32}^2} + \frac{\zeta_{32}^2}{2}$ are used in (6.69).

Step 4 Consider the input saturation, the control input signal, and the adaptive laws are designed as

$$\bar{u}_{i2} = -k_{61} S_{i6} - \frac{S_{i6} \hat{\chi}_{i41} \boldsymbol{\varphi}_{i4}^T \boldsymbol{\varphi}_{i4}}{2\zeta_{41}^2} - \frac{S_{i6} \hat{\chi}_{i42}}{2\zeta_{42}^2} - b_{i6} |S_{i6}|^{r_i} \operatorname{sign}(S_{i6}) + k_{62} \eta_i \tag{6.70}$$

$$\dot{\hat{\chi}}_{i41} = g_{41} \left(\frac{S_{i6}^2 \boldsymbol{\varphi}_{i4}^T \boldsymbol{\varphi}_{i4}}{2\zeta_{41}^2} - h_{41} \hat{\chi}_{i41} \right) \tag{6.71}$$

$$\dot{\hat{\chi}}_{i42} = g_{42} \left(\frac{S_{i6}^2}{2\zeta_{42}^2} - h_{42} \hat{\chi}_{i42} \right) \tag{6.72}$$

where $k_{61}, k_{62}, \zeta_{41}, \zeta_{42}, g_{41}, g_{42}, h_{41}, h_{42}$, and b_{i6} are positive design parameters and $\boldsymbol{\varphi}_{i4}$ is the basis function vector. η_{i2} is an auxiliary signal constructed to compensate the input saturation. $\hat{\chi}_{i41}$ and $\hat{\chi}_{i42}$ are the estimations of χ_{i41} and χ_{i42}, respectively.

Choose the Lyapunov function candidate at this step as

$$L_{i6} = \frac{1}{2}S_{i6}^2 + \frac{1}{2g_{41}}\tilde{\chi}_{i41}^2 + \frac{1}{2g_{42}}\tilde{\chi}_{i42}^2 + \frac{1}{2}\eta_{i2}^2 \tag{6.73}$$

where $\tilde{\chi}_{i41} = \chi_{i41} - \hat{\chi}_{i41}$ and $\tilde{\chi}_{i42} = \chi_{i42} - \hat{\chi}_{i42}$ are the estimation errors.

Taking the time derivative of (6.73) yields

$$\begin{aligned}\dot{L}_{i6} =& S_{i6}\left[f_{i4}(x_{i4}, x_{i5}, x_{i6}, u_{i2f}) + \varepsilon_{i3} + u_{i2} - \dot{\bar{x}}_{i6d} + b_{i5}r_i|S_{i5}|^{r_i-1}\dot{S}_{i5}\right] \\ &+ \frac{1}{g_{41}}\tilde{\chi}_{i41}\dot{\tilde{\chi}}_{i41} + \frac{1}{g_{42}}\tilde{\chi}_{i42}\dot{\tilde{\chi}}_{i42} + \eta_{i2}\dot{\eta}_{i2}\end{aligned} \tag{6.74}$$

Define $F_{i4}(x_{i2}, x_{i4}, x_{i5}, x_{i6}, \dot{\bar{x}}_{i6d}, u_{i2f}) = f_{i4}(x_{i4}, x_{i5}, x_{i6}, u_{i2f}) + b_{i5}r_i|S_{i5}|^{r_i-1}$ $\dot{S}_{i5} - \dot{\bar{x}}_{i6d}$. By using RBFNN, $F_{i4}(x_{i2}, x_{i4}, x_{i5}, x_{i6}, \dot{\bar{x}}_{i6d}, u_{i2f}) = \boldsymbol{\theta}_{i4}^{*T}\boldsymbol{\varphi}_{i4}(x_{i2}, x_{i4}, x_{i5}, x_{i6}, \dot{\bar{x}}_{i6d}, u_{i2f}) + w_{i4}$. Similar to (6.42), the FOSMD is used to calculate $\dot{\bar{x}}_{i6d}$ in $\boldsymbol{\varphi}_{i4}(\cdot)$.

Then, one has

$$\begin{aligned}&F_{i4}(x_{i2}, x_{i4}, x_{i5}, x_{i6}, \dot{\bar{x}}_{i6d}, u_{i2f}) \\ &= \boldsymbol{\theta}_{i4}^{*T}\boldsymbol{\varphi}_{i4}(x_{i2}, x_{i4}, x_{i5}, x_{i6}, \dot{\bar{x}}_{i6ds}, u_{i2f}) + \xi_{i4} + w_{i4}\end{aligned} \tag{6.75}$$

where ξ_{i4} is bounded by $|\xi_{i4}| \le \bar{\xi}_{i4}$ and $\dot{\bar{x}}_{i6ds}$ is obtained by passing $\bar{x}_{i6d}$ to the FOSMD.

Similar to (6.45), the input signal under the input limitations can be described as

$$u_{i2} = \text{sat}(\bar{u}_{i2}) = \begin{cases} u_{i2\max}, & \bar{u}_{i2} \ge u_{i2\max} \\ \bar{u}_{i2}, & u_{i2\min} < \bar{u}_{i2} < u_{i2\max} \\ u_{i2\min}, & \bar{u}_{i2} \le u_{i2\min} \end{cases} \tag{6.76}$$

where $i \in F$. $u_{i2\min} = -\delta_{ie0\max}$, $u_{i2\max} = -\delta_{ie0\min}$. $\delta_{ie0\max}$ and $\delta_{ie0\min}$ are the maximum and minimum elevator deflection angles, respectively.

Then, one has

$$\begin{aligned}\dot{L}_{i6} \le& -k_{61}S_{i6}^2 + S_{i6}\Delta u_{i2} - b_{i6}S_{i6}|S_{i6}|^{r_i}\text{sign}(S_{i6}) \\ &+ k_{62}\eta_{i2}S_{i6} + \eta_{i2}\dot{\eta}_{i2} - \frac{h_{41}}{2}\tilde{\chi}_{i41}^2 - \frac{h_{42}}{2}\tilde{\chi}_{i42}^2 \\ &+ \frac{h_{41}}{2}\chi_{i41}^2 + \frac{h_{42}}{2}\chi_{i42}^2 + \frac{\zeta_{41}^2}{2} + \frac{\zeta_{42}^2}{2}\end{aligned} \tag{6.77}$$

where $\Delta u_{i2} = u_{i2} - \bar{u}_{i2}$ is the error between the designed control input signal and the actual control signal, $\chi_{i41} = \boldsymbol{\theta_{i4}}^{*T}\boldsymbol{\theta_{i4}}^{*}$, $\chi_{i42} = (\bar{\varepsilon}_{i3} + \bar{\xi}_{i4} + \bar{w}_{i4})^2$. $S_{i6}\boldsymbol{\theta}_{i4}^{*T}\boldsymbol{\varphi}_{i4} \leq \frac{S_{i6}^2\chi_{i41}\boldsymbol{\varphi}_{i4}^T\boldsymbol{\varphi}_{i4}}{2\zeta_{41}^2} + \frac{\zeta_{41}^2}{2}$ and $S_{i6}(\varepsilon_{i3} + \xi_{i4} + w_{i4}) \leq \frac{S_{i6}^2\chi_{i42}}{2\zeta_{42}^2} + \frac{\zeta_{42}^2}{2}$ are used in (6.77).

Again, to handle the input saturation, an auxiliary system for each follower UAV is defined as

$$\dot{\eta}_{i2} = \begin{cases} -k_{62}\eta_{i2} - \frac{|S_{i6}\Delta u_{i2}| + 0.5\Delta u_{i2}^2}{|\eta_{i2}|^2}\eta_{i2} + \Delta u_{i2} & |\eta_{i2}| \geq \nu_{i2} \\ 0 & |\eta_{i2}| < \nu_{i2} \end{cases} \tag{6.78}$$

The treatment mentioned in Remark 6.6 can also be applied once $\Delta u_{i2} \neq 0$ when $|\eta_{i2}| < \nu_{i2}$. Then, (6.77) is given by

$$\begin{aligned} \dot{L}_{i6} \leq &- \left(k_{61} - \frac{1}{2}k_{62}\right)S_{i6}^2 - \frac{h_{41}}{2}\tilde{\chi}_{i41}^2 - \frac{h_{42}}{2}\tilde{\chi}_{i42}^2 \\ &- \left(\frac{k_{62}}{2} - \frac{1}{2}\right)\eta_{i2}^2 - b_{i6}S_{i6}|S_{i6}|^{r_i}\operatorname{sign}(S_{i6}) + \varsigma_{i4} \end{aligned} \tag{6.79}$$

where $\varsigma_{i4} = \frac{h_{41}}{2}\chi_{i41}^2 + \frac{h_{42}}{2}\chi_{i42}^2 + \frac{\zeta_{41}^2}{2} + \frac{\zeta_{42}^2}{2}$.

The overall control architecture is illustrated in Fig. 6.2.

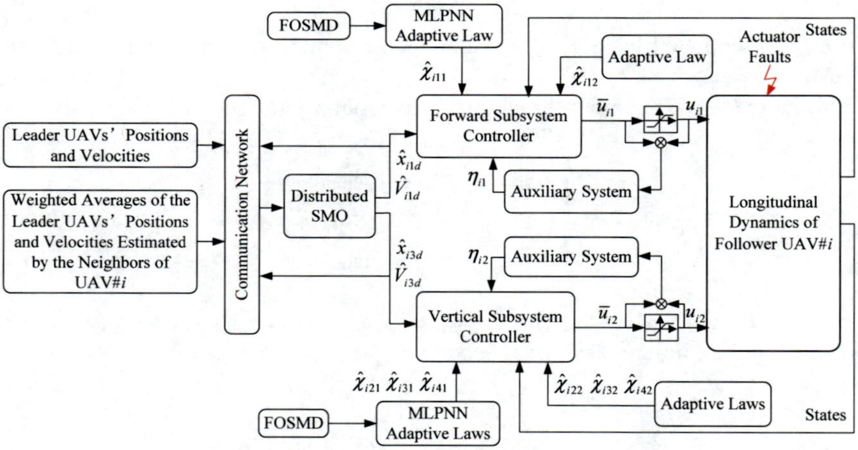

Fig. 6.2 Overall control architecture

6.3.4 Stability Analysis

Theorem 6.1 *Consider a group of N follower UAVs and M leader UAVs with Assumptions 6.1–6.4 satisfied, and if the distributed finite-time SMO is designed as (6.22), (6.23), the control laws are designed as (6.34), (6.37), (6.53), (6.56), (6.63), (6.70), the adaptive laws are designed as (6.38), (6.39), (6.57), (6.58), (6.64), (6.65), (6.71), (6.72), the auxiliary systems are constructed as (6.46), (6.78), then the forward and vertical positions of all follower UAVs converge into the convex hull spanned by the forward and vertical positions of all leader UAVs in the longitudinal plane, and the overall tracking errors $E_{i1} = S_{i1} + \tilde{x}_{i1d}$, $E_{i3} = S_{i3} + \tilde{x}_{i3d}$, $i \in F$, of each follower UAV will be convergent in finite time.*

Proof Consider the following Lyapunov function candidate as

$$L_i = L_{i1} + L_{i2} + L_{i3} + L_{i4} + L_{i5} + L_{i6} \tag{6.80}$$

By applying (6.36), (6.47), (6.55), (6.62), (6.69), and (6.79), the time derivative of (6.80) is given by

$$\begin{aligned}
\dot{L}_i \leq & -\left(k_{11} - \frac{1}{2\zeta_{01}^2}\right) S_{i1}^2 - \left(k_{21} - \frac{k_{22}}{2} - \frac{\zeta_{01}^2}{2}\right) S_{i2}^2 \\
& - \frac{1}{\bar{V}_i^2}\left(k_{31} - \frac{1}{2\zeta_{02}^2}\right) S_{i3}^2 - \left(k_{41} - \frac{1}{2} - \frac{\zeta_{02}^2}{2}\right) S_{i4}^2 \\
& - (k_{51} - 1) S_{i5}^2 - \left(k_{61} - \frac{k_{62}}{2} - \frac{1}{2}\right) S_{i6}^2 \\
& - \frac{h_{11}}{2}\tilde{\chi}_{i11}^2 - \frac{h_{12}}{2}\tilde{\chi}_{i12}^2 - \frac{h_{21}}{2}\tilde{\chi}_{i21}^2 - \frac{h_{22}}{2}\tilde{\chi}_{i22}^2 \\
& - \frac{h_{31}}{2}\tilde{\chi}_{i31}^2 - \frac{h_{32}}{2}\tilde{\chi}_{i32}^2 - \frac{h_{41}}{2}\tilde{\chi}_{i41}^2 - \frac{h_{42}}{2}\tilde{\chi}_{i42}^2 \\
& - g_{ix} b_{i1} S_{i1} |S_{i1}|^{r_i} \operatorname{sign}(S_{i1}) - b_{i2} S_{i2} |S_{i2}|^{r_i} \operatorname{sign}(S_{i2}) \\
& - \frac{V_i}{\bar{V}_i^2} b_{i3} S_{i3} |S_{i3}|^{r_i} \operatorname{sign}(S_{i3}) - b_{i4} S_{i4} |S_{i4}|^{r_i} \operatorname{sign}(S_{i4}) \\
& - b_{i5} S_{i5} |S_{i5}|^{r_i} \operatorname{sign}(S_{i5}) - b_{i6} S_{i6} |S_{i6}|^{r_i} \operatorname{sign}(S_{i6}) \\
& - \left(\frac{k_{22}}{2} - \frac{1}{2}\right) \eta_{i1}^2 - \left(\frac{k_{62}}{2} - \frac{1}{2}\right) \eta_{i2}^2 + \varsigma_i
\end{aligned} \tag{6.81}$$

where $\varsigma_i = \varsigma_{i1} + \varsigma_{i2} + \varsigma_{i3} + \varsigma_{i4}$.

Then, one has

$$
\begin{aligned}
\dot{L}_i \leq & -\left(k_{11}-\frac{1}{2\zeta_{01}^2}\right)S_{i1}^2-\left(k_{21}-\frac{k_{22}}{2}-\frac{\zeta_{01}^2}{2}\right)S_{i2}^2 \\
& -\frac{1}{\bar{V}_i^2}\left(k_{31}-\frac{1}{2\zeta_{02}^2}\right)S_{i3}^2-\left(k_{41}-\frac{1}{2}-\frac{\zeta_{02}^2}{2}\right)S_{i4}^2-\frac{h_{12}}{2}\tilde{\chi}_{i12}^2 \\
& -(k_{51}-1)S_{i5}^2-\left(k_{61}-\frac{k_{62}}{2}-\frac{1}{2}\right)S_{i6}^2-\frac{h_{11}}{2}\tilde{\chi}_{i11}^2 \\
& -\frac{h_{21}}{2}\tilde{\chi}_{i21}^2-\frac{h_{22}}{2}\tilde{\chi}_{i22}^2-\frac{h_{31}}{2}\tilde{\chi}_{i31}^2-\frac{h_{32}}{2}\tilde{\chi}_{i32}^2-\frac{h_{41}}{2}\tilde{\chi}_{i41}^2 \\
& -\frac{h_{42}}{2}\tilde{\chi}_{i42}^2-\left(\frac{k_{22}}{2}-\frac{1}{2}\right)\eta_{i1}^2-\left(\frac{k_{62}}{2}-\frac{1}{2}\right)\eta_{i2}^2+\varsigma_i \\
\leq & -\upsilon_i L_i+\varsigma_i
\end{aligned}
\tag{6.82}
$$

where υ_i is expressed as

$$
\upsilon_i=\min\left\{\begin{array}{c}
2\left(k_{11}-\frac{1}{2\zeta_{01}^2}\right),\ 2\left(k_{21}-\frac{k_{22}}{2}-\frac{\zeta_{01}^2}{2}\right), \\
2\left(k_{31}-\frac{1}{2\zeta_{02}^2}\right),\ 2\left(k_{41}-\frac{1}{2}-\frac{\zeta_{02}^2}{2}\right), \\
2(k_{51}-1),\ 2\left(k_{61}-\frac{k_{62}}{2}-\frac{1}{2}\right), \\
h_{11},\ h_{12},\ h_{21},\ h_{22}, h_{31},\ h_{32}, \\
h_{41},\ h_{42},\ k_{22}-1,\ k_{62}-1
\end{array}\right\}>0
\tag{6.83}
$$

According to the boundedness theorem, S_{i1}, S_{i2}, S_{i3}, S_{i4}, S_{i5}, S_{i6}, η_{i1}, η_{i2}, $\tilde{\chi}_{i11}$, $\tilde{\chi}_{i12}$, $\tilde{\chi}_{i21}$, $\tilde{\chi}_{i22}$, $\tilde{\chi}_{i31}$, $\tilde{\chi}_{i32}$, $\tilde{\chi}_{i41}$, and $\tilde{\chi}_{i42}$ are UUB. It can be further concluded that the control input signals u_{i1}, u_{i2} are bounded, and the errors Δu_{i1}, Δu_{i2} are bounded. Therefore, one can assume that $|\eta_{i1}| \leq \eta_{i1m}$, $|\eta_{i2}| \leq \eta_{i2m}$, $|\tilde{\chi}_{i11}| \leq \chi_{i11m}$, $|\tilde{\chi}_{i12}| \leq \chi_{i12m}$, $|\tilde{\chi}_{i21}| \leq \chi_{i21m}$, $|\tilde{\chi}_{i22}| \leq \chi_{i22m}$, $|\tilde{\chi}_{i31}| \leq \chi_{i31m}$, $|\tilde{\chi}_{i32}| \leq \chi_{i32m}$, $|\tilde{\chi}_{i41}| \leq \chi_{i41m}$, $|\tilde{\chi}_{i42}| \leq \chi_{i42m}$, where η_{i1m}, η_{i2m}, χ_{i11m}, χ_{i12m}, χ_{i21m}, χ_{i22m}, χ_{i31m}, χ_{i32m}, χ_{i41m}, and χ_{i42m} are positive constants.

Subsequently, another Lyapunov function candidate is used to analyze the finite-time stability.

$$
L_{if}=\frac{1}{2}S_{i1}^2+\frac{1}{2}S_{i2}^2+\frac{1}{2\bar{V}_i^2}S_{i3}^2+\frac{1}{2}S_{i4}^2+\frac{1}{2}S_{i5}^2+\frac{1}{2}S_{i6}^2
\tag{6.84}
$$

Taking the time derivative of L_{if} gives

$$
\begin{aligned}
\dot{L}_{if} \leq & -\left(k_{11} - \frac{1}{2\zeta_{01}^2}\right) S_{i1}^2 - g_{ix} b_{i1} S_{i1} |S_{i1}|^{r_i} \operatorname{sign}(S_{i1}) \\
& - \left(k_{21} - \frac{1}{2} - \frac{\zeta_{01}^2}{2} - \frac{k_{22}}{2} - \frac{|\chi_{i11m}| \boldsymbol{\varphi}_{i1}^T \boldsymbol{\varphi}_{i1}}{2\zeta_{11}^2} - \frac{|\chi_{i12m}|}{2\zeta_{12}^2}\right) S_{i2}^2 \\
& - b_{i2} S_{i2} |S_{i2}|^{r_i} \operatorname{sign}(S_{i2}) - \frac{1}{\bar{V}_i^2} \left(k_{31} - \frac{1}{2\zeta_{02}^2}\right) S_{i3}^2 \\
& - \frac{V_i}{\bar{V}_i^2} b_{i3} S_{i3} |S_{i3}|^{r_i} \operatorname{sign}(S_{i3}) \\
& - \left(k_{41} - \frac{1}{2} - \frac{\zeta_{02}^2}{2} - \frac{|\chi_{i21m}| \boldsymbol{\varphi}_{i2}^T \boldsymbol{\varphi}_{i2}}{2\zeta_{21}^2} - \frac{|\chi_{i22m}|}{2\zeta_{22}^2}\right) S_{i4}^2 \\
& - b_{i4} S_{i4} |S_{i4}|^{r_i} \operatorname{sign}(S_{i4}) \\
& - \left(k_{51} - 1 - \frac{|\chi_{i31m}| \boldsymbol{\varphi}_{i3}^T \boldsymbol{\varphi}_{i3}}{2\zeta_{31}^2} - \frac{|\chi_{i32m}|}{2\zeta_{32}^2}\right) S_{i5}^2 - b_{i5} S_{i5} |S_{i5}|^{r_i} \operatorname{sign}(S_{i5}) \\
& - \left(k_{61} - 1 - \frac{k_{62}}{2} - \frac{|\chi_{i41m}| \boldsymbol{\varphi}_{i4}^T \boldsymbol{\varphi}_{i4}}{2\zeta_{41}^2} - \frac{|\chi_{i42m}|}{2\zeta_{42}^2}\right) S_{i6}^2 \\
& - b_{i6} S_{i6} |S_{i6}|^{r_i} \operatorname{sign}(S_{i6}) + \sigma_i
\end{aligned} \tag{6.85}
$$

where $\sigma_i = \frac{k_{22}}{2}\eta_{i1}^2 + \frac{\Delta u_{i1}^2}{2} + \frac{\zeta_{11}^2}{2} + \frac{\zeta_{12}^2}{2} + \frac{\zeta_{21}^2}{2} + \frac{\zeta_{22}^2}{2} + \frac{\zeta_{31}^2}{2} + \frac{\zeta_{32}^2}{2} + \frac{\zeta_{41}^2}{2} + \frac{\zeta_{42}^2}{2} + \frac{k_{42}}{2}\eta_{i2}^2 + \frac{\Delta u_{i2}^2}{2}$.

Then, by using Lemma 6.5, one has

$$
\begin{aligned}
\dot{L}_{if} \leq & -\lambda_{\min}(\boldsymbol{P}_i) \cdot L_{if} - \lambda_{\min}(\boldsymbol{Q}_i) \cdot m_{ir} \\
& \cdot \left[\left(\frac{1}{2} S_{i1}^2\right)^{\frac{r_i+1}{2}} + \left(\frac{1}{2} S_{i2}^2\right)^{\frac{r_i+1}{2}} + \left(\frac{1}{2\bar{V}_i^2} S_{i3}^2\right)^{\frac{r_i+1}{2}} \right. \\
& \left. + \left(\frac{1}{2} S_{i4}^2\right)^{\frac{r_i+1}{2}} + \left(\frac{1}{2} S_{i5}^2\right)^{\frac{r_i+1}{2}} + \left(\frac{1}{2} S_{i6}^2\right)^{\frac{r_i+1}{2}} \right] + \sigma_i \\
\leq & -\lambda_{\min}(\boldsymbol{P}_i) \cdot L_{if} - \lambda_{\min}(\boldsymbol{Q}_i) \cdot m_{ir} \cdot L_{if}^{\frac{r_i+1}{2}} + \sigma_i
\end{aligned} \tag{6.86}
$$

where $m_{ir} = \min\left\{2^{\frac{r_i-1}{2}}, (2\bar{V}_i^2)^{\frac{r_i-1}{2}}\right\}$. $\boldsymbol{P}_i = \mathrm{diag}\Big[2\Big(k_{11}-\frac{1}{2\zeta_{01}^2}\Big), 2\Big(k_{21}-\frac{1}{2}-\frac{\zeta_{01}^2}{2}-\frac{k_{22}}{2}-\frac{|\chi_{i11m}|\boldsymbol{\varphi}_{i1}^T\boldsymbol{\varphi}_{i1}}{2\zeta_{11}^2}-\frac{|\chi_{i12m}|}{2\zeta_{12}^2}\Big), 2\Big(k_{31}-\frac{1}{2\zeta_{02}^2}\Big), 2\Big(k_{41}-\frac{1}{2}-\frac{\zeta_{02}^2}{2}-\frac{|\chi_{i21m}|\boldsymbol{\varphi}_{i2}^T\boldsymbol{\varphi}_{i2}}{2\zeta_{21}^2}-\frac{|\chi_{i22m}|}{2\zeta_{22}^2}\Big), 2\Big(k_{51}-1-\frac{|\chi_{i31m}|\boldsymbol{\varphi}_{i3}^T\boldsymbol{\varphi}_{i3}}{2\zeta_{31}^2}-\frac{|\chi_{i32m}|}{2\zeta_{32}^2}\Big), 2\Big(k_{61}-1-\frac{k_{62}}{2}-\frac{|\chi_{i41m}|\boldsymbol{\varphi}_{i4}^T\boldsymbol{\varphi}_{i4}}{2\zeta_{41}^2}-\frac{|\chi_{i42m}|}{2\zeta_{42}^2}\Big)\Big]$, $\boldsymbol{Q}_i = \mathrm{diag}[2g_{ix}b_{i1}, 2b_{i2}, 2V_ib_{i3}, 2b_{i4}, 2b_{i5}, 2b_{i6}]$.

According to Lemma 6.4, if the parameters satisfy $k_{11} > \frac{1}{2\zeta_{01}^2}$, $k_{21} > \frac{k_{22}}{2}+\frac{1}{2}+\frac{\zeta_{01}^2}{2}+\frac{|\chi_{i11m}|\boldsymbol{\varphi}_{i1}^T\boldsymbol{\varphi}_{i1}}{2\zeta_{11}^2}+\frac{|\chi_{i12m}|}{2\zeta_{12}^2}$, $k_{31} > \frac{1}{2\zeta_{02}^2}$, $k_{41} > \frac{1}{2}+\frac{\zeta_{02}^2}{2}+\frac{|\chi_{i21m}|\boldsymbol{\varphi}_{i2}^T\boldsymbol{\varphi}_{i2}}{2\zeta_{21}^2}+\frac{|\chi_{i22m}|}{2\zeta_{22}^2}$, $k_{51} > 1+\frac{|\chi_{i31m}|\boldsymbol{\varphi}_{i3}^T\boldsymbol{\varphi}_{i3}}{2\zeta_{31}^2}+\frac{|\chi_{i32m}|}{2\zeta_{32}^2}$, $k_{61} > 1+\frac{k_{62}}{2}-\frac{|\chi_{i41m}|\boldsymbol{\varphi}_{i4}^T\boldsymbol{\varphi}_{i4}}{2\zeta_{41}^2}+\frac{|\chi_{i42m}|}{2\zeta_{42}^2}$, the finite-time boundedness can be guaranteed, and the tracking errors S_{i1}, S_{i2}, S_{i3}, S_{i4}, S_{i5}, S_{i6} will converge to the set $\Omega_i = \min\{\Omega_{i1}, \Omega_{i2}\}$ in finite time $T_{i3} = \max\{T_{i3f1}, T_{i3f2}\}$, where Ω_{i1}, Ω_{i2}, T_{i3f1}, and T_{i3f2} are given by

$$\begin{cases} \Omega_{i1} = \left\{S_{ij}, j = 1, 2, \ldots, 6 : L_{if} \le \frac{\sigma_i}{(1-\vartheta_i)\lambda_{\min}(\boldsymbol{P}_i)}\right\} \\ \Omega_{i2} = \left\{S_{ij}, j = 1, 2, \ldots, 6 : L_{if}^{\frac{r_i+1}{2}} \le \frac{\sigma_i}{(1-\vartheta_i)\lambda_{\min}(\boldsymbol{Q}_i)m_{ir}}\right\} \\ T_{i3f1} \le \frac{1}{\vartheta_i\lambda_{\min}(\boldsymbol{P}_i)(\frac{1-r_i}{2})}\ln\frac{\vartheta_i\lambda_{\min}(\boldsymbol{P}_i)L_{if}^{\frac{1-r_i}{2}}(T_2)+\lambda_{\min}(\boldsymbol{Q}_i)m_{ir}}{\lambda_{\min}(\boldsymbol{Q}_i)m_{ir}} \\ T_{i3f2} \le \frac{1}{\lambda_{\min}(\boldsymbol{P}_i)(\frac{1-r_i}{2})}\ln\frac{\lambda_{\min}(\boldsymbol{P}_i)L_{if}^{\frac{1-r_i}{2}}(T_2)+\vartheta_i\lambda_{\min}(\boldsymbol{Q}_i)m_{ir}}{\vartheta_i\lambda_{\min}(\boldsymbol{Q}_i)m_{ir}} \end{cases} \tag{6.87}$$

where $0 < \vartheta_i < 1$ is a constant and $L_{if}(T_2)$ is the value of L_{if} at the time T_2.

Furthermore, by the combination of Lemma 6.6, the forward position tracking error $E_{i1} = S_{i1} + \tilde{x}_{i1d} = (x_{i1} - \hat{x}_{i1d}) + (\hat{x}_{i1d} - x_{i1d})$ and the vertical position tracking error $E_{i3} = S_{i3} + \tilde{x}_{i3d} = (x_{i3} - \hat{x}_{i3d}) + (\hat{x}_{i3d} - x_{i3d})$ of each follower UAV will be convergent in finite time, and the settling time is $T_i = T_2 + T_{i3}$. This ends the proof. □

Remark 6.7 From (6.87), it can be seen that S_{i1}, S_{i2}, S_{i3}, S_{i4}, S_{i5}, and S_{i6} can converge to arbitrarily small neighborhoods containing zero by setting $\lambda_{\min}(\boldsymbol{P}_i)$ and $\lambda_{\min}(\boldsymbol{Q}_i)$ sufficiently large. This can be achieved by choosing large enough k_{11}, k_{21}, k_{31}, k_{41}, k_{51}, k_{61}, b_{i1}, b_{i2}, b_{i3}, b_{i4}, b_{i5}, and b_{i6}. From the relationship between E_{ij} and S_{ij}, $j = 1, 3$, the tracking errors E_{i1} and E_{i3} can be reduced by properly increasing these parameters. However, very large values of these parameters may lead to second damage of the control surfaces in the faulty UAVs. Therefore, the values of these parameters are chosen by extensive tests and trials until a good performance is obtained.

Remark 6.8 Define $S_{i1} = e_{i1}$, $S_{i2} = e_{i2}$, $S_{i3} = e_{i3}$, $S_{i4} = e_{i4}$, $S_{i5} = e_{i5}$, $S_{i6} = e_{i6}$, and remove the terms $b_{i2}|S_{i2}|^{r_i}\mathrm{sign}(S_{i2})$, $b_{i6}|S_{i6}|^{r_i}\mathrm{sign}(S_{i6})$ from (6.37), (6.70), respectively, and then the control scheme will reduce to the regular infinite-time stability control problem if the control laws and the adaptive laws are designed as (6.34), (6.37), (6.38), (6.39), (6.53), (6.56), (6.57), (6.58), (6.63), (6.64), (6.65), (6.70), (6.71), and (6.72). The remaining proof of infinite-time stability is similar to that in (6.80), which is omitted here.

6.4 Simulation Results

In the simulation, to verify the effectiveness of the proposed finite-time FTCC scheme, a topology consisting of four follower UAVs (UAVs#1–4) and two leader UAVs (UAVs#5–6) and the corresponding weights are illustrated in Fig. 6.3. The structure parameters and coefficient values of the ith follower UAV, $i = 1, 2, 3, 4$, can be referred to Chap. 3. The input saturation limits of the ith follower UAV are listed in Table 6.1.

Two obstacle zones are introduced in the simulation to demonstrate the effectiveness of the proposed cooperative control scheme and support the claim in Remark 6.4, which states that follower UAVs may be safer and easier to avoid obstacles if the leader UAVs are equipped with high-precision sensors to detect the obstacles and the follower UAVs are in the convex hull spanned by the leader UAVs. The first obstacle zone is located at 2,500 m with respect to the x direction of the ground coordinate frame with a width of 2,000 m and a height of 430 m. The second obstacle zone is located at 9,000 m with respect to the x direction of the ground coordinate frame with a width of 3,000 m and a height of 450 m. The obstacles are illustrated in Fig. 6.3. When the obstacle zones are detected by the sensors on the leader UAVs, the leader UAVs will change their trajectories to fly over the obstacles. Then, the leader UAVs will decrease their heights once the sensors on the leader UAVs detect the termination of obstacles. In the simulation, it is assumed that the detection range of sensor is 1,000 m, which is in the detection range of laser/light detection and ranging system. For the sense/detection ranges of different sensors in the unmanned aircraft systems, interested readers can refer to [28] for more details. In the current study, the role of leader UAVs is to detect the obstacles and send their updated trajectories to the follower UAVs via the communication network.

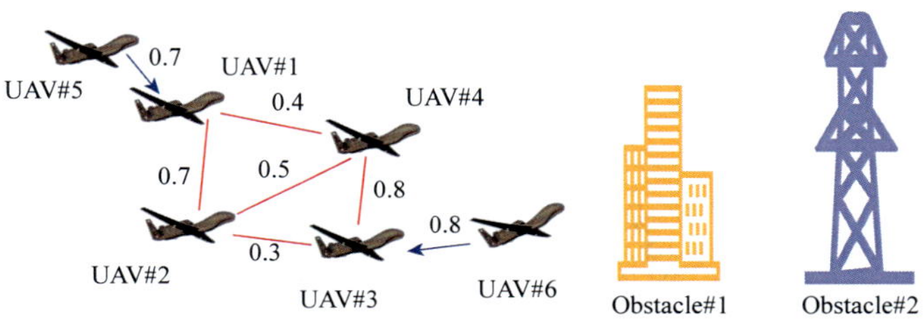

Fig. 6.3 Communication topology and obstacles

Table 6.1 Input saturation limits of the ith follower UAV

Lower limit	Upper limit
$T_{i\min} = 0$ N	$T_{i\max} = 60$ N
$\delta_{ie0\min} = -0.9075$ rad	$\delta_{ie0\max} = 0.9075$ rad

The elevator faults are assumed to be sequentially encountered by the follower UAV#1, UAV#2, and UAV#3 at $t = 40$ s, $t = 130$ s, and $t = 300$ s, respectively. Based on the fault model (6.7), the fault expressions ρ_i and h_{if} of UAVs#1–3 are chosen as $\rho_1 = 1$, $h_{1f} = 0$ for $t < 40$ s, $\rho_1 = 0.4e^{-0.8(t-40)} + 0.6$, $h_{1f} = -0.2e^{-0.7(t-40)} + 0.2$ for $t \geq 40$ s; $\rho_2 = 1$, $h_{2f} = 0$ for $t < 130$ s, $\rho_2 = 0.3e^{-0.8(t-130)} + 0.7$, $h_{2f} = -0.25e^{-0.7(t-130)} + 0.25$ for $t \geq 130$ s; $\rho_3 = 1$, $h_{3f} = 0$ for $t < 300$ s, $\rho_3 = 0.25e^{-0.6(t-300)} + 0.75$, $h_{3f} = -0.2e^{-0.6(t-300)} + 0.2$ for $t \geq 300$ s.

The initial values of follower UAVs are $x_1(0) = 10$ m, $z_1(0) = 468$ m, $x_2(0) = 50$ m, $z_2(0) = 450$ m, $x_3(0) = 100$ m, $z_3(0) = 425$ m, $x_4(0) = 120$ m, $z_4(0) = 442$ m, $V_i(0) = 30$ m, $\gamma_i(0) = 0$ rad, $\alpha_i(0) = 0.0322$ rad, $q_i(0) = 0$ rad/s, $i = 1, 2, 3, 4$. The forward and vertical positions of leader UAV#5 at $t = 0$ s are 30 m and 500 m, respectively. The forward and vertical positions of leader UAV#6 at $t = 0$ s are 150 m and 400 m, respectively. The design parameters for the control scheme of the ith follower UAV, $i = 1, 2, 3, 4$, are shown in Table 6.2. The initial values of the distributed SMO (6.22), (6.23) have the same values as the initial values of follower UAVs. The initial values of the adaptive laws (6.38), (6.39), (6.57), (6.58), (6.64), (6.65), (6.71), (6.72) are set as zeros, and the initial values of the auxiliary systems (6.46), (6.78) are set as 0.01. The BLPFs $B_{iu_1}(s)$, $B_{i5}(s)$, $B_{iu_2}(s)$ in (6.14) are chosen as $1/(s^2 + 1.414s + 1)$ [37]. Since sign function is used in the developed control scheme, the chattering may be caused due to the discontinuous design. To reduce the chattering phenomenon, a smooth hyperbolic tangent function $\tanh(\cdot/\kappa)$ is utilized to replace $\text{sign}(\cdot)$ in the simulation, where $\kappa = 0.02$. In the simulation, the engines and the elevator control surfaces are assumed to behave as a first-order system $20/(s + 20)$.

The trajectories of follower UAVs and leader UAVs are illustrated in Fig. 6.4. When the sensors on the leader UAV#6 detect the Obstacle#1 at $x_{610} = 1{,}500$ m ($t = 44.9$ s), the obstacle information is sent to the leader UAV#5, and then the

Table 6.2 Design parameters

Parameter	Value
$\beta_{1\iota}, \beta_{2\iota}, \beta_{3\iota}, \beta_{4\iota}, a$	0.5, 0.3, 0.2, 3, 0.4
$b_{i1}, b_{i2}, b_{i3}, b_{i4}, b_{i5}, b_{i6}, r_i$	0.2, 0.4, 0.05, 0.33, 0.1, 0.2, 0.8
$k_{11}, k_{21}, k_{22}, k_{31}, k_{41}, k_{51}, k_{61}, k_{62}$	0.2, 2, 1.1, 0.8, 3, 2.06, 1.35, 1.2
$\zeta_{11}, \zeta_{12}, \zeta_{21}, \zeta_{22}, \zeta_{31}, \zeta_{32}, \zeta_{41}, \zeta_{42}$	0.6, 0.15, 0.45, 0.3, 1.34, 0.4, 1.8, 0.49
$g_{11}, g_{12}, g_{21}, g_{22}, g_{31}, g_{32}, g_{41}, g_{42}$	15, 13, 15, 12, 20, 30, 8.2, 10.3
$h_{11}, h_{12}, h_{21}, h_{22}, h_{31}, h_{32}, h_{41}, h_{42}$	0.001, 0.002, 0.008, 0.05, 0.02, 0.03, 0.05, 0.8
ν_{i1}, ν_{i2}	0.005, 0.005

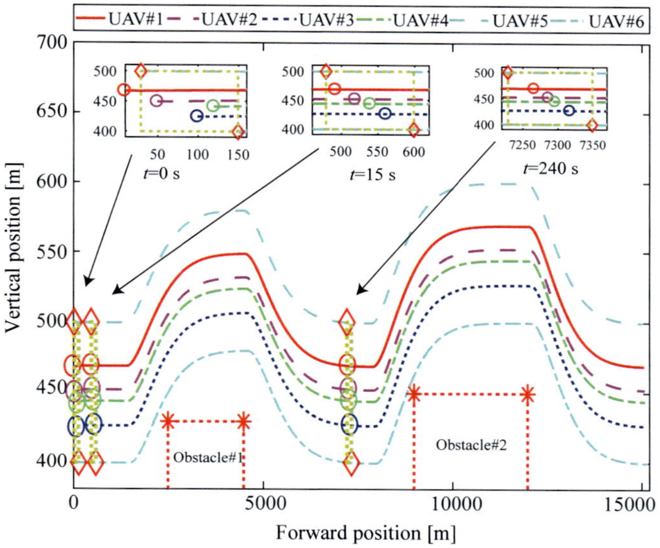

Fig. 6.4 Trajectories of follower UAVs (UAVs#1–4) and leader UAVs (UAVs#5–6)

leader UAV#5 and UAV#6, respectively, increase their heights to 580 and 480 m to avoid the obstacles and form a safe convex hull for the follower UAVs. When the sensors on the leader UAV#5 detect the termination of Obstacle#1 at $x_{510} = 4{,}500$ m ($t = 149$ s), the information is sent to the leader UAV#6, and then the leader UAV#5 and UAV#6 decrease their heights to 500 m and 400 m, respectively. When the higher Obstacle#2 is detected by the leader UAV#6 at $x_{610} = 8{,}000$ m ($t = 261.6$ s), the heights of leader UAVs#5–6 are respectively increased to 600 and 500 m to avoid the Obstacle#2. Then, the leader UAVs#5–6, respectively, decrease their heights to 500 and 400 m once the termination of Obstacle#2 is sensed by the leader UAV#5 at $x_{510} = 12{,}000$ m ($t = 395.3$ s). It is observed from Fig. 6.4 that the follower UAV#1 is outside the convex hull spanned by the forward and vertical positions of the leader UAVs#5–6 in the longitudinal plane at $t = 0$ s. Then, at $t = 15$ s and $t = 240$ s, it can be seen that all follower UAVs converge into the convex hull formed by the leader UAVs. With such a strategy, all follower UAVs can avoid the obstacles.

The tracking errors S_{i1} and S_{i3}, $i = 1, 2, 3, 4$, of follower UAVs with respect to the estimations of the weighted averages of leader UAVs' forward and vertical positions are demonstrated in Fig. 6.5. It can be observed that the tracking errors are convergent even when UAV#1, UAV#2, and UAV#3 encounter elevator faults at $t = 40$ s, $t = 130$ s, and $t = 300$ s, respectively. Figure 6.6 shows the forward position tracking errors of follower UAVs with respect to x_{i1d} and the vertical position tracking errors of follower UAVs with respect to x_{i3d}, $i = 1, 2, 3, 4$. It is indicated from Fig. 6.6 that the tracking errors converge into the small region containing

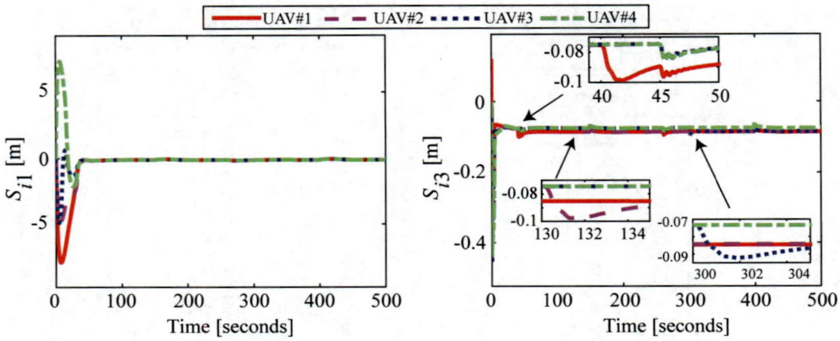

Fig. 6.5 Tracking errors S_{i1}, S_{i3} of four follower UAVs with respect to the estimations $\hat{x}_{i1d}$, $\hat{x}_{i3d}$

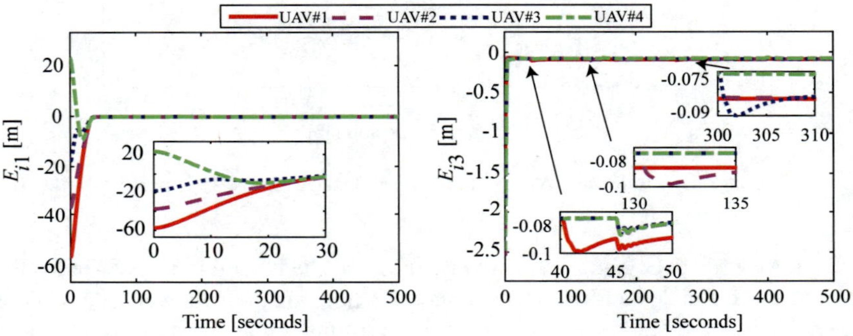

Fig. 6.6 Tracking errors E_{i1}, E_{i3} of four follower UAVs with respect to the weighted averages of the leader UAVs' forward and vertical positions

zero. Moreover, it can also be seen that when UAVs#1, 2, 3 are encountered by the elevator faults at $t = 40$ s, $t = 130$ s, and $t = 300$ s, respectively, slight degradation of tracking performances is induced. Then, under the proposed control scheme, the tracking errors are pulled back into the region containing zero. According to Figs. 6.5 and 6.6, one can find that the estimation errors $\tilde{x}_{i1d}$, $\tilde{x}_{i3d}$ are convergent since $\tilde{x}_{i1d} = E_{i1} - S_{i1}$, $\tilde{x}_{i3d} = E_{i3} - S_{i3}$.

Figure 6.7 displays the thrust inputs and elevator deflection angles of UAV#1-UAV#4. To react to the actuator faults encountered by UAVs#1, 2, 3, the elevator deflection angles are regulated to compensate for the faults in a timely manner and stabilize the system. By observing the time responses of thrust inputs, it is seen that the thrusts of follower UAVs reach the limits at the beginning of the simulation.

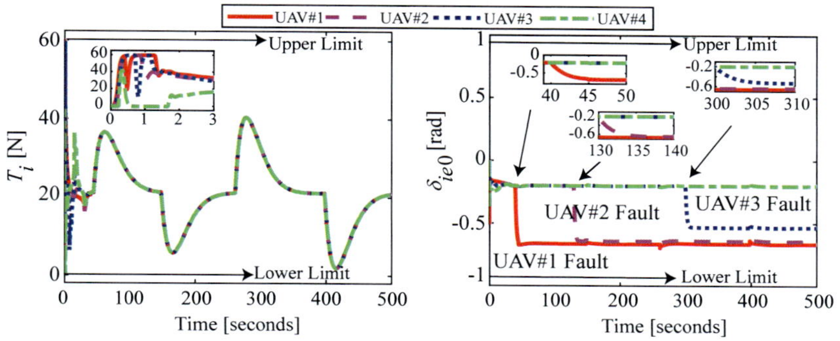

Fig. 6.7 Thrust inputs and elevator deflection angles

Then, the auxiliary system (6.46) starts working to pull the saturated control input signals T_i, $i = 1, 2, 3, 4$, back into the unsaturated region, such that persistent saturation is avoided. Figure 6.8 shows the estimated values, which are bounded by using the adaptive laws (6.38), (6.39), (6.57), (6.58), (6.64), (6.65), (6.71), and (6.72).

6.5 Conclusions

The distributed finite-time FTCC problem has been studied in this chapter for multi-UAVs with multiple leaders in the presence of actuator faults and input saturation. By using the distributed finite-time SMO, the forward and vertical position references can be estimated for each follower UAV in a distributed manner. Then, based on the estimated references, a finite-time control scheme is designed by combining a new set of error variables, NNs, and adaptive mechanism. Moreover, to reduce the computational burden, MLPNN and FOSMD techniques have been exploited to facilitate the finite-time control design. Lyapunov stability analysis has demonstrated that the forward and vertical positions of all follower UAVs can converge into the convex hull spanned by the forward and vertical positions of all leader UAVs in finite time. Simulation results have verified the effectiveness of the proposed control scheme.

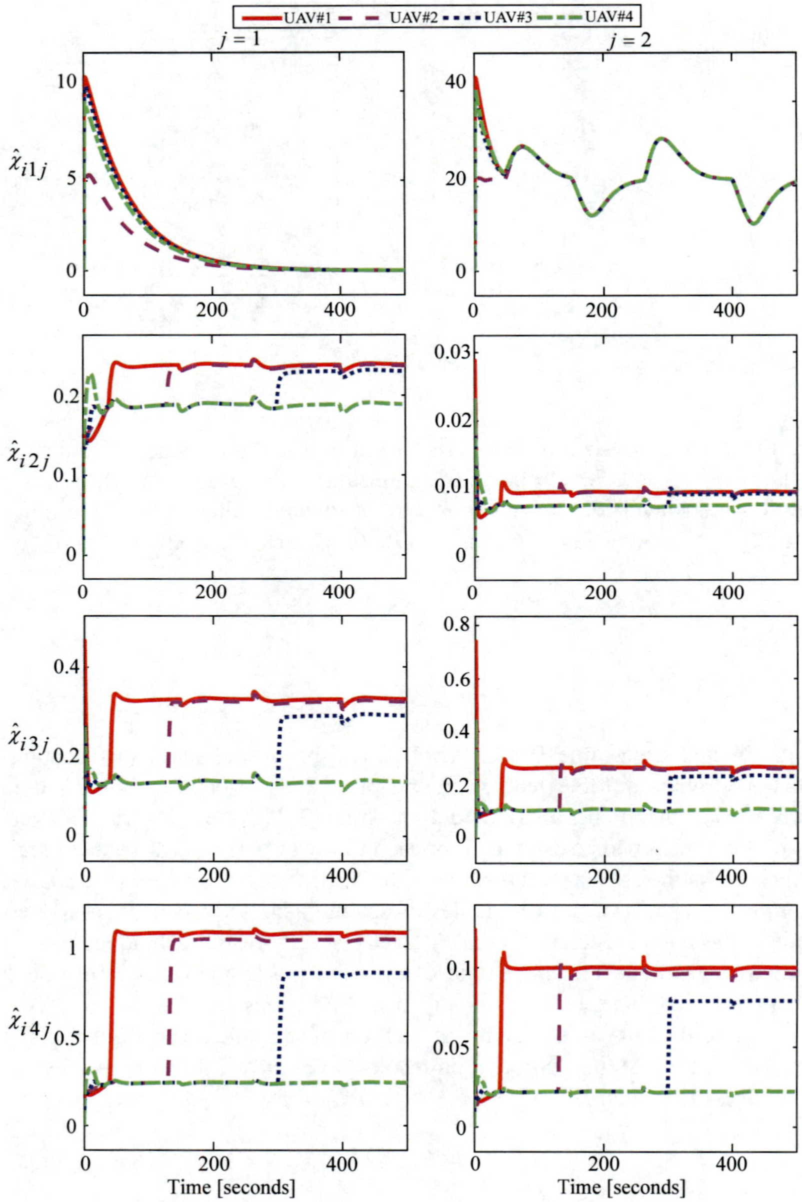

Fig. 6.8 Adaptive parameters updated by using (6.38), (6.39), (6.57), (6.58), (6.64), (6.65), (6.71), and (6.72)

References

1. Q. Ali, N. Gageik, S. Montenegro, A review on distributed control of cooperating mini UAVs. Int. J. Artif. Intell. Appl. **5**(4), 1–13 (2014)
2. S.P. Bhat, D.S. Bernstein, Continuous finite-time stabilization of the translational and rotational double integrators. IEEE Trans. Autom. Control **43**(5), 678–682 (1998)
3. J.D. Boskovic, S. Bergstrom, R.K. Mehra, Robust integrated flight control design under failures, damage, and state-dependent disturbances. J. Guid. Control Dynam. **28**(5), 902–917 (2005)
4. Y.C. Cao, W. Ren, Z.Y. Meng, Decentralized finite-time sliding mode estimators and their applications in decentralized finite-time formation tracking. Syst. Control Lett. **59**(9), 522–529 (2010)
5. M. Chen, Z.S. Ge, B.B. Ren, Adaptive tracking control of uncertain MIMO nonlinear systems with input constraints. Automatica **47**(3), 452–465 (2011)
6. H.B. Du, S.H. Li, Finite-time attitude stabilization for a spacecraft using homogeneous method. J. Guid. Control Dynam. **35**(3), 740–748 (2012)
7. A.A. Federal, Introduction to TCAS II. Version 7.1. Technical report, U.S. Department of Transportation (2011)
8. S.I. Han, H. Ha, J.M. Lee, Fuzzy finite-time dynamic surface control for nonlinear large-scale systems. Int. J. Fuzzy Syst. **18**(4), 570–584 (2016)
9. T. Han, Z.H. Guan, Y.H. Wu, D.F. Zheng, X.H. Zhang, J.W. Xiao, Three-dimensional containment control for multiple unmanned aerial vehicles. J. Frankl. Inst. **353**(13), 2929–2942 (2016)
10. T. Han, M. Chi, Z.H. Guan, B. Hu, J.W. Xiao, Y.H. Huang, Distributed three-dimensional formation containment control of multiple unmanned aerial vehicle systems. Asian J. Control **19**(3), 1103–1113 (2017)
11. A. Levant, Higher-order sliding modes, differentiation and output-feedback control. Int. J. Control **76**(9–10), 924–941 (2003)
12. S.H. Li, X.Y. Wang, Finite-time consensus and collision avoidance control algorithms for multiple AUVs. Automatica **49**(11), 3359–3367 (2013)
13. J.H. Li, Q.L. Zhang, A linear switching function approach to sliding mode control and observation of descriptor systems. Automatica **95**, 112–121 (2018)
14. D.Y. Li, G.F. Ma, C.J. Li, W. He, J. Mei, S.Z. Ge, Distributed attitude coordinated control of multiple spacecraft with attitude constraints via state and output feedback. IEEE Trans. Aerosp. Electron. Syst. **54**(5), 2233–2245 (2018)
15. C.J. Liu, W.H. Chen, Disturbance rejection flight control for small fixed-wing unmanned aerial vehicles. J. Guid. Control Dyn. **39**(12), 2810–2819 (2016)
16. H. Ma, H.J. Liang, Q. Zhou, C.K. Ahn, Adaptive dynamic surface control design for uncertain nonlinear strict-feedback systems with unknown control direction and disturbances. IEEE Trans. Syst. Man Cybern. Syst. **49**(3), 506–515 (2018)
17. Z.Y. Meng, W. Ren, Z. You, Distributed finite-time attitude containment control for multiple rigid bodies. Automatica **46**(12), 2092–2099 (2010)
18. J. Seo, Y. Kim, S. Kim, A. Tsourdos, Collision avoidance strategies for unmanned aerial vehicles in formation flight. IEEE Trans. Aerosp. Electron. Syst. **53**(6), 2718–2734 (2017)
19. C.P. Tan, X.H. Yu, Z.H. Man, Terminal sliding mode observers for a class of nonlinear systems. Automatica **46**(8), 1401–1404 (2010)
20. D. Wang, J. Huang, Adaptive neural network control for a class of uncertain nonlinear systems in pure-feedback form. Automatica **38**(8), 1365–1372 (2002)
21. X.Y. Wang, S.H. Li, Composite finite-time containment control for disturbed second-order multi-agent systems, in *Complex Systems and Networks* (Springer, Berlin, 2016), pp. 207–228
22. Y.J. Wang, Y.D. Song, Fraction dynamic-surface-based neuroadaptive finite-time containment control of multiagent systems in nonaffine pure-feedback form. IEEE Trans. Neural Netw. Learn. Syst. **28**(3), 678–689 (2017)

23. Y.J. Wang, Y.D. Song, W. Ren, Distributed adaptive finite-time approach for formation-containment control of networked nonlinear systems under directed topology. IEEE Trans. Neural Netw. Learn. Syst. **29**(7), 3164–3175 (2018)
24. B. Xu, Robust adaptive neural control of flexible hypersonic flight vehicle with dead-zone input nonlinearity. Nonlinear Dyn. **80**(3), 1509–1520 (2015)
25. Q. Xu, H. Yang, B. Jiang, D.H. Zhou, Y.M. Zhang, Fault tolerant formations control of UAVs subject to permanent and intermittent faults. J. Intell. Robot. Syst. **73**(1–4), 589–602 (2014)
26. B. Xu, C.G. Yang, Y.P. Pan, Global neural dynamic surface tracking control of strict-feedback systems with application to hypersonic flight vehicle. IEEE Trans. Neural Netw. Learn. Syst. **26**(10), 2563–2575 (2015)
27. B. Xu, D.W. Wang, Y.M. Zhang, Z.K. Shi, DOB based neural control of flexible hypersonic flight vehicle considering wind effects. IEEE Trans. Ind. Electron. **64**(11), 8676–8685 (2017)
28. X. Yu, Y.M. Zhang, Sense and avoid technologies with applications to unmanned aircraft systems: Review and prospects. Prog. Aeosp. Sci. **74**, 152–166 (2015)
29. S.H. Yu, X.H. Yu, B. Shirinzadeh, Z.H. Man, Continuous finite-time control for robotic manipulators with terminal sliding mode. Automatica **41**(11), 1957–1964 (2005)
30. X. Yu, Z.X. Liu, Y.M. Zhang, Fault-tolerant formation control of multiple UAVs in the presence of actuator faults. Int. J. Robust Nonlinear Control **26**(12), 2668–2685 (2016)
31. X. Yu, Z.X. Liu, Y.M. Zhang, Fault-tolerant flight control design with finite-time adaptation under actuator stuck failures. IEEE Trans. Control Syst. Technol. **25**(4), 1431–1440 (2017)
32. D. Zhai, L.W. An, J.X. Dong, Q.L. Zhang, Output feedback adaptive sensor failure compensation for a class of parametric strict feedback systems. Automatica **97**, 48–57 (2018)
33. Y.M. Zhang, J. Jiang, Bibliographical review on reconfigurable fault-tolerant control systems. Ann. Rev. Control **32**(2), 229–252 (2008)
34. K. Zhao, Y.D. Song, C.Y. Wen, Computationally inexpensive fault tolerant control of uncertain nonlinear systems with non-smooth asymmetric input saturation and undetectable actuation failures. IET Contr. Theory Appl. **10**(15), 1866–1873 (2016)
35. K. Zhao, Y.D. Song, Z.X. Shen, Neuroadaptive fault-tolerant control of nonlinear systems under output constraints and actuation faults. IEEE Trans. Neural Netw. Learn. Syst. **29**(2), 286–298 (2018)
36. Q. Zhou, S.Y. Zhao, H.Y. Li, R.Q. Lu, C.W. Wu, Adaptive neural network tracking control for robotic manipulators with dead zone. IEEE Trans. Neural Netw. Learn. Syst. **30**(12), 3611–3620 (2019)
37. A.M. Zou, Z.G. Hou, M. Tan, Adaptive control of a class of nonlinear pure-feedback systems using fuzzy backstepping approach. IEEE Trans. Fuzzy Syst. **16**(4), 886–897 (2008)
38. A.M. Zou, K.D. Kumar, Neural network-based distributed attitude coordination control for spacecraft formation flying with input saturation. IEEE Trans. Neural Netw. Learn. Syst. **23**(7), 1155–1162 (2012)

Chapter 7
Decentralized Finite-Time Attitude FTCC of Multi-UAVs with Prescribed Performance

7.1 Introduction

Recently, numerous results have been obtained on the cooperative control of multi-UAVs. In [32], a leader–follower configuration was adopted to navigate the UAVs to follow certain trajectories, while maintaining a fixed geometrical formation. Specifically, the leading UAV tried to fly on some predefined trajectories and each follower UAV maintained its position relative to the leading UAV, and this cooperative control manner can be seen as the centralized method. With such a method, all follower UAVs need to communicate with the leading UAV and may lose stability if the communication links between follower UAVs and the only one leading UAV encounter breakdowns. In order to improve the communication robustness in formation flight, decentralized control methods are employed in the cooperative control of multi-UAVs. Such methods are previously investigated on the cooperative control of MASs, especially for MASs represented by first-order and second-order dynamics [7, 11, 12, 21, 23]. Recently, an output-feedback formation control scheme was proposed for a group of UAVs to track a desired trajectory by introducing a state observer [9]. With the help of differential game theory, a distributed formation control design approach was presented for multi-UAVs [14]. In the system, each UAV was only able to exchange its information with its neighboring UAVs and each UAV tried to minimize its terminal formation errors and control efforts. A decentralized cohesive motion control scheme was proposed in [1] for a swarm of autonomous flight vehicles. In the proposed scheme, the formation team moved from its initial setting to a final desired setting, and the formation geometry was maintained all the time. More recently, to improve the robustness, a full-order SMC framework was studied in [22] for the robust synchronized formation flight of multi-UAVs with system uncertainties. A full-order sliding-mode surface, including the individual position tracking error and the synchronized formation error, was designed in a decentralized manner to facilitate the controller design. Although some results have been obtained for the decentralized control of

Z. Yu et al., *Fault-Tolerant Cooperative Control of Unmanned Aerial Vehicles*,
https://doi.org/10.1007/978-981-99-7661-4_7

multi-UAVs, more investigations should be conducted to improve flight safety in the presence of actuator faults.

Overall system performance degradation or collisions in the formation team may be induced if the actuator faults are not handled in a timely manner [27, 31]. Recently, inspired by the emerging cooperative control researches on multi-UAVs, a few results on the FTCC of multi-UAVs have been reported. By using the inner–outer-loop control architecture, the formation control problem of multi-UAVs was investigated in the presence of permanent and intermittent faults [24]. With a similar architecture, Yu et al. [25] further investigated the FTCC design methodology for multi-UAVs with the integration of reference generator technique. It should be stressed that the results reported in the aforementioned works [24, 25] are about the FTCC design for multi-UAVs in the leader–follower framework. Consider the diverse studies on the cooperative control of multi-UAVs in a complex communication network, FTCC schemes should be further investigated for multi-UAVs in a decentralized communication network.

Regarding the FTCC design for multi-UAVs in a decentralized communication network, actuator faults should be counteracted in a timely fashion. Failure to react to such in-flight faults within a limited amount of time may result in catastrophic consequence. Therefore, finite-time FTCC schemes for multi-UAVs need to be studied to reduce the time interval associated with the duration of the faults. Fortunately, finite-time methods can provide fast convergence rate, high accuracy, and good robustness [6, 8, 17]. With such advantages, the FTCC performance can be further improved if the finite-time feature is incorporated into the designed scheme for multi-UAVs in a decentralized communication network. Despite the finite-time FTC schemes for a single UAV [26] or multi-UAVs in a leader-following architecture [13], very few results have been reported on the FTCC design for multi-UAVs in a decentralized communication network. Moreover, to further improve the FTCC performance of multi-UAVs in a decentralized communication network, the integration of PPC in FTCC scheme needs to be further developed, so as to confine the transient performance when multi-UAVs encounter actuator faults abruptly.

Motivated by the aforementioned observations, this chapter proposes a decentralized finite-time FTCC scheme for multi-UAVs with prescribed attitude synchronization tracking performance. By using prescribed performance functions, a new set of synchronization tracking errors is first constructed. Meanwhile, the NNs are utilized to identify the unknown nonlinear functions, and the norms of the weight vectors are used for the estimation to reduce the computational burden. Finite-time differentiators are used to estimate the intermediate control signals and their first derivatives. Considering the estimation errors caused by the finite-time differentiators, auxiliary dynamic signals are also constructed. Based on the transformed synchronization tracking errors and the auxiliary dynamic signals, a finite-time FTCC scheme is finally developed for multi-UAVs with prescribed performance. The main contributions of this chapter are as follows:

(1) Compared with the works [15, 28], which designed the FTC schemes for a single UAV, this chapter further investigates the FTCC scheme to guarantee the

formation flight safety of multi-UAVs in the presence of actuator faults. Different from the results [24, 25], which present FTCC schemes for multi-UAVs in a leader–follower architecture. To enhance the communication robustness, the FTCC method in this chapter is developed for multi-UAVs in the decentralized communication network.

(2) Different from the existing results [1, 9], in which the closed-loop systems are asymptotically convergent, a finite-time synchronization tracking control scheme is presented in this chapter by utilizing the fractional power of the synchronized tracking errors and the finite-time differentiators. With such a finite-time FTCC scheme, the FTCC performance can be enhanced when multi-UAVs encounter actuator faults.

(3) Compared with the works [3, 4], in which the PPC method is investigated for a single UAV, by introducing the prescribed performance function into the synchronized tracking errors, this chapter further considers the prescribed performance-based decentralized FTCC of multi-UAVs against actuator faults. Furthermore, the norms of the weighting vectors are used for the estimation to reduce the computational burden, and the number of adaptive parameters is independent of neurons.

The rest of this chapter is organized as follows. Section 7.2 introduces the faulty UAV, basic graph theory, and control objective. Section 7.3 presents the synchronized tracking error transformation, controller design, and stability analysis. Simulation results are given in Sect. 7.4. Section 7.5 concludes this chapter.

(*Remark: The main control schemes and contents of this chapter are from the published journal paper "Z. Q. Yu, Y. M. Zhang, Z. X. Liu, Y. H. Qu, C.-Y. Su and B. Jiang. Decentralized finite-time adaptive fault-tolerant synchronization tracking control for multiple UAVs with prescribed performance, Journal of the Franklin Institute, 2020, 357(16): 11830–11862." The authors appreciate the permission from the Elsevier to reuse the results published in the relevant Journal.*)

7.2 Preliminaries and Problem Statement

7.2.1 *Faulty UAV Attitude Dynamics*

In this chapter, the attitude synchronization tracking control problem is considered. Consider a group of N identical UAVs and the UAV attitude model (2.13)–(2.14), and the following control-oriented attitude model of the ith UAV can be formulated as

$$\dot{\boldsymbol{X}}_{i1} = \boldsymbol{F}_{i1} + \boldsymbol{G}_{i1}\boldsymbol{X}_{i2} \tag{7.1}$$

$$\dot{\boldsymbol{X}}_{i2} = \boldsymbol{F}_{i2} + \boldsymbol{G}_{i2}\boldsymbol{U}_{i} \tag{7.2}$$

where $i = 1, 2, \ldots, N$, $\boldsymbol{X}_{i1} = [\mu_i, \alpha_i, \beta_i]^T$, $\boldsymbol{X}_{i2} = [p_i, q_i, r_i]^T$, $\boldsymbol{U}_i = [\delta_{ia}, \delta_{ie}, \delta_{ir}]^T$. $\boldsymbol{F}_{i1}$, $\boldsymbol{G}_{i1}$, $\boldsymbol{F}_{i2}$, and $\boldsymbol{G}_{i2}$ are given by

$$\boldsymbol{F}_{i1} = \begin{bmatrix} 0 & \sin\gamma_i + \cos\gamma_i \sin\mu_i \tan\beta_i & \cos\mu_i \tan\beta_i \\ 0 & -\frac{\cos\gamma_i \sin\mu_i}{\cos\beta_i} & -\frac{\cos\mu_i}{\cos\beta_i} \\ 0 & \cos\gamma_i \cos\mu_i & -\sin\mu_i \end{bmatrix}$$

$$\cdot \begin{bmatrix} \frac{-D_i + T_i \cos\alpha_i \cos\beta_i}{m_i} - g\sin\gamma_i \\ \frac{1}{m_i V_i \cos\gamma_i}[L_i \sin\mu_i + Y_i \cos\mu_i + T_i(\sin\alpha_i \sin\mu_i - \cos\alpha_i \sin\beta_i \cos\mu_i)] \\ \frac{1}{m_i V_i}[L_i \cos\mu_i - Y_i \sin\mu_i + T_i(\cos\alpha_i \sin\beta_i \sin\mu_i + \sin\alpha_i \cos\mu_i)] - \frac{g\cos\gamma_i}{V_i} \end{bmatrix}$$

$$\boldsymbol{G}_{i1} = \begin{bmatrix} \frac{\cos\alpha_i}{\cos\beta_i} & 0 & \frac{\sin\alpha_i}{\cos\beta_i} \\ -\cos\alpha_i \tan\beta_i & 1 & -\sin\alpha_i \tan\beta_i \\ \sin\alpha_i & 0 & -\cos\alpha_i \end{bmatrix}$$

$$\boldsymbol{F}_{i2} = [F_{i21}, F_{i22}, F_{i23}]^T$$

$$\begin{cases} F_{i21} = c_{i1} q_i r_i + c_{i2} p_i q_i + c_{i3} \bar{q}_i s_i b_i \left(C_{il0} + C_{il\beta}\beta_i + \frac{C_{ilp} b_i p_i}{2V_i} + \frac{C_{ilr} b_i r_i}{2V_i} \right) \\ \quad + C_{i4} \bar{q}_i s_i b_i \left(C_{in0} + C_{in\beta}\beta_i + \frac{C_{inp} b_i p_i}{2V_i} + \frac{C_{inr} b_i r_i}{2V_i} \right) \\ F_{i22} = c_{i5} p_i r_i - c_{i6}(p_i^2 - r_i^2) + c_{i7} \bar{q}_i s_i c_i \left(C_{im0} + C_{im\alpha}\alpha_i + \frac{C_{imq} c_i q_i}{2V_i} \right) \\ F_{i23} = c_{i8} p_i q_i - c_{i2} q_i r_i + c_{i4} \bar{q}_i s_i b_i \left(C_{il0} + C_{il\beta}\beta_i + \frac{C_{ilp} b_i p_i}{2V_i} + \frac{C_{ilr} b_i r_i}{2V_i} \right) \\ \quad + c_{i9} \bar{q}_i s_i b_i \left(C_{in0} + C_{in\beta}\beta_i + \frac{C_{inp} b_i p_i}{2V_i} + \frac{C_{inr} b_i r_i}{2V_i} \right) \end{cases}$$

$$\boldsymbol{G}_{i2} = \begin{bmatrix} G_{i211} & 0 & G_{i213} \\ 0 & G_{i222} & 0 \\ G_{i231} & 0 & G_{i233} \end{bmatrix}$$

$$\begin{cases} G_{i211} = c_{i3} \bar{q}_i s_i b_i C_{il\delta_a} + c_{i4} \bar{q}_i s_i b_i C_{in\delta_a} \\ G_{i213} = c_{i3} \bar{q}_i s_i b_i C_{il\delta_r} + c_{i4} \bar{q}_i s_i b_i C_{in\delta_r} \\ G_{i222} = c_{i7} \bar{q}_i s_i c_i C_{im\delta_e} \\ G_{i231} = c_{i4} \bar{q}_i s_i b_i C_{il\delta_a} + c_{i9} \bar{q}_i S_i b_i C_{in\delta_a} \\ G_{i233} = c_{i4} \bar{q}_i s_i b_i C_{il\delta_r} + c_{i9} \bar{q}_i s_i b_i C_{in\delta_r} \end{cases}$$

In practical engineering applications, it is often difficult to obtain the exact knowledge of aerodynamic parameters. Considering such uncertainties, $\boldsymbol{F}_{i1}$ in (7.1) and $\boldsymbol{F}_{i2}$, $\boldsymbol{G}_{i2}$ in (7.2) can be considered as unknown nonlinear functions. Despite this fact, rough values of aerodynamic parameters can be obtained for the controller design. It can be seen from the expression of $\boldsymbol{F}_{i1}$ that it is not easy to decompose $\boldsymbol{F}_{i1}$ into known nonlinear function and unknown nonlinear function due to the fact that $\boldsymbol{F}_{i1}$ is a highly nonlinear function. As a contrast, it is observed from the expressions of $\boldsymbol{F}_{i2}$ and $\boldsymbol{G}_{i2}$ that $\boldsymbol{F}_{i2}$ and $\boldsymbol{G}_{i2}$ can be decomposed into known $\boldsymbol{F}_{i20}$, $\boldsymbol{G}_{i20}$ and unknown $\Delta\boldsymbol{F}_{i2}$, $\Delta\boldsymbol{G}_{i2}$, respectively.

Then, the attitude model (7.1)–(7.2) can be described as

$$\dot{\boldsymbol{X}}_{i1} = \boldsymbol{F}_{i1} + \boldsymbol{G}_{i1}\boldsymbol{X}_{i2} \tag{7.3}$$

$$\dot{\boldsymbol{X}}_{i2} = \boldsymbol{F}_{i20} + \Delta\boldsymbol{F}_{i2} + (\boldsymbol{G}_{i20} + \Delta\boldsymbol{G}_{i2})\boldsymbol{U}_i \tag{7.4}$$

where $\boldsymbol{F}_{i1}$, $\Delta\boldsymbol{F}_{i2}$, and $\Delta\boldsymbol{G}_{i2}$ are unknown nonlinear functions caused by the uncertainties. $\boldsymbol{F}_{i20}$ and $\boldsymbol{G}_{i20}$ are known functions.

Assumption 7.1 The elements of the gain matrix $\boldsymbol{G}_{i1}$ are bounded.

Remark 7.1 Assumption 7.1 is reasonable since the considered system is a practical engineering system and the elements of $\boldsymbol{G}_{i1}$ cannot be infinite due to the physical limitations. Besides, $\det(\boldsymbol{G}_{i1}) = -\sec\beta_i$. Therefore, $\boldsymbol{G}_{i1}$ is invertible if $\beta_i \neq \pm\pi/2$. Regarding $\boldsymbol{G}_{i20}$, it can be regarded as the control allocation matrix, which is invertible in the flight envelopes.

The control-oriented attitude model (7.3)–(7.4) is presented in the absence of actuator faults. However, in practical engineering applications, actuator faults are often encountered by the highly complicated systems, which may lead to an unsatisfactory performance or even system instability if prompt actions are not adopted in a timely manner. Therefore, actuator faults should be considered in the controller design to maintain the safe flight of one UAV or the safe formation flight of multi-UAVs. As analyzed in previous chapters, loss-of-effectiveness and bias faults are the common faults. Similar to (3.6), the actuator fault model is given as [2]

$$\boldsymbol{U}_i = \boldsymbol{\rho}_i\boldsymbol{U}_{i0} + \boldsymbol{U}_{if} \tag{7.5}$$

where $\boldsymbol{U}_i = [\delta_{ia}, \delta_{ie}, \delta_{ir}]^T$ is the applied control signal by the aileron, elevator, and rudder actuators, and $\boldsymbol{U}_{i0} = [\delta_{ia0}, \delta_{ie0}, \delta_{ir0}]^T$ is the control signal commanded by the controller. The aileron, elevator, and rudder control input vectors of all UAVs are defined as $\boldsymbol{\delta}_{a0} = [\delta_{1a0}, \delta_{2a0}, \ldots, \delta_{Na0}]^T$, $\boldsymbol{\delta}_{e0} = [\delta_{1e0}, \delta_{2e0}, \ldots, \delta_{Ne0}]^T$, and $\boldsymbol{\delta}_{r0} = [\delta_{1r0}, \delta_{2r0}, \ldots, \delta_{Nr0}]^T$, respectively. $\boldsymbol{\rho}_i = \mathrm{diag}\{\rho_{i1}, \rho_{i2}, \rho_{i3}\}$ denotes the loss-of-effectiveness fault, and $\boldsymbol{U}_{if} = [u_{if1}, u_{if2}, u_{if3}]^T$ represents the bounded bias fault.

By substituting the actuator fault model (7.5) into (7.4), then the attitude dynamic model in the presence of actuator faults can be formulated as

$$\begin{aligned} \dot{\boldsymbol{X}}_{i2} &= \boldsymbol{F}_{i20} + \Delta \boldsymbol{F}_{i2} + (\boldsymbol{G}_{i20} + \Delta \boldsymbol{G}_{i2})(\boldsymbol{\rho}_i \boldsymbol{U}_{i0} + \boldsymbol{U}_{if}) \\ &= \boldsymbol{F}_{i20} + \boldsymbol{G}_{i20}\boldsymbol{U}_{i0} + \left[\boldsymbol{G}_{i20}(\boldsymbol{\rho}_i - \boldsymbol{I}) + \Delta \boldsymbol{G}_{i2}\boldsymbol{\rho}_i\right]\boldsymbol{U}_{i0} + \Delta \boldsymbol{F}_{i2} \\ &\quad + \boldsymbol{G}_{i20}\boldsymbol{U}_{if} + \Delta \boldsymbol{G}_{i2}\boldsymbol{U}_{if} \\ &= \boldsymbol{F}_{i20} + \boldsymbol{G}_{i20}\boldsymbol{U}_{i0} + \boldsymbol{\Delta}_i \end{aligned} \tag{7.6}$$

where $\boldsymbol{\Delta}_i = \Delta \boldsymbol{F}_{i2} + \left[\boldsymbol{G}_{i20}(\boldsymbol{\rho}_i - \boldsymbol{I}) + \Delta \boldsymbol{G}_{i2}\boldsymbol{\rho}_i\right]\boldsymbol{U}_{i0} + \boldsymbol{G}_{i20}\boldsymbol{U}_{if} + \Delta \boldsymbol{G}_{i2}\boldsymbol{U}_{if}$ is the lumped uncertainty induced by the actuator faults and system uncertainties.

Assumption 7.2 It is assumed that $\left\|\left[\boldsymbol{G}_{i20}(\boldsymbol{\rho}_i - \boldsymbol{I}) + \Delta \boldsymbol{G}_{i2}\boldsymbol{\rho}_i\right]\boldsymbol{G}_{i20}^{-1}\right\|_\infty < 1$ and $\frac{\partial \boldsymbol{\Delta}_i}{\partial \boldsymbol{U}_{i0}} + \boldsymbol{G}_{i20} \neq \boldsymbol{0}$ holds.

Remark 7.2 Assumption 7.2 is a controllability condition for (7.6), which guarantees the existence of controller $\boldsymbol{U}_{i0}$ [30, 34] and similar to Assumption 6.2 in Chap. 6. Moreover, the control signal $\boldsymbol{G}_{i20}\boldsymbol{U}_{i0}$ dominates the fault vector $[\boldsymbol{G}_{i20}(\boldsymbol{\rho}_i - \boldsymbol{I}) + \Delta \boldsymbol{G}_{i2}\boldsymbol{\rho}_i]\boldsymbol{U}_{i0}$, and the faulty UAVs can be stabilized by using the faulty actuators with the help of the constructed control scheme [26].

From (7.6), it can be seen that the control signal $\boldsymbol{U}_{i0}$ is involved in the unknown nonlinear function $\boldsymbol{\Delta}_i$. The algebraic loops will be introduced into the control scheme if NNs are used to approximate the function $\boldsymbol{\Delta}_i$. To break the algebraic loops, BLPF technique is employed to filter the signal $\boldsymbol{U}_{i0}$ involved in the NN approximator [19, 34]. Then, one has $\boldsymbol{\Delta}_i = \boldsymbol{\Delta}_{if} + \boldsymbol{h}_{if}$, where $\boldsymbol{\Delta}_{if} = \Delta \boldsymbol{F}_{i2} + \left[\boldsymbol{G}_{i20}(\boldsymbol{\rho}_i - \boldsymbol{I}) + \Delta \boldsymbol{G}_{i2}\boldsymbol{\rho}_i\right]\boldsymbol{U}_{i0f} + \boldsymbol{G}_{i20}\boldsymbol{U}_{if} + \Delta \boldsymbol{G}_{i2}\boldsymbol{U}_{if}$, $\boldsymbol{U}_{i0f}$ is the filtered signal, and $\boldsymbol{h}_{if} = [h_{if1}, h_{if2}, h_{if3}]^T$ is the bounded filter error [30].

By considering the system uncertainties and actuator faults, the control-oriented model is given by

$$\dot{\boldsymbol{X}}_{i1} = \boldsymbol{F}_{i1} + \boldsymbol{G}_{i1}\boldsymbol{X}_{i2} \tag{7.7}$$

$$\dot{\boldsymbol{X}}_{i2} = \boldsymbol{F}_{i20} + \boldsymbol{G}_{i20}\boldsymbol{U}_{i0} + \boldsymbol{\Delta}_{if} + \boldsymbol{h}_{if} \tag{7.8}$$

7.2.2 Basic Graph Theory

In the formation flight of multi-UAVs, the motion synchronization depends on the information exchange among UAVs in the communication network. The topology of the information flow between UAVs is described by a weighted undirected graph $\mathcal{G} = (\Upsilon, \mathcal{E}, \mathcal{A})$, where $\Upsilon = \{r_1, r_2, \ldots, r_N\}$ is the set of UAVs, $\mathcal{E} \subseteq \Upsilon \times \Upsilon$ is the set of edges, and $\mathcal{A} = [a_{ij}] \in R^{N\times N}$ is the weighted adjacency matrix of topology $\mathcal{G}$ with nonnegative elements. The edge in $\mathcal{G}$ is denoted by an unordered pair (r_i, r_j).

$(r_i, r_j) \in \mathcal{E}$ if and only if there is an information exchange between the ith UAV and the jth UAV, i.e., $(r_i, r_j) \in \mathcal{E} \leftrightarrow (r_j, r_i) \in \mathcal{E}$. The set of neighbors of the UAV#i is denoted as $N_i = \{r_j | (r_i, r_j) \in \mathcal{E}, i \neq j\}$. A path from UAV r_{i_1} to UAV r_{i_l} is a sequence of edges (r_{i_1}, r_{i_2}), (r_{i_2}, r_{i_3}), ..., $(r_{i_{l-1}}, r_{i_l})$. For any two UAVs i and j, if there exists a path between them, then $\mathcal{G}$ is called a connected graph. The degree matrix is defined as $\mathcal{D} = \text{diag}\{d_i\} \in R^{N\times N}$, where $d_i = \sum_{j=1}^{N} a_{ij},\ j \in N_i$. The Laplacian matrix $\mathcal{L} \in R^{N\times N}$ is given by

$$\mathcal{L} = \mathcal{D} - \mathcal{A} \tag{7.9}$$

Assumption 7.3 The communication topology $\mathcal{G}$ involving N UAVs is connected.

Lemma 7.1 *If communication topology $\mathcal{G}$ is undirected and connected and $\mathbf{\Omega} \in R^N$ is a nonzero nonnegative vector, then the matrices $\mathcal{L}$, $\mathcal{L}+\text{diag}(\mathbf{\Omega})$ are symmetric positive definite matrices [18, 22].*

Remark 7.3 To design the controller for a UAV, it is usually to decompose the kinematics and dynamics of a UAV into position loop and attitude loop. Based on the decomposition, a position-loop controller and an attitude-loop controller are constructed to provide a good flight performance. Regarding the formation flying of multi-UAVs, numerous efforts are devoted to the investigations on the position loop of UAV, i.e., only the point-mass model of a UAV is considered. To further investigate the attitude coordination control of multi-UAVs, the synchronization tracking control of the angles of attack, bank angles, and sideslip angles is investigated in this chapter, which can be regarded as a supplement of existing results on the cooperative position-loop controller [14, 16].

To achieve the decentralized finite-time adaptive FTCC for multi-UAVs, the following lemmas are first given.

Lemma 7.2 *Let $\boldsymbol{K}$ be a real diagonal matrix with appropriate dimension. For the vectors $\boldsymbol{X} = [x_1, x_2, \ldots, x_n]^T$ and $\boldsymbol{Y} = [y_1, y_2, \ldots, y_n]^T$ satisfying $x_i y_i \geq 0$, there exist maximum and minimum eigenvalues of $\boldsymbol{K}$, such that [20]*

$$\lambda_{\min}(\boldsymbol{K})\boldsymbol{X}^T\boldsymbol{Y} \leq \boldsymbol{X}^T\boldsymbol{K}\boldsymbol{Y} \leq \lambda_{\max}(\boldsymbol{K})\boldsymbol{X}^T\boldsymbol{Y} \tag{7.10}$$

Lemma 7.3 *For $x_i \in R$, $i = 1, \ldots, n$, $0 < a \leq 1$, the following inequality can be maintained [10].*

$$\left(\sum_{i=1}^{n} |x_i|\right)^a \leq \sum_{i=1}^{n} |x_i|^a \leq n^{1-a}\left(\sum_{i=1}^{n} |x_i|\right)^a \tag{7.11}$$

7.2.3 Control Objective

Consider a group of UAVs in a decentralized communication network, the control objective is to develop a set of decentralized finite-time adaptive FTCC laws such that all signals of the closed-loop system are bounded and the synchronization tracking errors of all UAVs are finite-time stable. Furthermore, the synchronization tracking errors are confined within the prescribed bounds.

7.3 Main Results

In this section, a finite-time FTCC scheme is designed for multi-UAVs in the presence of uncertainties and actuator faults. To achieve the decentralized synchronization tracking control objective, the synchronization tracking error is first given. Then, a new set of errors is defined by introducing the prescribed performance functions. Based on the new set of errors, finite-time synchronization tracking control can be guaranteed by combining graph theory and robust exact differentiator.

7.3.1 Error Transformation

To achieve the synchronization tracking control of multi-UAVs, the synchronization tracking error in the decentralized form is expressed as

$$\boldsymbol{E}_{i1} = \lambda_{i1}\boldsymbol{e}_{i1} + \lambda_{i2} \sum_{j \in N_i} a_{ij}(\boldsymbol{e}_{i1} - \boldsymbol{e}_{j1}) \tag{7.12}$$

where $\boldsymbol{E}_{i1} = [E_{i11}, E_{i12}, E_{i13}]^T$, λ_{i1}, and λ_{i2} are positive constants, which are used to regulate the tracking and synchronization performances, respectively. $\boldsymbol{e}_{i1} = [e_{i11}, e_{i12}, e_{i13}]^T = \boldsymbol{X}_{i1} - \boldsymbol{X}_{i1d} = [\mu_i - \mu_{id}, \alpha_i - \alpha_{id}, \beta_i - \beta_{id}]^T$ are the tracking errors of the ith UAV with respect to its individual desired attitude angles $\boldsymbol{X}_{i1d} = [\mu_{id}, \alpha_{id}, \beta_{id}]^T$.

Remark 7.4 The first term in (7.12) denotes the attitude tracking error for each individual UAV, and the second term in (7.12) represents the attitude synchronization error among UAVs.

By rewriting (7.12), one has

$$\boldsymbol{E}_{i1} = (\lambda_{i1} + \lambda_{i2}\mathcal{L}_{ii})\,\boldsymbol{e}_{i1} + \lambda_{i2} \sum_{j \in N_i} \mathcal{L}_{ij}\boldsymbol{e}_{j1} \tag{7.13}$$

where $\mathcal{L}_{ij} = -a_{ij}$ and $\mathcal{L}_{ii} = d_i = \sum_{j=1}^{N} a_{ij}$ are the elements of Laplacian matrix $\mathcal{L}$ defined in (7.9).

Then, the synchronization errors for the N UAVs can be expressed as

$$\boldsymbol{E}_1 = [(\boldsymbol{\lambda}_1 + \boldsymbol{\lambda}_2 \mathcal{L}) \otimes \boldsymbol{I}_3]\, \boldsymbol{e}_1 \tag{7.14}$$

where $\boldsymbol{E}_1 = \left[\boldsymbol{E}_{11}^T, \ldots, \boldsymbol{E}_{N1}^T\right]^T$, $\boldsymbol{\lambda}_1 = \text{diag}\{\lambda_{11}, \ldots, \lambda_{N1}\}$, $\boldsymbol{\lambda}_2 = \text{diag}\{\lambda_{12}, \ldots, \lambda_{N2}\}$, and $\otimes$ denotes the Kronecker matrix product.

It yields $\|\boldsymbol{e}_1\| = \left\|\left[(\boldsymbol{\lambda}_1 + \boldsymbol{\lambda}_2 \mathcal{L})^{-1} \otimes \boldsymbol{I}_3\right] \boldsymbol{E}_1\right\| \leq \frac{1}{\underline{\sigma}(\boldsymbol{\lambda}_1 + \boldsymbol{\lambda}_2 \mathcal{L})} \|\boldsymbol{E}_1\|$ by recalling Lemma 7.1, where $\underline{\sigma}(\cdot)$ denotes the minimum singular value of a matrix. Therefore, the synchronization tracking control objective can be achieved when $\boldsymbol{E}_1$ converges into the small region containing zero.

Next, prescribed performance functions are imposed on the synchronization tracking error $\boldsymbol{E}_{i1}$ of the ith UAV, $i = 1, 2, \ldots, N$, to enhance the flight safety and guarantee steady-state stability in the pre-fault and post-fault phases, especially the transient stability in the duration of faults.

The prescribed performance-based FTCC designed for multi-UAVs means that the synchronization tracking error converges into the predefined residual set and the convergence rate cannot exceed the predefined value in the presence of actuator faults. The prescribed performance can be obtained if $E_{i1\tau}$ evolves strictly within predefined decaying bounds, given by

$$-\underline{k}_{i1\tau} \varepsilon_{i1\tau}(t) \leq E_{i1\tau} \leq \overline{k}_{i1\tau} \varepsilon_{i1\tau}(t) \tag{7.15}$$

where $\tau = 1, 2, 3$. $\underline{k}_{i1\tau}$ and $\overline{k}_{i1\tau}$ are positive design parameters. $\varepsilon_{i1\tau}(t)$ is a strictly decreasing smooth function with $\lim_{t\to\infty} \varepsilon_{i1\tau}(t) = \varepsilon_{i1\tau\infty}$, and $\varepsilon_{i1\tau\infty}$ represents the maximum allowable value of $E_{i1\tau}$ at steady state.

For each element of $\boldsymbol{E}_{i1} = [E_{i11}, E_{i12}, E_{i13}]^T$, the performance function $\varepsilon_{i1\tau}(t)$ is chosen as

$$\varepsilon_{i1\tau}(t) = (\varepsilon_{i1\tau 0} - \varepsilon_{i1\tau\infty})\, e^{-\nu_{i1\tau} t} + \varepsilon_{i1\tau\infty} \tag{7.16}$$

where $\tau = 1, 2, 3$. $\varepsilon_{i1\tau 0}$, $\varepsilon_{i1\tau\infty}$, and $\nu_{i1\tau}$ are positive constants, and $\varepsilon_{i1\tau 0} > \varepsilon_{i1\tau\infty}$. Note that $\varepsilon_{i1\tau 0}$, $\underline{k}_{i1\tau}$, and $\overline{k}_{i1\tau}$ should be properly chosen such that the initial error $E_{i1\tau}(0)$ is in the interval $\left[-\underline{k}_{i1\tau}\varepsilon_{i1\tau 0},\ \overline{k}_{i1\tau}\varepsilon_{i1\tau 0}\right]$.

Remark 7.5 From (7.15), it is observed that $-\underline{k}_{i1\tau}\varepsilon_{i1\tau}(t)$ and $\overline{k}_{i1\tau}\varepsilon_{i1\tau}(t)$ serve as the upper bound of overshoot and the lower bound of undershoot of $E_{i1\tau}$, respectively. The convergence rate of $E_{i1\tau}$ cannot exceed the decreasing rate of $\varepsilon_{i1\tau}(t)$, and the allowable steady-state value of $E_{i1\tau}$ is denoted by $\left[-\underline{k}_{i1\tau}\varepsilon_{i1\tau\infty},\ \overline{k}_{i1\tau}\varepsilon_{i1\tau\infty}\right]$.

To facilitate the prescribed performance controller design, the constraint (7.15) can be transformed into the following equality form:

$$E_{i1\tau} = \varepsilon_{i1\tau} \Gamma(\xi_{i1\tau}) \tag{7.17}$$

where $\tau = 1, 2, 3$, $\Gamma(\cdot)$ is a smooth and strictly increasing function, which has the following properties:

$$\begin{cases} \Gamma(0) = 0 \\ -\underline{k}_{i1\tau} < \Gamma(\xi_{i1\tau}) < \overline{k}_{i1\tau} \\ \lim_{\xi_{i1\tau}\to+\infty}\Gamma(\xi_{i1\tau}) = \overline{k}_{i1\tau},\ \lim_{\xi_{i1\tau}\to-\infty}\Gamma(\xi_{i1\tau}) = -\underline{k}_{i1\tau} \end{cases} \tag{7.18}$$

In this section, $\Gamma(\xi_{i1\tau})$ is chosen as

$$\Gamma(\xi_{i1\tau}) = \frac{\overline{k}_{i1\tau}e^{\xi_{i1\tau}+\kappa_{i1\tau}} - \underline{k}_{i1\tau}e^{-\xi_{i1\tau}-\kappa_{i1\tau}}}{e^{\xi_{i1\tau}+\kappa_{i1\tau}} + e^{-\xi_{i1\tau}-\kappa_{i1\tau}}} \tag{7.19}$$

where $\kappa_{i1\tau} = \frac{1}{2}\ln\frac{\underline{k}_{i1\tau}}{\overline{k}_{i1\tau}}$.

Then, the transformed error is given by

$$\xi_{i1\tau} = \Gamma^{-1}\left(\frac{E_{i1\tau}}{\varepsilon_{i1\tau}}\right) = \frac{1}{2}\ln\frac{\overline{k}_{i1\tau}\underline{k}_{i1\tau} + \overline{k}_{i1\tau}\sigma_{i1\tau}}{\overline{k}_{i1\tau}\underline{k}_{i1\tau} - \underline{k}_{i1\tau}\sigma_{i1\tau}} \tag{7.20}$$

where $\sigma_{i1\tau} = \frac{E_{i1\tau}}{\varepsilon_{i1\tau}}$.

By taking the time derivative of (7.20), one has

$$\dot{\xi}_{i1\tau} = \eta_{i1\tau}\left(\dot{E}_{i1\tau} - \frac{E_{i1\tau}\dot{\varepsilon}_{i1\tau}}{\varepsilon_{i1\tau}}\right) \tag{7.21}$$

where $\eta_{i1\tau} = \frac{1}{2\varepsilon_{i1\tau}}\left(\frac{1}{\underline{k}_{i1\tau}+\sigma_{i1\tau}} + \frac{1}{\overline{k}_{i1\tau}-\sigma_{i1\tau}}\right)$.

By rewriting (7.21) in matrix form, one can obtain

$$\dot{\boldsymbol{\xi}}_{i1} = \boldsymbol{\eta}_{i1}\left(\dot{\boldsymbol{E}}_{i1} - \boldsymbol{\varepsilon}_{i1}^{-1}\dot{\boldsymbol{\varepsilon}}_{i1}\boldsymbol{E}_{i1}\right) \tag{7.22}$$

where $\boldsymbol{\xi}_{i1} = [\xi_{i11}, \xi_{i12}, \xi_{i13}]^T$, $\dot{\boldsymbol{\xi}}_{i1} = [\dot{\xi}_{i11}, \dot{\xi}_{i12}, \dot{\xi}_{i13}]^T$, $\boldsymbol{\eta}_{i1} = \text{diag}\{\eta_{i11}, \eta_{i12}, \eta_{i13}\}$, $\dot{\boldsymbol{E}}_{i1} = [\dot{E}_{i11}, \dot{E}_{i12}, \dot{E}_{i13}]^T$, $\boldsymbol{\varepsilon}_{i1} = \text{diag}\{\varepsilon_{i11}, \varepsilon_{i12}, \varepsilon_{i13}\}$, $\dot{\boldsymbol{\varepsilon}}_{i1} = \text{diag}\{\dot{\varepsilon}_{i11}, \dot{\varepsilon}_{i12}, \dot{\varepsilon}_{i13}\}$.

The chosen function (7.19) has all required properties (7.18). By using (7.17) and (7.19), the prescribed synchronization tracking performance (7.15) can be achieved if the transformed error $\xi_{i1\tau}$ is bounded. Therefore, to achieve the finite-time prescribed synchronization tracking performance, the new error $\xi_{i1\tau}$ needs to be guaranteed bounded in finite time.

7.3.2 Controller Design

In this section, the finite-time FTCC scheme for multi-UAVs with prescribed performance is proposed for the formation flying of multi-UAVs. In the controller design for the attitude kinematic model, the intermediate control signal $\bar{\boldsymbol{X}}_{i2d}$ is developed as

$$\bar{\boldsymbol{X}}_{i2d} = \boldsymbol{G}_{i1}^{-1}\Bigg\{ - \boldsymbol{\eta}_{i1}^{-1}\boldsymbol{K}_{11}\boldsymbol{\xi}_{i1} - \boldsymbol{\eta}_{i1}^{-1}\boldsymbol{K}_{12}\text{sig}\left(\boldsymbol{\xi}_{i1} - \boldsymbol{\zeta}_{i1}\right)^{\upsilon} + \dot{\boldsymbol{X}}_{i1d} - \frac{\boldsymbol{\eta}_{i1}^{-1}\hat{\Phi}_{i1}\boldsymbol{\Xi}_{i11}}{2h_{11}^2} - \frac{\boldsymbol{\eta}_{i1}^{-1}\boldsymbol{\Xi}_{i12}}{2h_{12}^2} + \frac{\lambda_{i2}}{\Theta_i}\sum_{j\in N_i} a_{ij}\dot{\boldsymbol{e}}_{j1} + \frac{1}{\Theta_i}\boldsymbol{\varepsilon}_{i1}^{-1}\dot{\boldsymbol{\varepsilon}}_{i1}\boldsymbol{E}_{i1}\Bigg\} \tag{7.23}$$

where $\bar{\boldsymbol{X}}_{i2d} = [\bar{X}_{i2d1}, \bar{X}_{i2d2}, \bar{X}_{i2d3}]^T$. $\upsilon \in R$, $\boldsymbol{K}_{11} \in R^{3\times3}$, and $\boldsymbol{K}_{12} \in R^{3\times3}$ are real diagonal matrices to be designed by the engineer. $\hat{\Phi}_{i1} \in R$ is the estimation of Φ_{i1}. $\Theta_i = \lambda_{i1} + \lambda_{i2}\sum_{j\in N_i} a_{ij}$. Φ_{i1}, $\boldsymbol{\zeta}_{i1} = [\zeta_{i11}, \zeta_{i12}, \zeta_{i13}]^T$, $\Xi_{i11} = [\Xi_{i111}, \Xi_{i112}, \Xi_{i113}]^T$, and $\boldsymbol{\Xi}_{i12} = [\Xi_{i121}, \Xi_{i122}, \Xi_{i123}]^T$ will be defined later.

To obtain the overall control signal with the backstepping architecture, the first derivative of the intermediate control signal $\bar{\boldsymbol{X}}_{i2d}$ must be needed. However, the derivative of the intermediate control signal is not easily obtained due to the existence of actuator faults and disturbances in practical engineering. To effectively extract the first derivative of the intermediate control signal, the robust exact differentiator is employed as [5]

$$\begin{cases} \dot{X}_{i2d\tau} = -h_{13}\left[|\epsilon_{i1\tau}|^{\frac{1}{2}}\,\text{sign}(\epsilon_{i1\tau}) + h_{14}\,|\epsilon_{i1\tau}|^{\frac{3}{2}}\,\text{sign}(\epsilon_{i1\tau})\right] + \Psi_{i\tau} \\ \dot{\Psi}_{i\tau} = -h_{15}\left[\frac{1}{2}\text{sign}(\epsilon_{i1\tau}) + 2h_{14}\epsilon_{i1\tau} + \frac{3}{2}h_{14}^2\,|\epsilon_{i1\tau}|^2\,\text{sign}(\epsilon_{i1\tau})\right] \end{cases} \tag{7.24}$$

where $\tau = 1, 2, 3$. $X_{i2d\tau}$ and $\Psi_{i\tau}$ are the estimations of $\bar{X}_{i2d\tau}$ and $\dot{\bar{X}}_{i2d\tau}$, respectively. h_{13}, h_{14}, h_{15} are positive design parameters. $\epsilon_{i1\tau} = X_{i2d\tau} - \bar{X}_{i2d\tau}$ is the bounded estimation error, and $\Psi_{i\tau}$ can exactly converge to $\dot{\bar{X}}_{i2d\tau}$ in finite time [5].

Design the following auxiliary dynamic system:

$$\dot{\boldsymbol{\zeta}}_{i1} = \Theta_i\left[-\boldsymbol{K}_{11}\boldsymbol{\zeta}_{i1} + \boldsymbol{\eta}_{i1}\boldsymbol{G}_{i1}\boldsymbol{\epsilon}_{i1} + \boldsymbol{\eta}_{i1}\boldsymbol{G}_{i1}\boldsymbol{\zeta}_{i2} - \tau_{i1}\text{sign}(\boldsymbol{\zeta}_{i1})\right] \tag{7.25}$$

where $\boldsymbol{\zeta}_{i1} = [\zeta_{i11}, \zeta_{i12}, \zeta_{i13}]^T$, and $\boldsymbol{\zeta}_{i2} = [\zeta_{i21}, \zeta_{i22}, \zeta_{i23}]^T$ are the auxiliary dynamic signal vectors, and $\boldsymbol{\zeta}_{i2}$ will be constructed later in the controller design procedure. $\boldsymbol{\epsilon}_{i1} = [\epsilon_{i11}, \epsilon_{i12}, \epsilon_{i13}]^T$ is the estimation error vector. τ_{i1} is a positive design parameter.

Define the compensated synchronization tracking error and the angle rate tracking error as

$$\boldsymbol{S}_{i1} = \boldsymbol{\xi}_{i1} - \boldsymbol{\zeta}_{i1} \tag{7.26}$$

$$\boldsymbol{E}_{i2} = \boldsymbol{X}_{i2} - \boldsymbol{X}_{i2d} \tag{7.27}$$

where $\boldsymbol{S}_{i1} = [S_{i11}, S_{i12}, S_{i13}]^T$, $\boldsymbol{E}_{i2} = [E_{i21}, E_{i22}, E_{i23}]^T$, $\boldsymbol{X}_{i2d} = [X_{i2d1}, X_{i2d2}, X_{i2d3}]^T$.

Then, by recalling (7.7), (7.12), (7.22), and (7.25), one has

$$\begin{aligned} \boldsymbol{S}_{i1}^T\dot{\boldsymbol{S}}_{i1} &= \boldsymbol{S}_{i1}^T\left(\dot{\boldsymbol{\xi}}_{i1} - \dot{\boldsymbol{\zeta}}_{i1}\right) \\ &= \boldsymbol{S}_{i1}^T\boldsymbol{\eta}_{i1}\left(\dot{\boldsymbol{E}}_{i1} - \boldsymbol{\varepsilon}_{i1}^{-1}\dot{\boldsymbol{\varepsilon}}_{i1}\boldsymbol{E}_{i1}\right) - \boldsymbol{S}_{i1}^T\dot{\boldsymbol{\zeta}}_{i1} \\ &= \boldsymbol{S}_{i1}^T\boldsymbol{\eta}_{i1}\Theta_i\left(\boldsymbol{F}_{i1} + \boldsymbol{G}_{i1}\left(\bar{\boldsymbol{X}}_{i2d} + \boldsymbol{\epsilon}_{i1} + \boldsymbol{E}_{i2}\right) - \dot{\boldsymbol{X}}_{i1d}\right) \\ &\quad + \boldsymbol{S}_{i1}^T\boldsymbol{\eta}_{i1}\left(-\lambda_{i2}\sum\nolimits_{j\in N_i} a_{ij}\dot{\boldsymbol{e}}_{j1}(t) - \boldsymbol{\varepsilon}_{i1}^{-1}\dot{\boldsymbol{\varepsilon}}_{i1}\boldsymbol{E}_{i1}\right) - \boldsymbol{S}_{i1}^T\dot{\boldsymbol{\zeta}}_{i1} \end{aligned} \tag{7.28}$$

To facilitate the controller design, the RBFNN will be used to approximate $\boldsymbol{F}_{i1}$, which is given by

$$F_{i1\tau} = \boldsymbol{\theta}_{i1\tau}^{*T}\boldsymbol{\varphi}_{i1\tau} + w_{i1\tau} \tag{7.29}$$

where $\tau = 1, 2, 3$, $F_{i1\tau}$ is the element of nonlinear function vector $\boldsymbol{F}_{i1} = [F_{i11}, F_{i12}, F_{i13}]$. $\boldsymbol{\theta}_{i1\tau}^*$, $\boldsymbol{\varphi}_{i1\tau}$, and $w_{i1\tau}$ are the optimal weighting vector, basis function vector, and minimum approximation error, respectively.

Therefore, one has

$$\begin{aligned} \boldsymbol{S}_{i1}^T\boldsymbol{\eta}_{i1}\boldsymbol{F}_{i1} &= \boldsymbol{S}_{i1}^T\boldsymbol{\eta}_{i1}\left[\boldsymbol{\theta}_{i11}^{*T}\boldsymbol{\varphi}_{i11} + w_{i11},\ \boldsymbol{\theta}_{i12}^{*T}\boldsymbol{\varphi}_{i12} + w_{i12},\ \boldsymbol{\theta}_{i13}^{*T}\boldsymbol{\varphi}_{i13} + w_{i13}\right]^T \\ &= \sum_{\tau=1}^{3} S_{i1\tau}\eta_{i1\tau}\boldsymbol{\theta}_{i1\tau}^{*T}\boldsymbol{\varphi}_{i1\tau} + \sum_{\tau=1}^{3} S_{i1\tau}\eta_{i1\tau}w_{i1\tau} \end{aligned} \tag{7.30}$$

By using Young's inequality, one has

$$\begin{aligned} S_{i1\tau}\eta_{i1\tau}\boldsymbol{\theta}_{i1\tau}^{*T}\boldsymbol{\varphi}_{i1\tau} &\le \frac{S_{i1\tau}^2\eta_{i1\tau}^2\Phi_{i1}\boldsymbol{\varphi}_{i1\tau}^T\boldsymbol{\varphi}_{i1\tau}}{2h_{11}^2} + \frac{h_{11}^2}{2}, \\ S_{i1\tau}\eta_{i1\tau}w_{i1\tau} &\le \frac{S_{i1\tau}^2\eta_{i1\tau}^2}{2h_{12}^2} + \frac{h_{12}^2\bar{w}_{i1\tau}^2}{2} \end{aligned} \tag{7.31}$$

where $\Phi_{i1} = \max\left\{\boldsymbol{\theta}_{i11}^{*T}\boldsymbol{\theta}_{i11}, \boldsymbol{\theta}_{i12}^{*T}{}_{i12}, \boldsymbol{\theta}_{i13}^{*T}\boldsymbol{\theta}_{i13}\right\}$, $w_{i1\tau} \le \bar{w}_{i1\tau}$.

Define $\boldsymbol{\Xi}_{i11} = [S_{i11}\eta_{i11}^2\boldsymbol{\varphi}_{i11}^T\boldsymbol{\varphi}_{i11}, S_{i12}\eta_{i12}^2\boldsymbol{\varphi}_{i12}^T\boldsymbol{\varphi}_{i12}, S_{i13}\eta_{i13}^2\boldsymbol{\varphi}_{i13}^T\boldsymbol{\varphi}_{i13}]^T$, $\boldsymbol{\Xi}_{i12} = \left[S_{i11}\eta_{i11}^2, S_{i12}\eta_{i12}^2, S_{i13}\eta_{i13}^2\right]^T$, and then (7.28) can be formulated as

$$\begin{aligned}
&\boldsymbol{S}_{i1}^T\dot{\boldsymbol{S}}_{i1}\\
&\leq \boldsymbol{S}_{i1}^T\boldsymbol{\eta}_{i1}\Theta_i\left[\boldsymbol{G}_{i1}\bar{\boldsymbol{X}}_{i2d} + \boldsymbol{G}_{i1}\boldsymbol{E}_{i2} - \dot{\boldsymbol{X}}_{i1d} + \frac{\boldsymbol{\eta}_{i1}^{-1}\Phi_{i1}\boldsymbol{\Xi}_{i11}}{2h_{11}^2} + \frac{\boldsymbol{\eta}_{i1}^{-1}\boldsymbol{\Xi}_{i12}}{2h_{12}^2} + \boldsymbol{G}_{i1}\boldsymbol{\epsilon}_{i1}\right]\\
&- \lambda_{i2}\boldsymbol{S}_{i1}^T\boldsymbol{\eta}_{i1}\sum_{j\in N_i} a_{ij}\dot{\boldsymbol{e}}_{j1} - \boldsymbol{S}_{i1}^T\boldsymbol{\eta}_{i1}\boldsymbol{\varepsilon}_{i1}^{-1}\dot{\boldsymbol{\varepsilon}}_{i1}\boldsymbol{E}_{i1} - \boldsymbol{S}_{i1}^T\dot{\boldsymbol{\zeta}}_{i1} + \Theta_i\sum_{\tau=1}^{3}\frac{h_{11}^2 + h_{12}^2\bar{w}_{i1\tau}^2}{2}
\end{aligned} \tag{7.32}$$

By substituting the intermediate control signal (7.23) and the auxiliary signal (7.25) into (7.32), the following expression can be obtained:

$$\begin{aligned}
&\boldsymbol{S}_{i1}^T\dot{\boldsymbol{S}}_{i1}\\
&\leq \boldsymbol{S}_{i1}^T\boldsymbol{\eta}_{i1}\Theta_i\left[-\boldsymbol{\eta}_{i1}^{-1}\boldsymbol{K}_{11}\boldsymbol{\xi}_{i1} - \boldsymbol{\eta}_{i1}^{-1}\boldsymbol{K}_{12}\mathrm{sig}\left(\boldsymbol{\xi}_{i1} - \boldsymbol{\zeta}_{i1}\right)^{\upsilon}\right] + \Theta_i\sum_{\tau=1}^{3}\frac{h_{11}^2 + h_{12}^2\bar{w}_{i1\tau}^2}{2}\\
&+ \boldsymbol{S}_{i1}^T\boldsymbol{\eta}_{i1}\Theta_i\left(\boldsymbol{G}_{i1}\boldsymbol{E}_{i2} + \frac{\boldsymbol{\eta}_{i1}^{-1}\tilde{\Phi}_{i1}\boldsymbol{\Xi}_{i11}}{2h_{11}^2} + \boldsymbol{G}_{i1}\boldsymbol{\epsilon}_{i1} - \frac{1}{\lambda_{i1} + \lambda_{i2}\sum_{j\in N_i} a_{ij}}\boldsymbol{\eta}_{i1}^{-1}\dot{\boldsymbol{\zeta}}_{i1}\right)\\
&= \Theta_i\boldsymbol{S}_{i1}^T\left[-\boldsymbol{K}_{11}\boldsymbol{S}_{i1} - \boldsymbol{K}_{12}\mathrm{sig}(\boldsymbol{S}_{i1})^{\upsilon} + \frac{\tilde{\Phi}_{i1}\boldsymbol{\Xi}_{i11}}{2h_{11}^2} + \boldsymbol{\eta}_{i1}\boldsymbol{G}_{i1}\boldsymbol{S}_{i2} + \tau_{i1}\mathrm{sign}(\boldsymbol{\zeta}_{i1})\right]\\
&+ \Theta_i\sum_{\tau=1}^{3}\frac{h_{11}^2 + h_{12}^2\bar{w}_{i1\tau}^2}{2}
\end{aligned} \tag{7.33}$$

Design the adaptive law as

$$\dot{\hat{\Phi}}_{i1} = k_{13}\left(-k_{14}\hat{\Phi}_{i1} + \frac{\boldsymbol{S}_{i1}^T\boldsymbol{\Xi}_{i11}}{2h_{11}^2}\right) \tag{7.34}$$

where k_{13} and k_{14} are positive design parameters.

Then, choose the following first Lyapunov function candidate as

$$L_{i1} = \frac{1}{2}\boldsymbol{S}_{i1}^T\boldsymbol{S}_{i1} + \frac{\lambda_{i1} + \lambda_{i2}\sum_{j\in N_i} a_{ij}}{2k_{13}}\tilde{\Phi}_{i1}^2 \tag{7.35}$$

where $\tilde{\Phi}_{i1} = \Phi_{i1} - \hat{\Phi}_{i1}$ is the estimation error.

By taking the time derivative of (7.35) along the trajectories of (7.33) and (7.34), one has

$$
\begin{aligned}
\dot{L}_{i1} \leq & \Theta_i \boldsymbol{S}_{i1}^T \left[-\boldsymbol{K}_{11}\boldsymbol{S}_{i1} - \boldsymbol{K}_{12}\mathrm{sig}(\boldsymbol{S}_{i1})^{\upsilon} + \boldsymbol{\eta}_{i1}\boldsymbol{G}_{i1}\boldsymbol{S}_{i2} + \tau_{i1}\mathrm{sign}(\boldsymbol{\zeta}_{i1}) \right] \\
& + \Theta_i \tilde{\Phi}_{i1} \left[\frac{\boldsymbol{S}_{i1}^T \boldsymbol{\Xi}_{i11}}{2h_{11}^2} - \frac{1}{k_{13}} \dot{\hat{\Phi}}_{i1} \right] + \Theta_i \sum_{\tau=1}^{3} \frac{h_{11}^2 + h_{12}^2 \bar{w}_{i1\tau}^2}{2} \\
= & \Theta_i \left[-\boldsymbol{S}_{i1}^T \boldsymbol{K}_{11}\boldsymbol{S}_{i1} - \boldsymbol{S}_{i1}^T \boldsymbol{K}_{12}\mathrm{sig}(\boldsymbol{S}_{i1})^{\upsilon} + \boldsymbol{S}_{i1}^T \boldsymbol{\eta}_{i1}\boldsymbol{G}_{i1}\boldsymbol{S}_{i2} + \tau_{i1}\boldsymbol{S}_{i1}^T\mathrm{sign}(\boldsymbol{\zeta}_{i1}) \right] \\
& + \Theta_i k_{14} \tilde{\Phi}_{i1} \hat{\Phi}_{i1} + \Theta_i \sum_{\tau=1}^{3} \frac{h_{11}^2 + h_{12}^2 \bar{w}_{i1\tau}^2}{2}
\end{aligned}
\tag{7.36}
$$

Next, by using the tracking error $\boldsymbol{E}_{i2} = \boldsymbol{X}_{i2} - \boldsymbol{X}_{i2d}$ defined in (7.27), the overall controller is given by

$$
\begin{aligned}
\boldsymbol{U}_{i0} = & \boldsymbol{G}_{i20}^{-1} \Big[-\boldsymbol{K}_{21}\boldsymbol{E}_{i2} - \boldsymbol{F}_{i20} + \boldsymbol{\Psi}_i - \boldsymbol{K}_{22}\mathrm{sig}\left(\boldsymbol{E}_{i2} - \boldsymbol{\zeta}_{i2}\right)^{\upsilon} \\
& - \frac{\hat{\Phi}_{i2}\boldsymbol{\Xi}_{i2}}{2h_{21}^2} - \frac{\boldsymbol{E}_{i2} - \boldsymbol{\zeta}_{i2}}{2h_{22}^2} \Big]
\end{aligned}
\tag{7.37}
$$

where $\boldsymbol{K}_{21}$ and $\boldsymbol{K}_{22}$ are positive real diagonal matrices. $\hat{\Phi}_{i2}$ is the estimation of Φ_{i2}. Φ_{i2}, $\boldsymbol{\zeta}_{i2} = [\zeta_{i21}, \zeta_{i22}, \zeta_{i23}]^T$, $\boldsymbol{\Xi}_{i2} = [\Xi_{i21}, \Xi_{i22}, \Xi_{i23}]^T$ will be defined later.

Design the following auxiliary dynamic system:

$$
\dot{\boldsymbol{\zeta}}_{i2} = -\boldsymbol{K}_{21}\boldsymbol{\zeta}_{i2} + \Theta_i \boldsymbol{G}_{i1}^T \boldsymbol{\eta}_{i1}\boldsymbol{S}_{i1} - \tau_{i2}\mathrm{sign}(\boldsymbol{\zeta}_{i2})
\tag{7.38}
$$

where τ_{i2} is a positive control parameter. $\boldsymbol{\zeta}_{i2} = [\zeta_{i21}, \zeta_{i22}, \zeta_{i23}]^T$.

Define the compensated angular rate tracking error as

$$
\boldsymbol{S}_{i2} = \boldsymbol{E}_{i2} - \boldsymbol{\zeta}_{i2}
\tag{7.39}
$$

where $\boldsymbol{S}_{i2} = [S_{i21}, S_{i22}, S_{i23}]^T$.

By recalling the robust exact differentiator in (7.24) and taking the time derivative of $\boldsymbol{S}_{i2}$ along the trajectory of (7.8), one has

$$
\boldsymbol{S}_{i2}^T \dot{\boldsymbol{S}}_{i2} = \boldsymbol{S}_{i2}^T \left(\boldsymbol{F}_{i20} + \boldsymbol{G}_{i20}\boldsymbol{U}_{i0} + \boldsymbol{\Delta}_{if} + \boldsymbol{h}_{if} - \Psi_i - \dot{\boldsymbol{\zeta}}_{i2} \right)
\tag{7.40}
$$

The RBFNN will be utilized to approximate the nonlinear function $\boldsymbol{\Delta}_{if}$, which is given by

$$
\Delta_{if\tau} = \boldsymbol{\theta}_{i2\tau}^{*T} \boldsymbol{\varphi}_{i2\tau} + w_{i2\tau}
\tag{7.41}
$$

where $\tau = 1, 2, 3$, $\Delta_{if\tau}$ is the element of nonlinear function vector $\boldsymbol{\Delta}_{if} = [\Delta_{if1}, \Delta_{if2}, \Delta_{if3}]$. $\boldsymbol{\theta}_{i2\tau}^{*}$, $\boldsymbol{\varphi}_{i2\tau}$, and $w_{i2\tau}$ are the optimal weighting vector, basis function vector, and minimum approximation error, respectively.

Then, one has

$$\begin{aligned} \boldsymbol{S}_{i2}^{T}(\boldsymbol{\Delta}_{if} + \boldsymbol{h}_{if}) &= \boldsymbol{S}_{i2}^{T}\left[\boldsymbol{\theta}_{i21}^{*T}\boldsymbol{\varphi}_{i21} + w_{i21l},\ \boldsymbol{\theta}_{i22}^{*T}\boldsymbol{\varphi}_{i22} + w_{i22l},\ \boldsymbol{\theta}_{i23}^{*T}\boldsymbol{\varphi}_{i23} + w_{i23l}\right]^{T} \\ &= \sum_{\tau=1}^{3} S_{i2\tau}\boldsymbol{\theta}_{i2\tau}^{*T}\boldsymbol{\varphi}_{i2\tau} + \sum_{\tau=1}^{3} S_{i2\tau} w_{i2\tau l} \end{aligned} \tag{7.42}$$

where $w_{i2\tau l} = w_{i2\tau} + h_{if\tau}$, satisfying $|w_{i2\tau l}| \leq \bar{w}_{i2\tau l}$, $\tau = 1, 2, 3$.

By using Young's inequality, one has

$$S_{i2\tau}\boldsymbol{\theta}_{i2\tau}^{*T}\boldsymbol{\varphi}_{i2\tau} \leq \frac{S_{i2\tau}^{2}\Phi_{i2}\boldsymbol{\varphi}_{i2\tau}^{T}\boldsymbol{\varphi}_{i2\tau}}{2h_{21}^{2}} + \frac{h_{21}^{2}}{2},\quad S_{i2\tau}w_{i2\tau l} \leq \frac{S_{i2\tau}^{2}}{2h_{22}^{2}} + \frac{h_{22}^{2}\bar{w}_{i2\tau l}^{2}}{2} \tag{7.43}$$

where $\Phi_{i2} = \max\left\{\boldsymbol{\theta}_{i21}^{*T}\boldsymbol{\theta}_{i21}, \boldsymbol{\theta}_{i22}^{*T}\boldsymbol{\theta}_{i22}, \boldsymbol{\theta}_{i23}^{*T}\boldsymbol{\theta}_{i23}\right\}$.

Define $\boldsymbol{\Xi}_{i2} = \left[S_{i21}\boldsymbol{\varphi}_{i21}^{T}\boldsymbol{\varphi}_{i21}, S_{i21}\boldsymbol{\varphi}_{i22}^{T}\boldsymbol{\varphi}_{i22}, S_{i21}\boldsymbol{\varphi}_{i23}^{T}\boldsymbol{\varphi}_{i23}\right]^{T}$, by substituting the controller (7.37) and the auxiliary dynamic signal (7.38) into (7.40), and then the following expression can be derived:

$$\begin{aligned} \boldsymbol{S}_{i2}^{T}\dot{\boldsymbol{S}}_{i2} \leq& \boldsymbol{S}_{i2}^{T}\left[\boldsymbol{F}_{i20} + \boldsymbol{G}_{i20}\boldsymbol{U}_{i0} + \frac{\Phi_{i2}\boldsymbol{\Xi}_{i2}}{2h_{21}^{2}} + \frac{\boldsymbol{S}_{i2}}{2h_{22}^{2}} - \Psi_{i} - \dot{\boldsymbol{\zeta}}_{i2}\right] \\ &+ \sum_{\tau=1}^{3}\frac{h_{21}^{2} + h_{22}^{2}\bar{w}_{i2\tau l}^{2}}{2} \\ =& \boldsymbol{S}_{i2}^{T}\left[-\boldsymbol{K}_{21}\boldsymbol{E}_{i2} - \boldsymbol{K}_{22}\text{sig}(\boldsymbol{S}_{i2})^{\upsilon} + \frac{\tilde{\Phi}_{i2}\boldsymbol{\Xi}_{i2}}{2h_{21}^{2}} - \dot{\boldsymbol{\zeta}}_{i2}\right] \\ &+ \sum_{\tau=1}^{3}\frac{h_{21}^{2} + h_{22}^{2}\bar{w}_{i2\tau l}^{2}}{2} \\ =& \boldsymbol{S}_{i2}^{T}\Bigg[-\boldsymbol{K}_{21}\boldsymbol{S}_{i2} - \boldsymbol{K}_{22}\text{sig}(\boldsymbol{S}_{i2})^{\upsilon} - \Theta_{i}\boldsymbol{G}_{i1}^{T}\boldsymbol{\eta}_{i1}\boldsymbol{S}_{i1} \\ &+ \frac{\tilde{\Phi}_{i2}\boldsymbol{\Xi}_{i2}}{2h_{21}^{2}} + \tau_{i2}\text{sign}(\boldsymbol{\zeta}_{i2})\Bigg] + \sum_{\tau=1}^{3}\frac{h_{21}^{2} + h_{22}^{2}\bar{w}_{i2\tau l}^{2}}{2} \end{aligned} \tag{7.44}$$

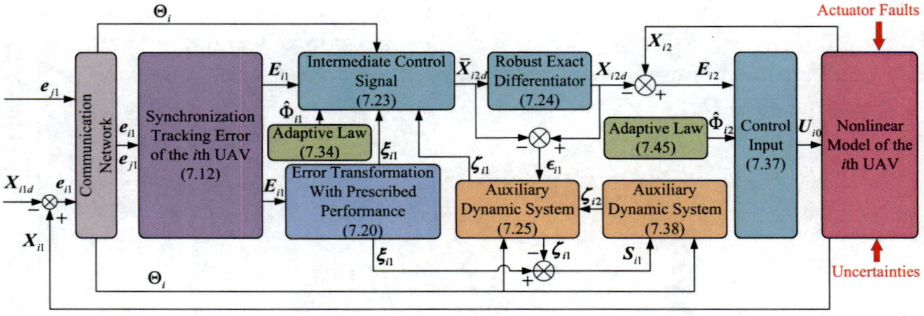

Fig. 7.1 Overall control scheme for the ith UAV

Design the adaptive law as

$$\dot{\hat{\Phi}}_{i2} = k_{23}\left(-k_{24}\hat{\Phi}_{i2} + \frac{\boldsymbol{S}_{i2}^T\boldsymbol{\Xi}_{i2}}{2h_{21}^2}\right) \tag{7.45}$$

where k_{23} and k_{24} are positive design parameters.

Choose the second Lyapunov function candidate as

$$L_{i2} = \frac{1}{2}\boldsymbol{S}_{i2}^T\boldsymbol{S}_{i2} + \frac{1}{2k_{23}}\tilde{\Phi}_{i2}^2 \tag{7.46}$$

where $\tilde{\Phi}_{i2} = \Phi_{i2} - \hat{\Phi}_{i2}$ is the estimation error.

By taking the time derivative of (7.46), one has

$$\begin{aligned}\dot{L}_{i2} =& \boldsymbol{S}_{i2}^T\dot{\boldsymbol{S}}_{i2} + \frac{1}{k_{23}}\tilde{\Phi}_{i2}\dot{\tilde{\Phi}}_{i2}\\ \leq& -\boldsymbol{S}_{i2}^T\boldsymbol{K}_{21}\boldsymbol{S}_{i2} - \boldsymbol{S}_{i2}^T\boldsymbol{K}_{22}\text{sig}(\boldsymbol{S}_{i2})^{\upsilon} - \Theta_i\boldsymbol{S}_{i2}^T\boldsymbol{G}_{i1}^T\boldsymbol{\eta}_{i1}\boldsymbol{S}_{i1}\\ &+ \tau_{i2}\boldsymbol{S}_{i2}^T\text{sign}(\boldsymbol{\zeta}_{i2}) + k_{24}\tilde{\Phi}_{i2}\hat{\Phi}_{i2} + \sum_{\tau=1}^{3}\frac{h_{21}^2 + h_{22}^2\bar{w}_{i2\tau l}^2}{2}\end{aligned} \tag{7.47}$$

The overall control architecture of the ith UAV is illustrated in Fig. 7.1.

7.3.3 *Stability Analysis*

Theorem 7.1 *Consider a group of N UAVs in the connected communication network, and if the control laws are designed as (7.23), (7.37), the adaptive laws are constructed as (7.34), (7.45), the auxiliary dynamic systems are designed as (7.25), (7.38), and the robust exact differentiator is employed as (7.24), then the attitude synchronization tracking errors of all UAVs can converge into the* *region containing zero in finite time in the presence of uncertainties and actuator faults (7.5). Furthermore, the attitude synchronization tracking errors are confined within the predefined bounds specified by (7.15).*

Proof Choose the following Lyapunov function candidate as

$$L = \sum_{i=1}^{N} L_i = \sum_{i=1}^{N} (L_{i1} + L_{i2}) \tag{7.48}$$

where L_{i1} and L_{i2} are defined in (7.35) and (7.46), respectively.

By recalling (7.36) and (7.47), the time derivative of L_i is given by

$$\begin{aligned}\dot{L}_i \leq & \Theta_i \left[-\boldsymbol{S}_{i1}^T \boldsymbol{K}_{11} \boldsymbol{S}_{i1} - \boldsymbol{S}_{i1}^T \boldsymbol{K}_{12} \text{sig}(\boldsymbol{S}_{i1})^{\upsilon} + \boldsymbol{S}_{i1}^T \boldsymbol{\eta}_{i1} \boldsymbol{G}_{i1} \boldsymbol{S}_{i2} + \tau_{i1} \boldsymbol{S}_{i1}^T \text{sign}(\boldsymbol{\zeta}_{i1}) \right] \\ & + \Theta_i k_{14} \tilde{\Phi}_{i1} \hat{\Phi}_{i1} + \Theta_i \sum_{\tau=1}^{3} \frac{h_{11}^2 + h_{12}^2 \bar{w}_{i1\tau}^2}{2} - \boldsymbol{S}_{i2}^T \boldsymbol{K}_{21} \boldsymbol{S}_{i2} - \boldsymbol{S}_{i2}^T \boldsymbol{K}_{22} \text{sig}(\boldsymbol{S}_{i2})^{\upsilon} \\ & - \Theta_i \boldsymbol{S}_{i2}^T \boldsymbol{G}_{i1}^T \boldsymbol{\eta}_{i1} \boldsymbol{S}_{i1} + \tau_{i2} \boldsymbol{S}_{i2}^T \text{sign}(\boldsymbol{\zeta}_{i2}) + k_{24} \tilde{\Phi}_{i2} \hat{\Phi}_{i2} + \sum_{\tau=1}^{3} \frac{h_{21}^2 + h_{22}^2 \bar{w}_{i2\tau l}^2}{2}\end{aligned} \tag{7.49}$$

By using Young's inequality, the following inequalities can be obtained

$$\tau_{i1} \boldsymbol{S}_{i1}^T \text{sign}(\boldsymbol{\zeta}_{i1}) \leq \frac{\tau_{i1} ||\boldsymbol{S}_{i1}||^2}{2} + \frac{\tau_{i1} ||\text{sign}(\boldsymbol{\zeta}_{i1})||^2}{2} = \frac{\tau_{i1} \boldsymbol{S}_{i1}^T \boldsymbol{S}_{i1}}{2} + \frac{3\tau_{i1}}{2} \tag{7.50}$$

$$\tau_{i2} \boldsymbol{S}_{i2}^T \text{sign}(\boldsymbol{\zeta}_{i2}) \leq \frac{\tau_{i2} ||\boldsymbol{S}_{i2}||^2}{2} + \frac{\tau_{i2} ||\text{sign}(\boldsymbol{\zeta}_{i2})||^2}{2} = \frac{\tau_{i2} \boldsymbol{S}_{i2}^T \boldsymbol{S}_{i2}}{2} + \frac{3\tau_{i2}}{2} \tag{7.51}$$

Then, one has

$$\begin{aligned}\dot{L}_i \leq & - \Theta_i \left[\lambda_{\min}(\boldsymbol{K}_{11}) - \frac{\tau_{i1}}{2} \right] \boldsymbol{S}_{i1}^T \boldsymbol{S}_{i1} - \Theta_i \lambda_{\min}(\boldsymbol{K}_{12}) \boldsymbol{S}_{i1}^T \text{sig}(\boldsymbol{S}_{i1})^{\upsilon} + \sigma_{i1} \\ & - \left[\lambda_{\min}(\boldsymbol{K}_{21}) - \frac{\tau_{i2}}{2} \right] \boldsymbol{S}_{i2}^T \boldsymbol{S}_{i2} - \lambda_{\min}(\boldsymbol{K}_{22}) \boldsymbol{S}_{i2}^T \text{sig}(\boldsymbol{S}_{i2})^{\upsilon} - \frac{k_{14}}{2} \tilde{\Phi}_{i1}^2 - \frac{k_{24}}{2} \tilde{\Phi}_{i2}^2 \\ = & - \Theta_i \left[2\lambda_{\min}(\boldsymbol{K}_{11}) - \tau_{i1} \right] \frac{1}{2} \boldsymbol{S}_{i1}^T \boldsymbol{S}_{i1} - \Theta_i \lambda_{\min}(\boldsymbol{K}_{12}) 2^{\frac{\upsilon+1}{2}} \left(\frac{1}{2} \boldsymbol{S}_{i1}^T \boldsymbol{S}_{i1} \right)^{\frac{\upsilon+1}{2}} \\ & - \left[2\lambda_{\min}(\boldsymbol{K}_{21}) - \tau_{i2} \right] \frac{1}{2} \boldsymbol{S}_{i2}^T \boldsymbol{S}_{i2} - \lambda_{\min}(\boldsymbol{K}_{22}) 2^{\frac{\upsilon+1}{2}} \left(\frac{1}{2} \boldsymbol{S}_{i2}^T \boldsymbol{S}_{i2} \right)^{\frac{\upsilon+1}{2}} - \frac{k_{14}}{2} \tilde{\Phi}_{i1}^2 \\ & - k_{23} k_{24} \frac{1}{2k_{23}} \tilde{\Phi}_{i2}^2 + \sigma_{i1}\end{aligned} \tag{7.52}$$

where $\sigma_{i1} = \Theta_i \left(\frac{3h_{11}^2}{2} + \frac{3\tau_{i1}}{2} + \sum_{\tau=1}^{3} \frac{h_{12}^2 \bar{w}_{i1\tau}^2}{2} \right) + \frac{3h_{21}^2}{2} + \frac{3\tau_{i2}}{2} + \sum_{\tau=1}^{3} \frac{h_{22}^2 \bar{w}_{i2\tau l}^2}{2} + \frac{k_{14}\Phi_{i1}^2}{2} + \frac{k_{24}\Phi_{i2}^2}{2}$.

Therefore, (7.52) can be rewritten as

$$\dot{L}_i \leq -l_{i1} L_i + \sigma_{i1} \tag{7.53}$$

where

$$l_{i1} = \min \left\{ -\Theta_i \left[2\lambda_{\min}(\boldsymbol{K}_{11}) - \tau_{i1} \right], \left[2\lambda_{\min}(\boldsymbol{K}_{21}) - \tau_{i2} \right], \frac{k_{14}}{2}, k_{23}k_{24} \right\} \tag{7.54}$$

According to Lyapunov's stability theory, it can be seen from (7.53) that the compensated synchronization tracking error $\boldsymbol{S}_{i1}$, the compensated attitude rate tracking error $\boldsymbol{S}_{i2}$, and the adaptive estimation errors $\tilde{\Phi}_{i1}$, $\tilde{\Phi}_{i2}$ are UUB. Therefore, one can assume that $|\tilde{\Phi}_{i1}| \leq \Phi_{i1m}$, $|\tilde{\Phi}_{i2}| \leq \Phi_{i2m}$, $|\eta_{i1\tau}| \leq \eta_{i1\tau m}$, where Φ_{i1m}, Φ_{i2m} are positive constants.

Next, another Lyapunov function candidate is utilized to analyze the finite-time stability.

$$L_f = \sum_{i=1}^{N} L_{if} = \sum_{i=1}^{N} \left(\frac{1}{2} \boldsymbol{S}_{i1}^T \boldsymbol{S}_{i1} + \frac{1}{2} \boldsymbol{S}_{i2}^T \boldsymbol{S}_{i2} \right) \tag{7.55}$$

Taking the time derivative of L_{if} yields

$$\begin{aligned} \dot{L}_{if} \leq & -\Theta_i \left[2\lambda_{\min}(\boldsymbol{K}_{11}) - \tau_{i1} + \frac{h_{13}}{h_{11}^2} \right] \frac{1}{2} \boldsymbol{S}_{i1}^T \boldsymbol{S}_{i1} \\ & - \left[2\lambda_{\min}(\boldsymbol{K}_{21}) - \tau_{i2} + \frac{h_{23}}{h_{21}^2} \right] \frac{1}{2} \boldsymbol{S}_{i2}^T \boldsymbol{S}_{i2} \\ & - \Theta_i \lambda_{\min}(\boldsymbol{K}_{12}) 2^{\frac{\upsilon+1}{2}} \left(\frac{1}{2} \boldsymbol{S}_{i1}^T \boldsymbol{S}_{i1} \right)^{\frac{\upsilon+1}{2}} - \lambda_{\min}(\boldsymbol{K}_{22}) 2^{\frac{\upsilon+1}{2}} \left(\frac{1}{2} \boldsymbol{S}_{i2}^T \boldsymbol{S}_{i2} \right)^{\frac{\upsilon+1}{2}} + \sigma_{i2} \\ \leq & - l_{i2} L_{if} - l_{i3} L_{if}^{\frac{\upsilon+1}{2}} + \sigma_{i2} \end{aligned} \tag{7.56}$$

where

$$\begin{cases} l_{i2} = \min\left\{\Theta_i\left[2\lambda_{\min}(\boldsymbol{K}_{11}) - \tau_{i1} + \frac{h_{13}}{h_{11}^2}\right], \left[2\lambda_{\min}(\boldsymbol{K}_{21}) - \tau_{i2} + \frac{h_{23}}{h_{21}^2}\right]\right\} \\ l_{i3} = \min\left\{\Theta_i\lambda_{\min}(\boldsymbol{K}_{12})2^{\frac{\upsilon+1}{2}},\ \lambda_{\min}(\boldsymbol{K}_{22})2^{\frac{\upsilon+1}{2}}\right\} \\ \sigma_{i2} = \Theta_i\left(\frac{3h_{11}^2}{2} + \frac{3\tau_{i1}}{2} + \sum_{\tau=1}^{3}\frac{h_{12}^2\bar{w}_{i1\tau}^2}{2}\right) + \frac{3h_{21}^2}{2} + \frac{3\tau_{i2}}{2} + \sum_{\tau=1}^{3}\frac{h_{22}^2\bar{w}_{i2\tau l}^2}{2} \\ h_{13} \geq \max\left\{\left|\tilde{\Phi}_{i11}\eta_{i11}^2\boldsymbol{\varphi}_{i11}^T\boldsymbol{\varphi}_{i11}\right|, \left|\tilde{\Phi}_{i12}\eta_{i12}^2\boldsymbol{\varphi}_{i12}^T\boldsymbol{\varphi}_{i12}\right|, \left|\tilde{\Phi}_{i13}\eta_{i13}^2\boldsymbol{\varphi}_{i13}^T\boldsymbol{\varphi}_{i13}\right|\right\} \\ h_{23} \geq \max\left\{\left|\tilde{\Phi}_{i21}\eta_{i21}^2\boldsymbol{\varphi}_{i21}^T\boldsymbol{\varphi}_{i21}\right|, \left|\tilde{\Phi}_{i22}\eta_{i22}^2\boldsymbol{\varphi}_{i22}^T\boldsymbol{\varphi}_{i22}\right|, \left|\tilde{\Phi}_{i23}\eta_{i23}^2\boldsymbol{\varphi}_{i23}^T\boldsymbol{\varphi}_{i23}\right|\right\} \end{cases}$$

Then, one has

$$\dot{L}_f \leq -l_2 L_f - l_3 L_f^{\frac{\upsilon+1}{2}} + \sigma_2 \tag{7.57}$$

where $l_2 = \min\{l_{i2}\}$, $l_3 = \min\{l_{i3}\}$, $i = 1, 2, \ldots, N$, $\sigma_2 = \sum_{i=1}^{N}\sigma_{i2}$.

By recalling Lemma 6.4 in Chap. 6, if the control parameters are chosen such that $2\lambda_{\min}(\boldsymbol{K}_{11}) - \tau_{i1} + \frac{h_{13}}{h_{11}^2} > 0$, $2\lambda_{\min}(\boldsymbol{K}_{21}) - \tau_{i2} + \frac{h_{23}}{h_{21}^2} > 0$, then there exists a constant $0 < a_0 < 1$ such that

$$L_f \leq \min\left\{\frac{\sigma_2}{(1-a_0)l_2}, \left(\frac{\sigma_2}{(1-a_0)l_3}\right)^{\frac{2}{\upsilon+1}}\right\} \tag{7.58}$$

The setting time T_1 is given by

$$T_1 \leq \max\left\{\begin{array}{l} \frac{1}{a_0 l_1\left(1-\frac{1+\upsilon}{2}\right)}\ln\frac{a_0 l_2 L_f^{1-(1+\upsilon)/2}(0)+l_3}{l_3}, \\ \frac{1}{l_2(1-(\upsilon+1)/2)}\ln\frac{L_f^{1-(1+\upsilon)/2}(0)+a_0 l_3}{a_0 l_3} \end{array}\right\} \tag{7.59}$$

Then, from (7.58), it can be seen that $\boldsymbol{S}_{i1}$ and $\boldsymbol{S}_{i2}$ will converge to the region $\{\|\boldsymbol{S}_{i1}\|,\ \|\boldsymbol{S}_{i2}\|\} \leq \min\left\{\sqrt{\frac{2\sigma_2}{(1-a_0)l_2}}, \sqrt{2\left(\frac{\sigma_2}{(1-a_0)l_3}\right)^{\frac{2}{\upsilon+1}}}\right\}$ in finite time T_1.

It should be stressed that in the above analysis, the finite-time stability of the compensated tracking errors $\boldsymbol{S}_{i1} = \boldsymbol{\xi}_{i1} - \boldsymbol{\zeta}_{i1}$, $\boldsymbol{S}_{i2} = \boldsymbol{E}_{i2} - \boldsymbol{\zeta}_{i2}$ has been analyzed. To further analyze the finite-time stability of tracking errors $\boldsymbol{\xi}_{i1}$, $\boldsymbol{E}_{i2}$, the Lyapunov function of the auxiliary system is chosen as

$$L_a = \sum_{i=1}^{N} L_{ia} = \sum_{i=1}^{N}\left(\frac{1}{2}\boldsymbol{\zeta}_{i1}^T\boldsymbol{\zeta}_{i1} + \frac{1}{2}\boldsymbol{\zeta}_{i2}^T\boldsymbol{\zeta}_{i2}\right) \tag{7.60}$$

By recalling (7.24), $\boldsymbol{X}_{i2d}$ converges to $\bar{\boldsymbol{X}}_{i2d}$ in finite time T_{i2}. For $t \geq T_2 = \max\{T_{i2}\}$, $i = 1, 2, \ldots, N$, the time derivative of L_{ia} yields

$$\begin{aligned}
\dot{L}_{ia} =& \boldsymbol{\zeta}_{i1}^T \dot{\boldsymbol{\zeta}}_{i1} + \boldsymbol{\zeta}_{i2}^T \dot{\boldsymbol{\zeta}}_{i2} \\
=& \boldsymbol{\zeta}_{i1}^T \left\{ \Theta_i \left[-\boldsymbol{K}_{11} \boldsymbol{\zeta}_{i1} + \boldsymbol{\eta}_{i1} \boldsymbol{G}_{i1} \boldsymbol{\epsilon}_{i1} + \boldsymbol{G}_{i1} \boldsymbol{\eta}_{i1} \boldsymbol{\zeta}_{i2} - \tau_{i1} \mathrm{sign}(\boldsymbol{\zeta}_{i1}) \right] \right\} \\
& + \boldsymbol{\zeta}_{i2}^T \left\{ -\boldsymbol{K}_{21} \boldsymbol{\zeta}_{i2} + \Theta_i \boldsymbol{G}_{i1}^T \boldsymbol{\eta}_{i1} \boldsymbol{S}_{i1} - \tau_{i2} \mathrm{sign}(\boldsymbol{\zeta}_{i2}) \right\} \\
\leq& - \Theta_i \left[\lambda_{\min}(\boldsymbol{K}_{11}) - \frac{3}{2} \right] \boldsymbol{\zeta}_{i1}^T \boldsymbol{\zeta}_{i1} - \tau_{i1} \Theta_i \left(\boldsymbol{\zeta}_{i1}^T \boldsymbol{\zeta}_{i1} \right)^{\frac{1}{2}} - \tau_{i2} \left(\boldsymbol{\zeta}_{i2}^T \boldsymbol{\zeta}_{i2} \right)^{\frac{1}{2}} \\
& - \left[\lambda_{\min}(\boldsymbol{K}_{21}) - \frac{\Theta_i}{2} - \frac{\Theta_i \left\| \boldsymbol{G}_{i1} \boldsymbol{\eta}_{i1} \right\|_F}{2} \right] \boldsymbol{\zeta}_{i2}^T \boldsymbol{\zeta}_{i2} + \frac{\Theta_i h_{31}^2}{2} + \frac{\Theta_i h_{32}^2}{2} \\
\leq& - l_{i4} L_{ia} - l_{i5} L_{ia}^{\frac{1}{2}} + \sigma_{i3}
\end{aligned} \tag{7.61}$$

where

$$\begin{cases}
l_{i4} = \min \left\{ \Theta_i \left[2\lambda_{\min}(\boldsymbol{K}_{11}) - 3 \right], \left[\lambda_{\min}(\boldsymbol{K}_{21}) - \frac{\Theta_i}{2} - \frac{\Theta_i h_{33}}{2} \right] \right\} \\
l_{i5} = \min \{ \tau_{i1} \Theta_i, \ \tau_{i2} \}, \ \sigma_{i3} = \frac{\Theta_i h_{31}^2}{2} + \frac{\Theta_i h_{32}^2}{2} \\
h_{31} \geq \left\| \boldsymbol{\eta}_{i1} \boldsymbol{G}_{i1} \boldsymbol{\epsilon}_{i1} \right\|, h_{32} \geq \left\| \boldsymbol{G}_{i1}^T \boldsymbol{\eta}_{i1} \boldsymbol{S}_{i1} \right\|, h_{33} \geq \left\| \boldsymbol{G}_{i1} \boldsymbol{\eta}_{i1} \right\|_F
\end{cases}$$

Then, the following inequality can be obtained as

$$\dot{L}_a \leq -l_4 L_a - l_5 L_a^{\frac{1}{2}} + \sigma_3 \tag{7.62}$$

where $l_4 = \min\{l_{i4}\}$, $l_5 = \min\{l_{i5}\}$, $i = 1, 2, \ldots, N$, $\sigma_3 = \sum_{i=1}^N \sigma_{i3}$.

By recalling Lemma 6.4 in Chap. 6, one has

$$L_a \leq \min \left\{ \frac{\sigma_3}{(1 - a_1) l_4}, \left(\frac{\sigma_3}{(1 - a_1) l_5} \right)^2 \right\} \tag{7.63}$$

The setting time T_3 is given by

$$\begin{aligned}
T_3 \leq \max & \left\{ T_2 + \frac{1}{a_1 l_4 \left(1 - \frac{1}{2} \right)} \ln \frac{a_1 l_5 L_f^{1/2}(T_2) + l_5}{l_5}, \right. \\
& \left. T_2 + \frac{1}{l_4 (1/2)} \ln \frac{L_a^{1/2}(T_2) + a_1 l_5}{a_1 l_5} \right\}
\end{aligned} \tag{7.64}$$

It can be seen from (7.63) that ζ_{i1} and ζ_{i2} will converge to the region $\{\|\zeta_{i1}\|, \|\zeta_{i2}\|\} \le \min\left\{\sqrt{\frac{2\sigma_3}{(1-a_1)l_4}}, \sqrt{2\left(\frac{\sigma_3}{(1-a_1)l_5}\right)^2}\right\}$ in finite time T_3. By recalling (7.26) and (7.39), it can be concluded that for $t \ge T_4 = \max\{T_1, T_3\}$, the tracking errors $\boldsymbol{\xi}_{i1}$ and $\boldsymbol{E}_{i2}$ are finite-time bounded. With the error transformation (7.17), it can be further concluded that the tracking error $\boldsymbol{E}_{i1}$ is eventually confined within the prescribed performance bound (7.15). Moreover, it can be concluded that all signals in the closed-loop system are bounded. This ends the proof. □

Remark 7.6 The convergence region can be made very small by choosing large parameters $\boldsymbol{K}_{11}$, $\boldsymbol{K}_{12}$, $\boldsymbol{K}_{21}$, and $\boldsymbol{K}_{22}$, and small parameters $\tau_{i1}, \tau_{i2}, h_{11}, h_{12}, h_{21}, h_{22}$ to guarantee the small radii of region of tracking errors. Moreover, the parameters $\varepsilon_{i1\tau0}$, $\underline{k}_{i1\tau}$, and $\overline{k}_{i1\tau}$ in the performance functions should be properly chosen such that the initial error $E_{i1\tau}(0)$ is in the interval $\left[-\underline{k}_{i1\tau}\varepsilon_{i1\tau0}, \overline{k}_{i1\tau}\varepsilon_{i1\tau0}\right]$.

Remark 7.7 It should be stressed that the control objective is to design a fault-tolerant synchronization tracking control scheme, such that all UAVs in the formation team can track their individual attitude references in a synchronized manner. If only one attitude reference signal is available to the formation team and only a subset of UAVs has access to the common attitude reference, the SMO technique can be first used to estimate the attitude references for all UAVs with nearest neighbor rules [29, 33]. Therefore, the proposed control scheme is still applicable by combining the SMO technique when only one attitude reference signal is involved in the formation flying of multi-UAVs.

Remark 7.8 Since the sign function is used in (7.23), (7.24), (7.25), (7.37), and (7.38), the control chattering may be induced. To attenuate the chattering effect in the simulation, the sign function is replaced by a smooth hyperbolic tangent function $\tanh(\cdot/\phi)$.

7.4 Simulation Results

Simulations with networked four UAVs are conducted to verify the effectiveness of the proposed decentralized finite-time adaptive FTCC scheme with prescribed performance. The communication topology is illustrated in Fig. 7.2, and the element a_{ij} of the corresponding adjacency matrix $\mathcal{A}$ equals one if there exists information exchange between the ith UAV and the jth UAV, otherwise $a_{ij} = 0, i, j = 1, 2, 3, 4$. The structure parameters and all coefficient values of each UAV are referred to [30]. The initial flight conditions of four UAVs are listed in Table 7.1. The simulations are performed with an initial velocity of 30 m/s. The bank angle commands (μ_{1c}, μ_{2c}, μ_{3c}, μ_{4c}) of four UAVs are set to step from 0° to 10°, 9°, 10°, and 9° at $t = 2$ s, respectively, and then step to 0° at $t = 32$ s. The angle of attack commands (α_{1c}, α_{2c}, α_{3c}, α_{4c}) of four UAVs are set to step from 2° to 8°, 7°, 8°, and 7° at $t = 2$

Fig. 7.2 Communication topology

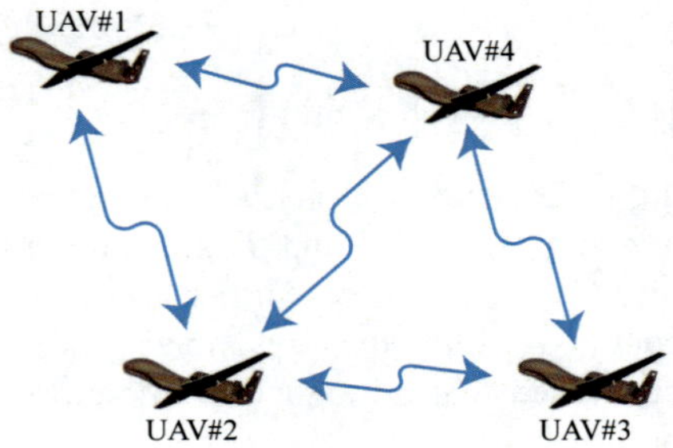

Table 7.1 Initial values of all UAVs

	$\mu_i(0)$ [deg]	$\alpha_i(0)$ [deg]	$\beta_i(0)$ [deg]	$p_i(0)$ [deg/s]	$q_i(0)$ [deg/s]	$r_i(0)$ [deg/s]
UAV#1	2.865	2.865	1.719	0	0	0
UAV#2	1.146	2.865	0.2865	0	0	0
UAV#3	0	1.834	2.865	0	0	0
UAV#4	1.719	2.292	1.146	0	0	0

s, respectively, and then step to 2° at $t = 32$ s. The sideslip angle commands (β_{1c}, β_{2c}, β_{3c}, β_{4c}) are 0° at the overall time interval. The commands mentioned above are shaped by the following filter to generate the smooth reference commands $\boldsymbol{X}_{i1d}$.

$$\frac{\boldsymbol{X}_{i1d}}{\boldsymbol{X}_{i1c}} = \frac{\omega_d^2}{s^2 + 2\xi_d\omega_d s + \omega_d^2} \tag{7.65}$$

where $\boldsymbol{X}_{i1c} = [\mu_{ic}, \alpha_{ic}, \beta_{ic}]$, $\omega_d = 0.9$, $\xi_d = 0.3$.

The following actuator faults encountered by UAV#1, UAV#2, UAV#3 and model uncertainties are considered in the simulations:

(1) Actuator faults:
UAV#1 fault ($t = 10$ s):

$$\rho_{11} = \begin{cases} 1, & 0 \le t < 10 \\ 0.55, & t \ge 10 \end{cases} \quad \rho_{12} = \begin{cases} 1, & 0 \le t < 10 \\ 0.8, & t \ge 10 \end{cases} \quad \rho_{13} = \begin{cases} 1, & 0 \le t < 10 \\ 0.65 & t \ge 10 \end{cases}$$

$$u_{1f1} = \begin{cases} 0^\circ, & 0 \le t < 10 \\ 20.055^\circ, & t \ge 10 \end{cases} \quad u_{1f2} = \begin{cases} 0^\circ, & 0 \le t < 10 \\ 2.865^\circ, & t \ge 10 \end{cases}$$

$$u_{1f3} = \begin{cases} 0^\circ, & 0 \le t < 10 \\ 17.19^\circ, & t \ge 10 \end{cases} \tag{7.66}$$

UAV#2 fault ($t = 20$ s):

$$\rho_{21} = \begin{cases} 1, & 0 \le t < 20 \\ 0.8, & t \ge 20 \end{cases} \quad \rho_{22} = \begin{cases} 1, & 0 \le t < 20 \\ 0.85, & t \ge 20 \end{cases} \quad \rho_{23} = \begin{cases} 1, & 0 \le t < 20 \\ 0.7 & t \ge 20 \end{cases}$$

$$u_{2f1} = \begin{cases} 0^\circ, & 0 \le t < 20 \\ 17.19^\circ, & t \ge 20 \end{cases} \quad u_{2f2} = \begin{cases} 0^\circ, & 0 \le t < 20 \\ 2.865^\circ, & t \ge 20 \end{cases}$$

$$u_{2f3} = \begin{cases} 0^\circ, & 0 \le t < 20 \\ 14.325^\circ, & t \ge 20 \end{cases} \tag{7.67}$$

UAV#3 fault ($t = 40$ s):

$$\rho_{31} = \begin{cases} 1, & 0 \le t < 40 \\ 0.7, & t \ge 40 \end{cases} \quad \rho_{32} = \begin{cases} 1, & 0 \le t < 40 \\ 0.75, & t \ge 40 \end{cases} \quad \rho_{33} = \begin{cases} 1, & 0 \le t < 40 \\ 0.4 & t \ge 40 \end{cases}$$

$$u_{3f1} = \begin{cases} 0^\circ, & 0 \le t < 40 \\ 11.46^\circ, & t \ge 40 \end{cases} \quad u_{3f2} = \begin{cases} 0^\circ, & 0 \le t < 40 \\ 2.865^\circ, & t \ge 40 \end{cases}$$

$$u_{3f3} = \begin{cases} 0^\circ, & 0 \le t < 40 \\ 17.19^\circ, & t \ge 40 \end{cases} \tag{7.68}$$

(2) The mismatch in the aerodynamic coefficients C_{iL0}, $C_{iL\alpha}$, C_{iD0}, $C_{iD\alpha}$, $C_{iD\alpha^2}$, C_{iY0}, $C_{iY\beta}$, C_{il0}, C_{im0}, C_{in0}, $C_{il\beta}$, $C_{il\delta_a}$, $C_{il\delta_r}$, C_{ilp}, C_{ilr}, $C_{im\alpha}$, $C_{im\delta_e}$, C_{ilq}, $C_{in\beta}$, $C_{in\delta_a}$, $C_{in\delta_r}$, C_{inp}, and C_{inr} is chosen as 20% upper uncertainties of the nominal values.

To quantitatively evaluate the fault-tolerant synchronization tracking performance, bank angle synchronization tracking error metric (BASTEM), angle of attack synchronization tracking error metric (AASTEM), and sideslip angle synchronization tracking error metric (SASTEM) are defined as

$$\text{BASTEM} = \sqrt{\sum_{i=1}^{4} |E_{i11}|^2}, \quad \text{AASTEM} = \sqrt{\sum_{i=1}^{4} |E_{i12}|^2},$$

$$\text{SASTEM} = \sqrt{\sum_{i=1}^{4} |E_{i13}|^2} \tag{7.69}$$

The control parameters of the proposed decentralized FTCC scheme are listed in Table 7.2. The initial values of the adaptive laws are set as $\Phi_{i1}(0) = 0$, $\Phi_{i2}(0) = 0$, $i = 1, 2, 3, 4$. The initial values of the robust exact differentiator are chosen as

Table 7.2 Controller parameters

Parameter	Value	Parameter	Value	Parameter	Value
λ_{i1}	1.2	λ_{i2}	1	$\underline{k}_{i1\tau}$	1.1
$\overline{k}_{i1\tau}$	1.1	$\varepsilon_{i1\tau0}$	22.92	$\varepsilon_{i1\tau\infty}$	8.595
$\nu_{i1\tau}$	0.5	$\boldsymbol{K}_{11}$	diag[3, 5, 1.5]	$\boldsymbol{K}_{12}$	diag[0.5, 0.25, 0.2]
$\boldsymbol{K}_{21}$	diag[10, 15, 8]	$\boldsymbol{K}_{22}$	diag[0.3, 2.5, 0.2]	υ	0.6
h_{11}	2.5	h_{12}	1.2	h_{13}	63.25
h_{14}	1	h_{15}	2×10^3	h_{21}	2
h_{22}	1.3	τ_{i1}	0.2	τ_{i2}	0.2
k_{13}	30	k_{14}	0.05	k_{23}	30
k_{24}	0.08	ϕ	0.002		

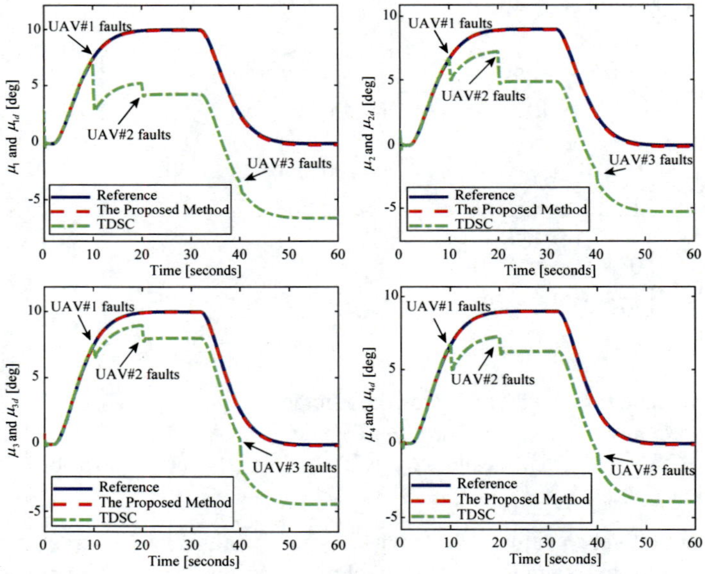

Fig. 7.3 Bank angles of four UAVs

$X_{i2d\tau}(0) = 0$, $\Psi_{i\tau}(0) = 0$, $\tau = 1, 2, 3$, and the initial values of the auxiliary dynamic systems (7.25), (7.38) are set as $\boldsymbol{\zeta}_{i1} = [0, 0, 0]^T$, $\boldsymbol{\zeta}_{i2} = [0, 0, 0]^T$, respectively.

Figure 7.3 shows the bank angles of four UAVs, and it can be seen that the bank angles of four UAVs can track their individual bank angle references before the actuator fault encountered by the UAV#1 at $t = 10$ s. Under the proposed decentralized finite-time FTCC scheme, the bank angles of four UAVs can track the bank angle references after the occurrences of the actuator faults at $t = 10$ s, $t = 20$ s, and $t = 40$ s. However, with traditional DSC (TDSC), the bank angles cannot track the bank angle references after the actuator fault at $t = 10$ s. At

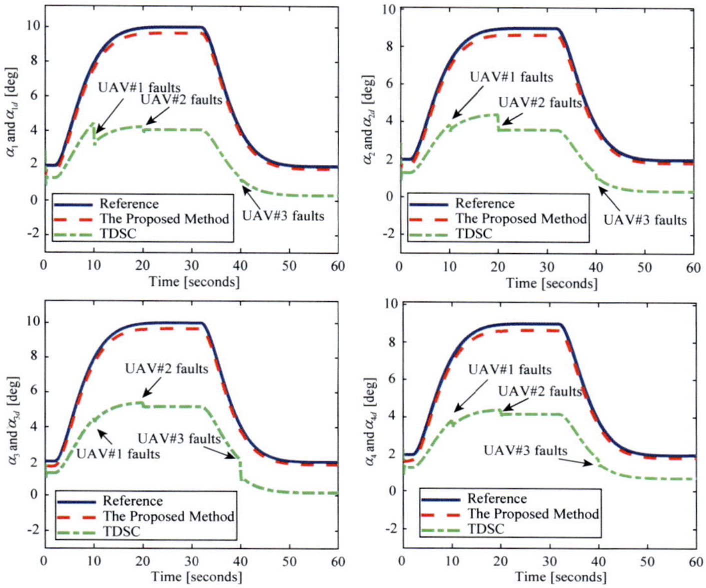

Fig. 7.4 Angles of attack of four UAVs

$t = 10$ s, the bank angle of UAV#1 has a larger drop than other UAVs, and this is induced by the actuator fault subjected by the UAV#1. Then, the information from the faulty UAV#1 is shared to UAV#2, UAV#3, and UAV#4, causing performance degradations of other UAVs. Figure 7.4 illustrates the angles of attack of all UAVs. It is observed from Fig. 7.4 that the proposed method has a better tracking performance than the TDSC, and the actuator faults in UAV#1, UAV#2, and UAV#3 have a significantly adverse influence on the tracking performance of TDSC, which may threaten the flight safety. Figure 7.5 gives the histories of sideslip angles, and it can be seen that the sideslip angles can be maintained at 0° with quite small bounded errors. The TDSC can maintain the sideslip angles of four UAVs at 0° from different initial values at the time interval [0 2] s. However, the sideslip angles cannot track the references once the actuator faults are encountered by UAV#1, UAV#2, and UAV#3. From Fig. 7.5, it is also observed that the sideslip angle of UAV#4 under TDSC cannot track its individual reference, and this can be explained by the fact that the UAV#4 is contaminated by the information from the faulty UAV#1, UAV#2, and UAV#3.

Figures 7.6, 7.7, and 7.8 show the bank angle, angle of attack, and sideslip angle tracking errors with respect to their individual references, respectively. From these curves, it is observed that the tracking performance by the presented control scheme outperforms the comparing TDSC scheme. Moreover, all the tracking errors are bounded. As can be seen from Fig. 7.7, the tracking errors of angle of attack under the proposed control scheme are significantly smaller than that of the TDSC scheme.

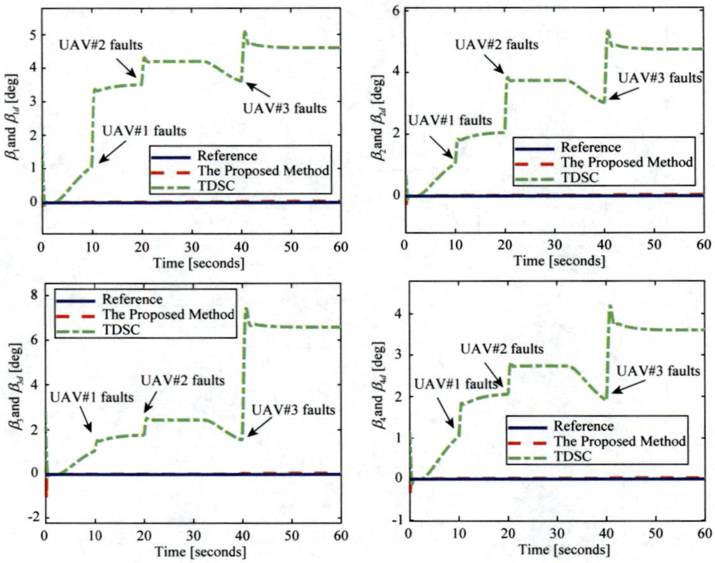

Fig. 7.5 Sideslip angles of four UAVs

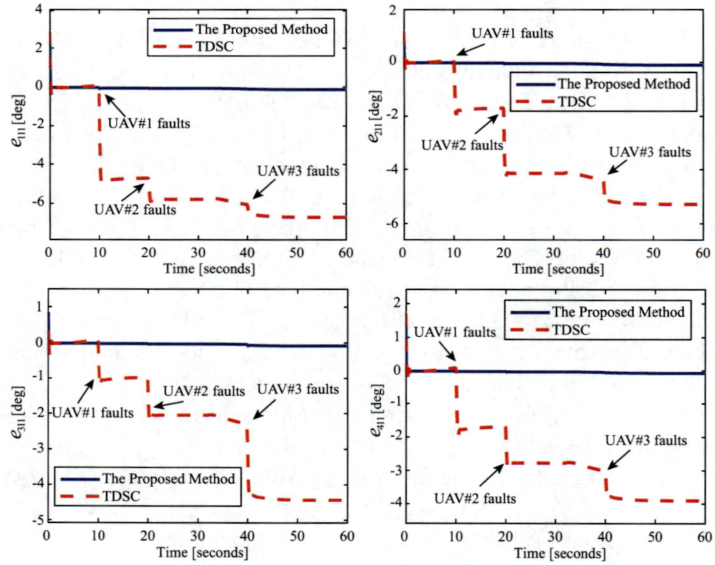

Fig. 7.6 Bank angle tracking errors of four UAVs with respect to their individual bank angle references

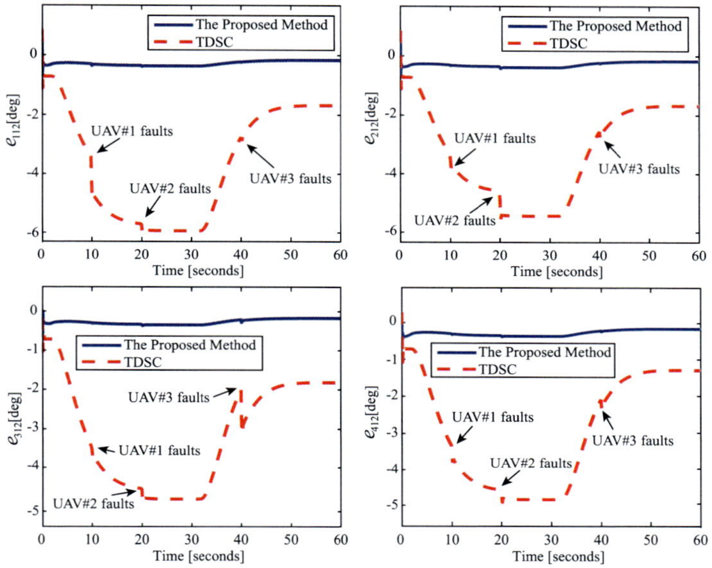

Fig. 7.7 Angle of attack tracking errors of four UAVs with respect to their individual angle of attack references

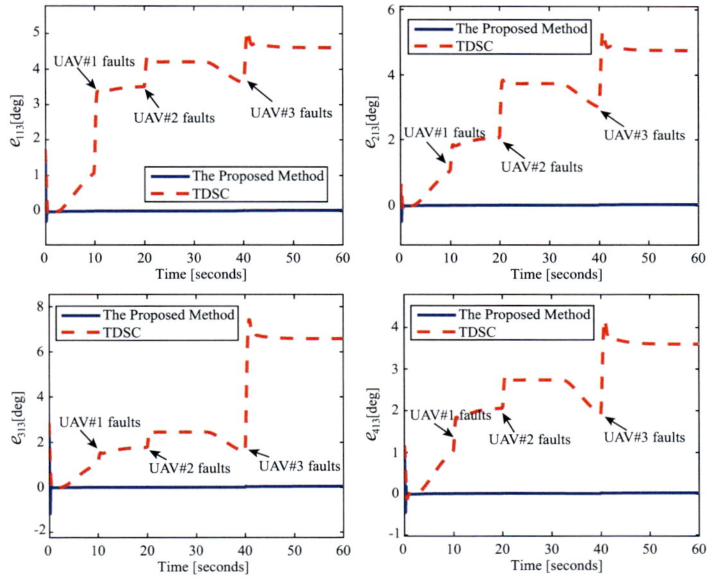

Fig. 7.8 Sideslip angle tracking errors of four UAVs with respect to their individual sideslip angle references

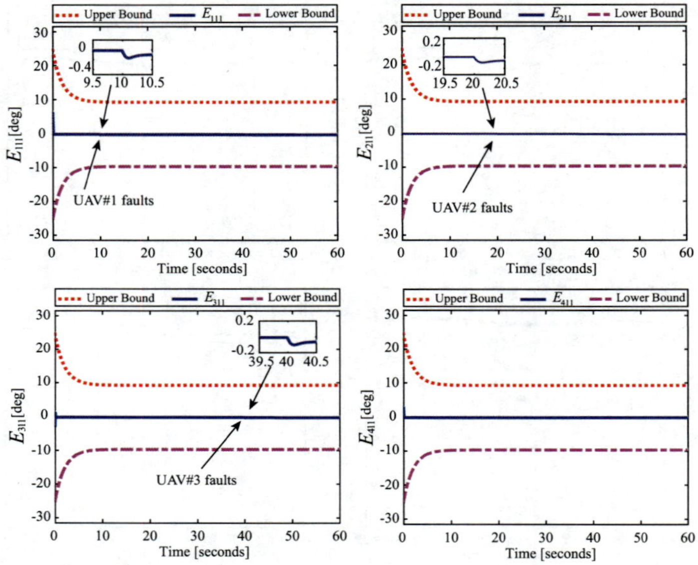

Fig. 7.9 Synchronized bank angle tracking errors of four UAVs

In terms of the curves in Fig. 7.8, the tracking errors of sideslip angles under TDSC deviate from the small region containing zero when four UAVs start following the changing reference signals of bank angles and angles of attack, which further shows that the nonlinear UAV model is highly coupled. As a contrast, the tracking errors of sideslip angles under the presented control scheme maintain within the small region containing zero even in the presence of actuator faults encountered by UAV#1, UAV#2, and UAV#3.

Figures 7.9, 7.10, and 7.11 illustrate the time responses of synchronized bank angle, angle of attack, and sideslip angle tracking errors. It is observed from Fig. 7.9 that the synchronized bank angle tracking error E_{111} of UAV#1 under the proposed control scheme achieves a good synchronization tracking performance. From the curves of E_{211}, E_{311}, and E_{411}, it can be seen that the synchronized tracking errors remain remarkably small. As can be seen from Fig. 7.10, the synchronized angle of attack tracking performances of four UAVs remains excellent in the case of the developed decentralized finite-time FTC scheme. It is illustrated in Fig. 7.11 that the synchronized sideslip angle tracking errors are significantly small, and the errors E_{113}, E_{213}, E_{313}, and E_{413} are constrained in the prescribed performance bounds.

The control input signals of four UAVs under the proposed control method are shown in Fig. 7.12, and it is observed from the time responses of $\boldsymbol{\delta}_{a0}$, $\boldsymbol{\delta}_{e0}$, and $\boldsymbol{\delta}_{r0}$ that the control input signals react to the actuator faults encountered by UAV#1 at $t = 10$ s, UAV#2 at $t = 20$ s, UAV#3 at $t = 40$ s in a timely manner. With such response modes, the actuator faults can be promptly compensated by adjusting the control input signals. Figure 7.13 illustrates the estimations of adaptive parameters

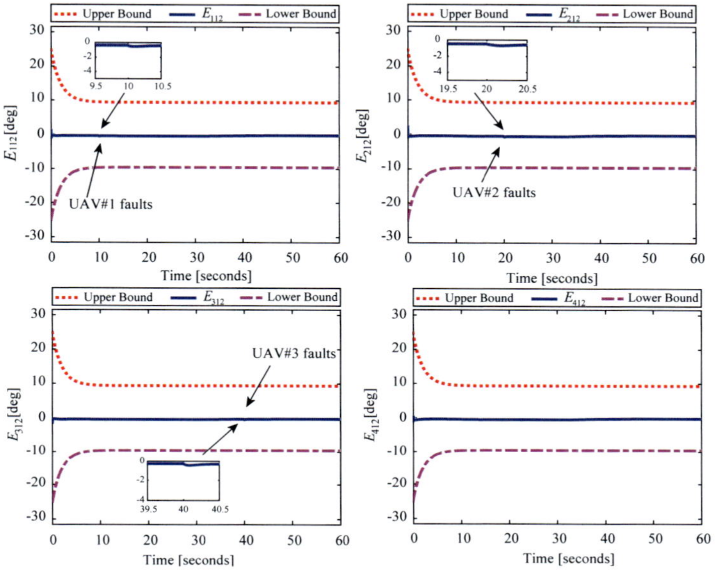

Fig. 7.10 Synchronized angle of attack tracking errors of four UAVs

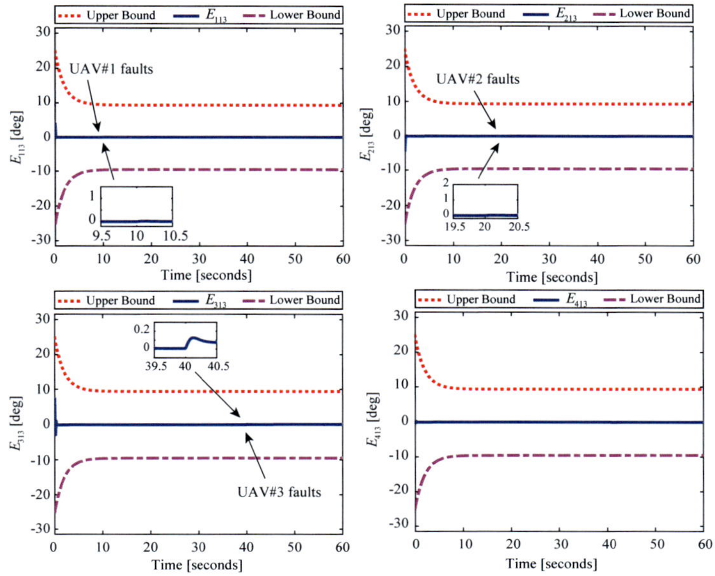

Fig. 7.11 Synchronized sideslip tracking errors of four UAVs

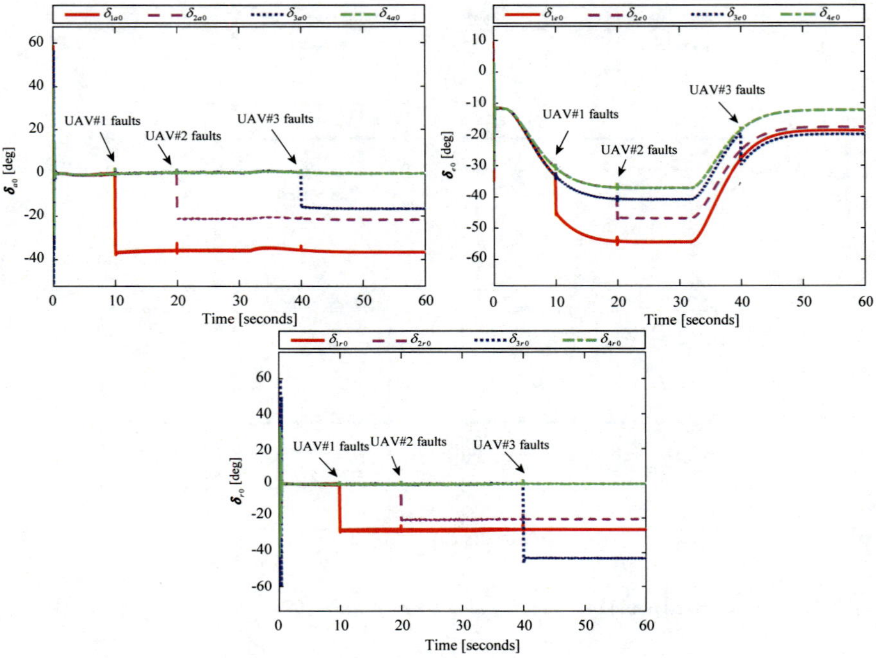

Fig. 7.12 Control input signals

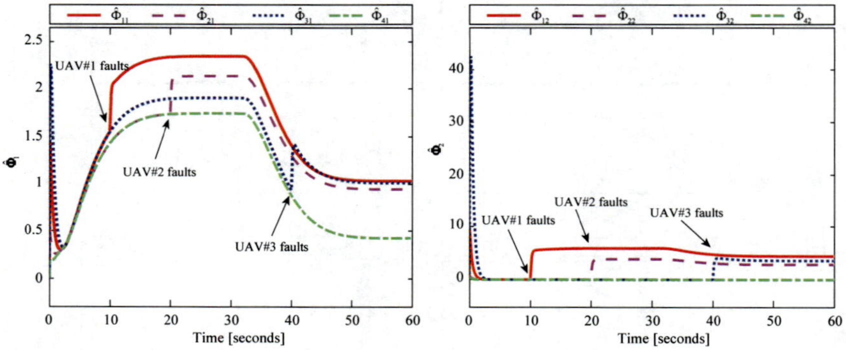

Fig. 7.13 Adaptive parameter estimations

Φ_{i1} and Φ_{i2}, and it can be seen from these curves that in order to attenuate the adverse effects caused by the actuator faults, the estimation values of Φ_{i1} and Φ_{i2} rapidly adjust to achieve satisfactory synchronization tracking performance.

Figure 7.14 presents the synchronization tracking error metrics, and it can be concluded from these curves that the presented control scheme performs very well, which further emphasizes that the decentralized FTCC scheme can ensure both

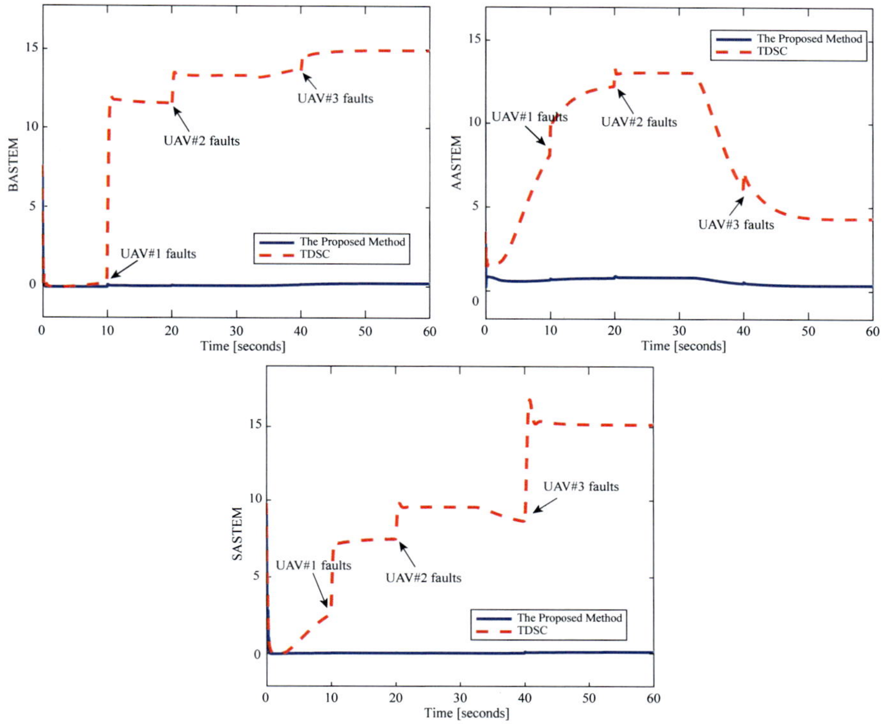

Fig. 7.14 BASTEM, AASTEM, and SASTEM

the formation flight safety and the good synchronization tracking performance. Furthermore, the presented control scheme can be used for the formation flight with faulty UAVs and the formation flight without faulty UAVs, since UAV#4 is healthy in the overall formation flight.

7.5 Conclusions

In this chapter, the decentralized finite-time adaptive FTCC scheme has been proposed for multi-UAVs in the presence of actuator faults. By introducing prescribed performance functions on the synchronized tracking errors, a new set of errors is first constructed to facilitate the decentralized synchronization tracking control design. Then, by using NNs and the norms of weight vectors, the number of adaptive parameters is significantly reduced and independent from the neurons. Moreover, the finite-time convergence feature has been incorporated into the control scheme to enhance the fault-tolerant performance. Lyapunov stability analysis has shown that all UAVs can track their individual attitude references in finite time, and the synchronization tracking errors are finite-time convergent and confined within the prescribed bounds. Comparative simulation results have demonstrated the effectiveness of the proposed control scheme.

References

1. I. Bayezit, B. Fidan, Distributed cohesive motion control of flight vehicle formations. IEEE Trans. Ind. Electron. **60**(12), 5763–5772 (2013)
2. J.D. Boskovic, S. Bergstrom, R.K. Mehra, Robust integrated flight control design under failures, damage, and state-dependent disturbances. J. Guid. Control Dyn. **28**(5), 902–917 (2005)
3. X.W. Bu, Guaranteeing prescribed output tracking performance for air-breathing hypersonic vehicles via non-affine back-stepping control design. Nonlinear Dyn. **91**(1), 525–538 (2018)
4. X.W. Bu, G.J. He, K. Wang, Tracking control of air-breathing hypersonic vehicles with non-affine dynamics via improved neural back-stepping design. ISA Trans. **75**, 88–100 (2018)
5. E. Cruz-Zavala, J.A. Moreno, L.M. Fridman, Uniform robust exact differentiator. IEEE Trans. Autom. Control **56**(11), 2727–2733 (2011)
6. H.B. Du, S.H. Li, Finite-time attitude stabilization for a spacecraft using homogeneous method. J. Guid. Control Dyn. **35**(3), 740–748 (2012)
7. S. Ghapani, W. Ren, F. Chen, Y.D. Song, Distributed average tracking for double-integrator multi-agent systems with reduced requirement on velocity measurements. Automatica **81**, 1–7 (2017)
8. M. Ghasemi, S.G. Nersesov, G. Clayton, Finite-time tracking using sliding mode control. J. Frankl. Inst. **351**(5), 2966–2990 (2014)
9. L. He, X.X. Sun, Y. Lin, Distributed output-feedback formation tracking control for unmanned aerial vehicles. Int. J. Syst. Sci. **47**(16), 3919–3928 (2016)
10. X.Q. Huang, W. Lin, B. Yang, Global finite-time stabilization of a class of uncertain nonlinear systems. Automatica **41**(5), 881–888 (2005)
11. J.Q. Li, K.D. Kumar, Decentralized fault-tolerant control for satellite attitude synchronization. IEEE Trans. Fuzzy Syst. **20**(3), 572–586 (2011)
12. J.S. Li, J.M. Li, Coordination control of multi-agent systems with second-order nonlinear dynamics using fully distributed adaptive iterative learning. J. Frankl. Inst. **352**(6), 2441–2463 (2015)
13. P. Li, X. Yu, X.Y. Peng, Z.Q. Zheng, Y.M. Zhang, Fault-tolerant cooperative control for multiple UAVs based on sliding mode techniques. Sci. China Inf. Sci. **60**(7), 070204 (2017)
14. W. Lin, Distributed UAV formation control using differential game approach. Aerosp. Sci. Technol. **35**, 54–62 (2014)
15. Z.X. Liu, C. Yuan, Y.M. Zhang, Active fault-tolerant control of unmanned quadrotor helicopter using linear parameter varying technique. J. Intell. Robot. Syst. **88**(2–4), 415–436 (2017)
16. Q.N. Luo, H.B. Duan, Distributed UAV flocking control based on homing pigeon hierarchical strategies. Aerosp. Sci. Technol. **70**, 257–264 (2017)
17. M.Y. Ou, H.B. Du, S.H. Li, Finite-time tracking control of multiple nonholonomic mobile robots. J. Frankl. Inst. **349**(9), 2834–2860 (2012)
18. Z.H. Qu, Matrix theory for cooperative systems, in *Cooperative Control of Dynamical Systems: Applications to Autonomous Vehicles* (Springer, Berlin, 2009), pp. 153–193
19. S.C. Tong, Y.M. Li, P. Shi, Observer-based adaptive fuzzy backstepping output feedback control of uncertain MIMO pure-feedback nonlinear systems. IEEE Trans. Fuzzy Syst. **20**(4), 771–785 (2012)
20. Y.J. Wang, Y.D. Song, F.L. Lewis. Robust adaptive fault-tolerant control of multiagent systems with uncertain nonidentical dynamics and undetectable actuation failures. IEEE Trans. Ind. Electron. **62**(6), 3978–3988 (2015)

21. B.L. Wu, D.W. Wang, E.K. Poh, Decentralized robust adaptive control for attitude synchronization under directed communication topology. J. Guid. Control Dyn. **34**(4), 1276–1282 (2011)
22. X.B. Xiang, C. Liu, H.S. Su, Q. Zhang, On decentralized adaptive full-order sliding mode control of multiple UAVs. ISA Trans. **71**, 196–205 (2017)
23. G.H. Xu, Z.H. Guan, D.X. He, M. Chi, Y.H. Wu, Distributed tracking control of second-order multi-agent systems with sampled data. J. Frankl. Inst. **351**(10), 4786–4801 (2014)
24. Q. Xu, H. Yang, B. Jiang, D.H. Zhou, Y.M. Zhang, Fault tolerant formations control of UAVs subject to permanent and intermittent faults. Int. J. Robust Nonlinear Control **73**(1–4), 589–602 (2014)
25. X. Yu, Z.X. Liu, Y.M. Zhang, Fault-tolerant formation control of multiple UAVs in the presence of actuator faults. Int. J. Robust Nonlinear Control **26**(12), 2668–2685 (2016)
26. X. Yu, P. Li, Y.M. Zhang, The design of fixed-time observer and finite-time fault-tolerant control for hypersonic gliding vehicles. IEEE Trans. Ind. Electron. **65**(5), 4135–4144 (2018)
27. Z.Q. Yu, Y.H. Qu, Y.M. Zhang, Distributed fault-tolerant cooperative control for multi-UAVs under actuator fault and input saturation. IEEE Trans. Control Syst. Technol. **27**(6), 2417–2429 (2018)
28. Z.Q. Yu, Y.H. Qu, Y.M. Zhang, Safe control of trailing UAV in close formation flight against actuator fault and wake vortex effect. Aerosp. Sci. Technol. **77**, 189–205 (2018)
29. Z.Q. Yu, Y.H. Qu, Y.M. Zhang, Fault-tolerant containment control of multiple unmanned aerial vehicles based on distributed sliding-mode observer. J. Intell. Robot. Syst. **93**(1–2), 163–177 (2019)
30. Z.Q. Yu, Y.M. Zhang, Z.X. Liu, Y.H. Qu, C.Y. Su, Distributed adaptive fractional-order fault-tolerant cooperative control of networked unmanned aerial vehicles via fuzzy neural networks. IET Contr. Theory Appl. **13**(17), 2917–2929 (2019)
31. Z.Q. Yu, Z.X. Liu, Y.M. Zhang, Y.H. Qu, C.Y. Su, Distributed finite-time fault-tolerant containment control for multiple unmanned aerial vehicles. IEEE Trans. Neural Netw. Learn. Syst. **31**(6), 2077–2091 (2020)
32. B. Yun, B.M. Chen, K.Y. Lum, T.H. Lee, A leader-follower formation flight control scheme for UAV helicopters, in *International Conference on Automation and Logistics*, Qingdao (2008)
33. L. Zhao, Y.M. Jia, Distributed adaptive containment control for second-order multi-agent systems via NTSM. J. Frankl. Inst. **352**(11), 5327–5341 (2015)
34. A.M. Zou, G. Zeng Z.G. Hou, M. Tan, Adaptive control of a class of nonlinear pure-feedback systems using fuzzy backstepping approach. IEEE Trans. Fuzzy Syst. **16**(4), 886–897 (2008)

Chapter 8
Decentralized Attitude FTCC of Multi-UAVs Under Directed Communication Topology

8.1 Introduction

The past decade has witnessed a great deal of research interests in cooperative control of multi-UAVs since the implementation of multi-UAVs instead of using a single UAV increases system reliability and efficiency, but reduces the overall mission cost [5, 6, 9, 12]. For the reduction of the communication and computation burden in cooperative control of multi-UAVs, decentralized control architecture is more efficient than the centralized control frame. It should be stressed that most existing results on the cooperative control of multi-UAVs are about the point-mass model [3, 4, 8, 11], i.e., outer-loop model. Regarding the attitude model, i.e., inner-loop model, few results have been obtained. However, to incorporate more practical situations into the research, such as actuator saturation, faults, and dead zone, the attitude model should be considered in the cooperative control of multi-UAVs [10, 13]. Furthermore, as analyzed in Chap. 7, investigation on the attitude synchronization tracking control of multi-UAVs can be seen as the replenishment of existing results about the point-mass UAV model.

This chapter further develops a decentralized FTCC scheme for multi-UAVs with prescribed attitude synchronization tracking performance in a directed communication network, such that all UAVs can synchronously track their attitude references in the presence of actuator faults. Specifically, to constrain the synchronization errors within the prescribed bounds, synchronization errors are first transformed into a new set of error variables by the prescribed performance functions. Then, NNs are used to approximate unknown nonlinear terms due to the nonlinearities inherent in the UAVs and the loss-of-effectiveness actuator faults. DOs are utilized to compensate for the NN approximation errors and the bias faults. Furthermore, prediction errors

Z. Yu et al., *Fault-Tolerant Cooperative Control of Unmanned Aerial Vehicles*,
https://doi.org/10.1007/978-981-99-7661-4_8

are incorporated into the NN adaptive laws and DOs to enhance the approximation abilities. The main contributions of this chapter are summarized as follows:

(1) The attitude synchronization tracking control of multi-UAVs is investigated in a decentralized communication network with directed information flows, rather than a centralized communication topology.
(2) Compared with numerous results on the FTC of a single UAV or the several studies on the FTCC of multi-UAVs in a leader–follower framework, an attitude synchronization FTCC scheme is further investigated for multi-UAVs in a decentralized communication network.
(3) Transient fault-tolerant synchronization tracking performance is considered by incorporating prescribed performance function.

The rest of this chapter is structured as follows. Section 8.2 provides the preliminaries and problem statement. The FTCC design and stability analysis are presented in Sect. 8.3. Numerical simulations are provided in Sect. 8.4 to demonstrate the effectiveness of the proposed control scheme, followed by conclusions in Sect. 8.5.

(*Remark: The main control schemes and contents of this chapter are from the published journal paper "Z. Q. Yu, Z. X. Liu, Y. M. Zhang, Y. H. Qu, and C.-Y Su. Decentralized fault-tolerant cooperative control of multiple UAVs with prescribed attitude synchronization tracking performance under directed communication topology, Frontiers of Information Technology & Electronic Engineering, 2019, 20: 685–700." The authors appreciate the permission from the Springer Nature to reuse the results published in the relevant Journal.*)

8.2 Preliminaries and Problem Statement

8.2.1 Faulty UAV Attitude Dynamics

In this chapter, it is assumed that there exist N UAVs in the formation team. By using the same control-oriented attitude model as (7.1)–(7.2) in Chap. 7, the following model is still used:

$$\dot{\boldsymbol{x}}_{i1} = \boldsymbol{f}_{i1} + \boldsymbol{g}_{i1}\boldsymbol{x}_{i2} \tag{8.1}$$

$$\dot{\boldsymbol{x}}_{i2} = \boldsymbol{f}_{i2} + \boldsymbol{g}_{i2}\boldsymbol{u}_{i} \tag{8.2}$$

where $i = 1, 2, \ldots, N$. $\boldsymbol{x}_{i1} = [\mu_i, \alpha_i, \beta_i]^T$, $\boldsymbol{x}_{i2} = [p_i, q_i, r_i]^T$, $\boldsymbol{u}_i = [\delta_{ia}, \delta_{ie}, \delta_{ir}]^T$. $\boldsymbol{f}_{i1}$, $\boldsymbol{f}_{i2}$, $\boldsymbol{g}_{i1}$, and $\boldsymbol{g}_{i2}$ have the same definitions of $\boldsymbol{F}_{i1}$, $\boldsymbol{F}_{i2}$, $\boldsymbol{G}_{i1}$, and $\boldsymbol{G}_{i2}$ in Chap. 7.

Assumption 8.1 Similar to the analysis in Chap. 7, the control gain function $\boldsymbol{g}_{i2}$ can be written as a known part $\boldsymbol{g}_{i2N}$ and an unknown part $\Delta\boldsymbol{g}_{i2}$.

To facilitate the FTCC design, the following actuator fault model is still used:

$$\boldsymbol{u}_i = \boldsymbol{\rho}_i \boldsymbol{u}_{i0} + \boldsymbol{u}_{if} \tag{8.3}$$

where $\boldsymbol{u}_i = [\delta_{ia},\ \delta_{ie},\ \delta_{ir}]^T$ is the applied signal, $\boldsymbol{u}_{i0} = [\delta_{ia0},\ \delta_{ie0},\ \delta_{ir0}]^T$ is the control signal commanded by the controller, $\boldsymbol{\rho}_i = \mathrm{diag}[\rho_{i1},\ \rho_{i2},\ \rho_{i3}]$ represents the remaining control effectiveness factor, and $\boldsymbol{u}_{if} = [u_{if1},\ u_{if2},\ u_{if3}]^T$ denotes the bounded bias fault.

By applying the actuator fault model (8.3) into (8.2), the attitude dynamic model with actuator faults is formulated as

$$\dot{\boldsymbol{x}}_{i2} = \boldsymbol{F}_{i2} + \boldsymbol{g}_{i2N}\boldsymbol{u}_{i0} + \boldsymbol{d}_i \tag{8.4}$$

where $\boldsymbol{F}_{i2} = \boldsymbol{f}_{i2} + \boldsymbol{g}_{i2N}\boldsymbol{\rho}_i\boldsymbol{u}_{i0} + \Delta\boldsymbol{g}_{i2}\boldsymbol{\rho}_i\boldsymbol{u}_{i0} - \boldsymbol{g}_{i2N}\boldsymbol{u}_{i0}$, $\boldsymbol{d}_i = (\boldsymbol{g}_{i2N} + \Delta\boldsymbol{g}_{i2})\boldsymbol{u}_{if}$.

Therefore, the attitude model in the presence of actuator faults is given by

$$\dot{\boldsymbol{x}}_{i1} = \boldsymbol{f}_{i1} + \boldsymbol{g}_{i1}\boldsymbol{x}_{i2} \tag{8.5}$$

$$\dot{\boldsymbol{x}}_{i2} = \boldsymbol{F}_{i2} + \boldsymbol{g}_{i2N}\boldsymbol{u}_{i0} + \boldsymbol{d}_i \tag{8.6}$$

It should be noted that the unknown nonlinear function $\boldsymbol{F}_{i2}$ involves the control input signal $\boldsymbol{u}_{i0}$. Algebraic loops will be introduced into the controller design if the RBFNN is employed to approximate the nonlinear function, since the input signal $\boldsymbol{u}_{i0}$ is directly fed into the Gaussian function of the RBFNN. Similar to Chaps. 3, 6, and 7, to break the algebraic loop, a BLPF is introduced as

$$\boldsymbol{u}_{i0f} = B_i(s)\boldsymbol{u}_{i0} \approx \boldsymbol{u}_{i0} \tag{8.7}$$

where $B_i(s)$ is a BLPF and $\boldsymbol{u}_{i0f}$ is the filtered signal.

Therefore, one has

$$\boldsymbol{\epsilon}_i = \boldsymbol{F}_{i2}(\boldsymbol{u}_{i0}) - \boldsymbol{F}_{i2b}(\boldsymbol{u}_{i0f}) \tag{8.8}$$

where $\boldsymbol{\epsilon}_i$ is the bounded filter error [14].

Considering the filter error $\boldsymbol{\epsilon}_i$, (8.6) can be further transformed as

$$\dot{\boldsymbol{x}}_{i2} = \boldsymbol{F}_{i2b} + \boldsymbol{g}_{i2N}\boldsymbol{u}_{i0} + \boldsymbol{d}_i + \boldsymbol{\epsilon}_i \tag{8.9}$$

Based on the control-oriented attitude model (8.5), (8.9), the decentralized FTCC scheme will be developed. Since the control scheme is to be developed for multi-UAVs in a decentralized communication network, basic graph theory is given in the subsequent section.

8.2.2 *Basic Graph Theory*

Assume that the information exchange of N UAVs is modeled by a weighted-directed graph $\mathcal{G} = (\mathcal{V}, \mathcal{E}, \mathcal{A})$, where $\mathcal{V} = \{v_1, v_2, \ldots, v_N\}$ is the set of nodes, $\mathcal{E} \subseteq \mathcal{V} \times \mathcal{V}$ is the set of edges, and $\mathcal{A} = [a_{ij}] \in R^{N \times N}$ is the weighted adjacency matrix of the graph $\mathcal{G}$. Node v_j can access information from node v_i if $(v_i, v_j) \in \mathcal{E}$. The set of neighbors of node v_i is denoted as $N_i = \{v_j \in \mathcal{V} | (v_j, v_i) \in \mathcal{E}\}$. A directed path from v_i to v_j is a sequence of edges of the form (v_i, v_{l_1}), (v_{l_1}, v_{l_2}), $\ldots$, $(v_{l_{k-1}}, v_{l_k})$, (v_{l_k}, v_j), where $v_{l_n} \in \mathcal{V}$ for $1 \leq l_n \leq l_k$. The element of $\mathcal{A}$ is defined as $a_{ij} > 0$ if $(v_j, v_i) \in \mathcal{E}$, and $a_{ij} = 0$, otherwise. The Laplacian matrix $\mathcal{L} = [l_{ij}] \in R^{N \times N}$ associated with the graph $\mathcal{G}$ is defined as $l_{ij} = \sum_{k=1}^{N} a_{ik}$ if $i = j$ and $l_{ij} = -a_{ij}$ if $i \neq j$.

Lemma 8.1 *For a weighted-directed graph $\mathcal{G}$ with N nodes, all the eigenvalues of the weighted Laplacian $\mathcal{L}$ have a nonnegative real part [7].*

Lemma 8.2 *Suppose that $\boldsymbol{X} \in R^{m \times m}$, $\boldsymbol{Y} \in R^{n \times n}$ and let $\lambda_{11}, \lambda_{12}, \ldots, \lambda_{1m}$ be the eigenvalues of $\boldsymbol{X}$ and $\lambda_{21}, \lambda_{22}, \ldots, \lambda_{2m}$ be the eigenvalues of $\boldsymbol{Y}$, respectively, then the eigenvalues of $\boldsymbol{X} \otimes \boldsymbol{Y}$ are $\lambda_{1i}\lambda_{2j}$, $i = 1, 2, \ldots, m$, $j = 1, 2, \ldots, n$, where $\otimes$ denotes the Kronecker product [7].*

8.2.3 *Control Objective*

The objective of this chapter is to design a set of decentralized FTCC laws for a group of UAVs in the presence of actuator faults and directed communication network, such that all attitudes of multi-UAVs can track their attitude references with prescribed synchronization tracking performance.

8.3 Main Results

In this section, the decentralized FTCC scheme is developed with the integration of PPC, NNs, and DOs. Auxiliary systems are also utilized in the proposed control scheme to compensate for the errors induced by the first-order filter and the actuator saturation.

8.3.1 Error Transformation

By defining the desired attitude reference for the ith UAV as $\boldsymbol{x}_{i1d} = [\mu_{id}, \alpha_{id}, \beta_{id}]^T$, the individual attitude tracking error of each UAV is denoted by $\tilde{\boldsymbol{x}}_{i1} = \boldsymbol{x}_{i1} - \boldsymbol{x}_{i1d}$. Define the attitude synchronization tracking error as

$$\boldsymbol{e}_{i1} = \lambda_1 \tilde{\boldsymbol{x}}_{i1} + \lambda_2 \sum_{j=1}^{N_i} a_{ij}(\tilde{\boldsymbol{x}}_{i1} - \tilde{\boldsymbol{x}}_{j1}) \tag{8.10}$$

where $\boldsymbol{e}_{i1} = [e_{i11}, e_{i12}, e_{i13}]^T$, and λ_1 and λ_2 are positive constants, which are determined by the controller designer to regulate the convergence rate of the state trajectory.

By using the Kronecker product, one has

$$\boldsymbol{e}_1 = [(\lambda_1 \boldsymbol{I}_N + \lambda_2 \mathcal{L}) \otimes \boldsymbol{I}_3]\tilde{\boldsymbol{x}}_1 \tag{8.11}$$

where $\boldsymbol{e}_1 = [\boldsymbol{e}_{11}^T, \ldots, \boldsymbol{e}_{N1}^T]^T$, $\tilde{\boldsymbol{x}}_1 = [\tilde{\boldsymbol{x}}_{11}^T, \ldots, \tilde{\boldsymbol{x}}_{N1}^T]^T$, and $\boldsymbol{I}_N$ and $\boldsymbol{I}_3$ are identity matrices with appropriate dimensions.

In view of Lemmas 8.1 and 8.2, one can conclude that $(\lambda_1 \boldsymbol{I}_N + \lambda_2 \mathcal{L}) \otimes \boldsymbol{I}_3$ has full rank. Therefore, it follows that $\tilde{\boldsymbol{x}}_1 \to \boldsymbol{0}$ when $\boldsymbol{e}_1 \to \boldsymbol{0}$.

Next, a new set of error variables is introduced to achieve the prescribed transient and steady performances, such that the fault-tolerant capability against actuator faults can be guaranteed. The PPC can be guaranteed if the following condition is always satisfied:

$$-\underline{k}_{i1\nu}\varepsilon_{i1\nu} \le e_{i1\nu} \le \overline{k}_{i1\nu}\varepsilon_{i1\nu}, \forall t \ge 0 \tag{8.12}$$

where $\nu = 1, 2, 3$, $\underline{k}_{i1\nu}$, and $\overline{k}_{i1\nu}$ are positive design parameters, $\varepsilon_{i1\nu}$ is a strictly decreasing smooth function, which is chosen as $\varepsilon_{i1\nu} = (\varepsilon_{i1\nu0} - \varepsilon_{i1\nu\infty})e^{-\iota_{i1\nu}t} + \varepsilon_{i1\nu\infty}$ with $\varepsilon_{i1\nu\infty}$ being the maximum allowable value of $e_{i1\nu}$ at steady state, $\varepsilon_{i1\nu0}$ being the initial value of $\varepsilon_{i1\nu}$. $\varepsilon_{i1\nu0}$, $\varepsilon_{i1\nu\infty}$, and $\iota_{i1\nu}$ are positive constants, which should be chosen to satisfy that $-\underline{k}_{i1\nu}\varepsilon_{i1\nu0} \le e_{i1\nu}(0) \le \overline{k}_{i1\nu}\varepsilon_{i1\nu0}$. $-\underline{k}_{i1\nu}\varepsilon_{i1\nu}$ and $\overline{k}_{i1\nu}\varepsilon_{i1\nu}$ are the allowable lower bound of the undershoot and the upper bound of the overshoot of $e_{i1\nu}$, respectively. The allowable steady-state value of $e_{i1\nu}$ is in the region $[-\underline{k}_{i1\nu}\varepsilon_{i1\nu\infty}, \overline{k}_{i1\nu}\varepsilon_{i1\nu\infty}]$.

To utilize the inequality (8.12) for the decentralized FTCC scheme design, (8.12) is transformed into the following equality:

$$e_{i1\nu} = \varepsilon_{i1\nu}\Theta(E_{i1\nu}) \tag{8.13}$$

where $\Theta(\cdot)$ is a smooth and strictly increasing function with the following properties:

$$\begin{cases} \Theta(0) = 0 \\ -\underline{k}_{i1\nu} \leq \Theta(E_{i1\nu}) \leq \overline{k}_{i1\nu} \\ \lim_{E_{i1\nu} \to +\infty} \Theta(E_{i1\nu}) = \overline{k}_{i1\nu} \\ \lim_{E_{i1\nu} \to -\infty} \Theta(E_{i1\nu}) = -\underline{k}_{i1\nu} \end{cases} \tag{8.14}$$

By using the error transformation (8.13), the controller design for the synchronized tracking error $e_{i1\nu}$ with prescribed performance can be converted into the control scheme design with bounded error $E_{i1\nu}$.

Similar to Chap. 7, $\Theta(E_{i1\nu})$ is chosen as

$$\Theta(E_{i1\nu}) = \frac{\overline{k}_{i1\nu} e^{E_{i1\nu}+\kappa_{i1\nu}} - \underline{k}_{i1\nu} e^{-E_{i1\nu}-\kappa_{i1\nu}}}{e^{E_{i1\nu}+\kappa_{i1\nu}} + e^{-E_{i1\nu}-\kappa_{i1\nu}}} \tag{8.15}$$

where $\kappa_{i1\nu} = \frac{1}{2}\ln\frac{\underline{k}_{i1\nu}}{\overline{k}_{i1\nu}}$.

Combining (8.13) and (8.15) yields

$$E_{i1\nu} = \Theta^{-1}\left(\frac{e_{i1\nu}}{\varepsilon_{i1\nu}}\right) = \frac{1}{2}\ln\frac{\overline{k}_{i1\nu}\underline{k}_{i1\nu} + \overline{k}_{i1\nu}\sigma_{i1\nu}}{\overline{k}_{i1\nu}\underline{k}_{i1\nu} - \underline{k}_{i1\nu}\sigma_{i1\nu}} \tag{8.16}$$

where $\sigma_{i1\nu} = \frac{e_{i1\nu}}{\varepsilon_{i1\nu}}$.

Taking the time derivative of (8.16) yields

$$\begin{aligned} \dot{E}_{i1\nu} =& \frac{1}{2\varepsilon_{i1\nu}}\left(\frac{1}{\underline{k}_{i1\nu} + \sigma_{i1\nu}} + \frac{1}{\overline{k}_{i1\nu} - \sigma_{i1\nu}}\right) \cdot \left(\dot{e}_{i1\nu} - \frac{e_{i1\nu}\dot{\varepsilon}_{i1\nu}}{\varepsilon_{i1\nu}}\right) \\ =& \eta_{i1\nu}\left(\dot{e}_{i1\nu} - \frac{e_{i1\nu}\dot{\varepsilon}_{i1\nu}}{\varepsilon_{i1\nu}}\right) \end{aligned} \tag{8.17}$$

where $\eta_{i1\nu} = \frac{1}{2\varepsilon_{i1\nu}}\left(\frac{1}{\underline{k}_{i1\nu}+\sigma_{i1\nu}} + \frac{1}{\overline{k}_{i1\nu}-\sigma_{i1\nu}}\right)$.

Then, the matrix form of (8.17) associated with the ith UAV is given by

$$\dot{\boldsymbol{E}}_{i1} = \boldsymbol{\eta}_{i1}(\dot{\boldsymbol{e}}_{i1} - \boldsymbol{\varepsilon}_{i1}^{-1}\dot{\boldsymbol{\varepsilon}}_{i1}\boldsymbol{e}_{i1}) \tag{8.18}$$

where $\boldsymbol{E}_{i1} = [E_{i11}, E_{i12}, E_{i13}]^T$, $\boldsymbol{\varepsilon}_{i1} = \text{diag}[\varepsilon_{i11}, \varepsilon_{i12}, \varepsilon_{i13}]$, $\boldsymbol{\eta}_{i1} = \text{diag}[\eta_{i11}, \eta_{i12}, \eta_{i13}]$.

8.3.2 Decentralized FTCC with Prescribed Attitude Synchronization Tracking Performance

Based on the transformed error variable (8.16), the attitude synchronization tracking controller with prescribed performance is developed with the integration of NNs and dOs. In view of (8.5) and (8.10), (8.18) can be derived as

$$\begin{aligned}\dot{\boldsymbol{E}}_{i1} &= \boldsymbol{\eta}_{i1}\left(\dot{\boldsymbol{e}}_{i1} - \boldsymbol{\varepsilon}_{i1}^{-1}\dot{\boldsymbol{\varepsilon}}_{i1}\boldsymbol{e}_{i1}\right)\\ &= \boldsymbol{\eta}_{i1}\left[\phi_i\dot{\bar{\boldsymbol{x}}}_{i1} - \lambda_2\sum_{j=1}^{N_i} a_{ij}\dot{\bar{\boldsymbol{x}}}_{j1} - \boldsymbol{\varepsilon}_{i1}^{-1}\dot{\boldsymbol{\varepsilon}}_{i1}\boldsymbol{e}_{i1}\right]\\ &= \boldsymbol{\eta}_{i1}\left[\phi_i\left(\boldsymbol{f}_{i1} + \boldsymbol{g}_{i1}\boldsymbol{x}_{i2} - \dot{\boldsymbol{x}}_{i1d}\right) - \lambda_2\sum_{j=1}^{N_i} a_{ij}\dot{\bar{\boldsymbol{x}}}_{j1} - \boldsymbol{\varepsilon}_{i1}^{-1}\dot{\boldsymbol{\varepsilon}}_{i1}\boldsymbol{e}_{i1}\right]\end{aligned} \tag{8.19}$$

where $\phi_i = \lambda_1 + \lambda_2\sum_{j=1}^{N_i} a_{ij}$.

By using RBFNN to approximate $\Gamma_1\boldsymbol{f}_{i1}$ with Γ_1 being the positive design parameter, one has

$$\begin{aligned}\dot{\boldsymbol{E}}_{i1} = \boldsymbol{\eta}_{i1}\Big[&\phi_i\left(\Gamma_1^{-1}\boldsymbol{W}_{i1}^{*T}\boldsymbol{\varphi}_{i1} + \boldsymbol{g}_{i1}\boldsymbol{x}_{i2} + \boldsymbol{D}_{i1} - \dot{\boldsymbol{x}}_{i1d}\right)\\ &- \lambda_2\sum_{j=1}^{N_i} a_{ij}\dot{\bar{\boldsymbol{x}}}_{j1} - \boldsymbol{\varepsilon}_{i1}^{-1}\dot{\boldsymbol{\varepsilon}}_{i1}\boldsymbol{e}_{i1}\Big]\end{aligned} \tag{8.20}$$

where $\boldsymbol{W}_{i1}^*$ is the optimal weight matrix and $\boldsymbol{\varphi}_{i1}$ is the Gaussian function vector. $\boldsymbol{D}_{i1} = \Gamma_1^{-1}\boldsymbol{\xi}_{i1}$ and $\boldsymbol{\xi}_{i1}$ is the bounded RBFNN approximation error.

Assumption 8.2 The disturbance $\boldsymbol{D}_{i1}$ is unknown but with bounded variation, and there exists an unknown positive constant $\bar{D}_{i1m}$ such that

$$\dot{\boldsymbol{D}}_{i1}^T\dot{\boldsymbol{D}}_{i1} \leq \bar{D}_{i1m} \tag{8.21}$$

To move on, the prediction error is introduced as

$$\begin{cases}\boldsymbol{\Upsilon}_{i1} = \boldsymbol{x}_{i1} - \hat{\boldsymbol{x}}_{i1}\\ \dot{\hat{\boldsymbol{x}}}_{i1} = \Gamma_1^{-1}\hat{\boldsymbol{W}}_{i1}^T\boldsymbol{\varphi}_{i1} + \boldsymbol{g}_{i1}\boldsymbol{x}_{i2} + \hat{\boldsymbol{D}}_{i1} + k_1\boldsymbol{\Upsilon}_{i1}\end{cases} \tag{8.22}$$

where $\hat{\boldsymbol{W}}_{i1}$ and $\hat{\boldsymbol{D}}_{i1}$ are the estimations of $\boldsymbol{W}_{i1}^*$ and $\boldsymbol{D}_{i1}$, respectively. $k_1 > 0$ is a positive design parameter.

By utilizing the prediction error (8.22), the DO is designed as

$$\begin{cases} \hat{\boldsymbol{D}}_{i1} = \boldsymbol{\pi}_{i1} + k_2 \boldsymbol{x}_{i1} \\ \dot{\boldsymbol{\pi}}_{i1} = -k_2 \boldsymbol{\pi}_{i1} + \boldsymbol{\phi}_i \boldsymbol{\eta}_{i1} \boldsymbol{E}_{i1} - k_2 \Big[\Gamma_1^{-1} \hat{\boldsymbol{W}}_{i1}^T \boldsymbol{\varphi}_{i1} + \boldsymbol{g}_{i1} \boldsymbol{x}_{i2} \\ \quad + k_2 \boldsymbol{x}_{i1} - k_2^{-1} (k_3 \phi_i \boldsymbol{\Upsilon}_{i1} + \boldsymbol{E}_{i1}) \Big] \end{cases} \tag{8.23}$$

where k_2 and k_3 are positive design parameters.

Taking the time derivative of (8.23) yields

$$\dot{\hat{\boldsymbol{D}}}_{i1} = k_2 \tilde{\boldsymbol{D}}_{i1} + k_2 \Gamma_1^{-1} \tilde{\boldsymbol{W}}_{i1}^T \boldsymbol{\varphi}_{i1} + k_3 \phi_i \boldsymbol{\Upsilon}_{i1} + \boldsymbol{E}_{i1} + \phi_i \boldsymbol{\eta}_{i1} \boldsymbol{E}_{i1} \tag{8.24}$$

where $\tilde{\boldsymbol{W}}_{i1} = \boldsymbol{W}_{i1}^* - \hat{\boldsymbol{W}}_{i1}$ and $\tilde{\boldsymbol{D}}_{i1} = \boldsymbol{D}_{i1} - \hat{\boldsymbol{D}}_{i1}$ are the estimation errors.

Taking the time derivative of $\tilde{\boldsymbol{D}}_{i1}$ gives

$$\dot{\tilde{\boldsymbol{D}}}_{i1} = \dot{\boldsymbol{D}}_{i1} - k_2 \tilde{\boldsymbol{D}}_{i1} - k_2 \Gamma_1^{-1} \tilde{\boldsymbol{W}}_{i1}^T \boldsymbol{\varphi}_{i1} - k_3 \phi_i \boldsymbol{\Upsilon}_{i1} - \boldsymbol{E}_{i1} - \phi_i \boldsymbol{\eta}_{i1} \boldsymbol{E}_{i1} \tag{8.25}$$

Design the intermittent control signal and the adaptive law as

$$\begin{aligned} \bar{\boldsymbol{x}}_{i2d} = & (\phi_i \boldsymbol{g}_{i1})^{-1} \Big[- \phi_i \Gamma_1^{-1} \hat{\boldsymbol{W}}_{i1}^T \boldsymbol{\varphi}_{i1} - \phi_i \hat{\boldsymbol{D}}_{i1} + \phi_i \dot{\boldsymbol{x}}_{i1d} \\ & + \boldsymbol{\varepsilon}_{i1}^{-1} \dot{\boldsymbol{\varepsilon}}_{i1} \boldsymbol{e}_{i1} - \boldsymbol{\eta}_{i1}^{-1} \boldsymbol{K}_4 \boldsymbol{E}_{i1} + k_5 \boldsymbol{\mu}_{i1} + \lambda_2 \sum_{j=1}^{N_i} a_{ij} \dot{\bar{\boldsymbol{x}}}_{j1} \Big] \end{aligned} \tag{8.26}$$

$$\dot{\hat{\boldsymbol{W}}}_{i1} = k_6 \Big[\boldsymbol{\varphi}_{i1} \phi_i \Gamma_1^{-1} (\boldsymbol{\eta}_{i1}^T \boldsymbol{E}_{i1} + k_3 \boldsymbol{\Upsilon}_{i1})^T - k_7 \hat{\boldsymbol{W}}_{i1} \Big] \tag{8.27}$$

where $\boldsymbol{K}_4$ is a diagonal matrix with positive elements and k_5 is a positive parameter. $\boldsymbol{\mu}_{i1}$ is an auxiliary signal to be designed later. k_6 and k_7 are positive design parameters.

In this chapter, the DSC technique is utilized to facilitate the controller design, which uses the first-order filter technique to obtain the virtual control signal $\boldsymbol{x}_{i2d}$ and its time derivative. The first-order filter is given by

$$\tau_i \dot{\boldsymbol{x}}_{i2d} + \boldsymbol{x}_{i2d} = \bar{\boldsymbol{x}}_{i2d}, \ \boldsymbol{x}_{i2d}(0) = \bar{\boldsymbol{x}}_{i2d}(0) \tag{8.28}$$

where τ_i is a positive design parameter. $\boldsymbol{x}_{i2d}$ is the virtual control signal, and the corresponding time derivative can be calculated as $\dot{\boldsymbol{x}}_{i2d} = (\bar{\boldsymbol{x}}_{i2d} - \boldsymbol{x}_{i2d})/\tau_i$.

To compensate for the filter error $\boldsymbol{\zeta}_{i1} = \boldsymbol{x}_{i2d} - \bar{\boldsymbol{x}}_{i2d}$, an auxiliary system, inspired by the work in [1], is designed as

$$\dot{\boldsymbol{\mu}}_{i1} = \begin{cases} -k_5\boldsymbol{\mu}_{i1} - \boldsymbol{\mu}_{i1}\Big[\phi_i(\boldsymbol{\mu}_{i1}^T\boldsymbol{\mu}_{i1})^{-1}\boldsymbol{E}_{i1}^T\boldsymbol{\eta}_{i1}\boldsymbol{g}_{i1}\boldsymbol{\zeta}_{i1} \\ \qquad +\frac{1}{2}(\boldsymbol{\mu}_{i1}^T\boldsymbol{\mu}_{i1})^{-1}\boldsymbol{\zeta}_{i1}^T\boldsymbol{\zeta}_{i1}\Big] + \boldsymbol{\zeta}_{i1} & \boldsymbol{\mu}_{i1}^T\boldsymbol{\mu}_{i1} \geq \mu_{i1b} \\ 0 & \boldsymbol{\mu}_{i1}^T\boldsymbol{\mu}_{i1} < \mu_{i1b} \end{cases} \tag{8.29}$$

where $\boldsymbol{\mu}_{i1} = [\mu_{i11}, \mu_{i12}, \mu_{i13}]^T$ and μ_{i1b} is a positive constant to be determined by the designer.

By using the auxiliary system (8.29) and following a similar analytical procedure to the one in [2], when $\boldsymbol{\mu}_{i1}^T\boldsymbol{\mu}_{i1} \geq \mu_{i1b}$, one has

$$\boldsymbol{\mu}_{i1}^T\dot{\boldsymbol{\mu}}_{i1} \leq -k_5\boldsymbol{\mu}_{i1}^T\boldsymbol{\mu}_{i1} - \phi_i\boldsymbol{E}_{i1}^T\boldsymbol{\eta}_{i1}\boldsymbol{g}_{i1}\boldsymbol{\zeta}_{i1} + \frac{1}{2}\boldsymbol{\mu}_{i1}^T\boldsymbol{\mu}_{i1} \tag{8.30}$$

When $\boldsymbol{\mu}_{i1}^T\boldsymbol{\mu}_{i1} < \mu_{i1b}$, one has

$$\boldsymbol{\mu}_{i1}^T\dot{\boldsymbol{\mu}}_{i1} = 0 \tag{8.31}$$

By defining the angular rate tracking error as $\boldsymbol{e}_{i2} = \boldsymbol{x}_{i2} - \boldsymbol{x}_{i2d}$ and using RBFNN to approximate $\Gamma_2\boldsymbol{F}_{i2b}$ with Γ_2 being a positive parameter, the time derivative of $\boldsymbol{e}_{i2}$ is given by

$$\begin{aligned} \dot{\boldsymbol{e}}_{i2} &= \dot{\boldsymbol{x}}_{i2} - \dot{\boldsymbol{x}}_{i2d} \\ &= \boldsymbol{F}_{i2b} + \boldsymbol{g}_{i2N}\boldsymbol{u}_{i0} + \boldsymbol{d}_i + \boldsymbol{\epsilon}_i - \dot{\boldsymbol{x}}_{i2d} \\ &= \Gamma_2^{-1}\boldsymbol{W}_{i2}^{*T}\boldsymbol{\varphi}_{i2} + \boldsymbol{g}_{i2N}\boldsymbol{u}_{i0} + \boldsymbol{D}_{i2} - \dot{\boldsymbol{x}}_{i2d} \end{aligned} \tag{8.32}$$

where $\boldsymbol{W}_{i2}^*$ is the optimal weight matrix and $\boldsymbol{\varphi}_{i2}$ is the Gaussian function vector. $\boldsymbol{D}_{i2} = \Gamma_2^{-1}\boldsymbol{\xi}_{i2} + \boldsymbol{\epsilon}_i + \boldsymbol{d}_i$ with $\boldsymbol{\xi}_{i2}$ being the RBFNN approximation error.

Assumption 8.3 For the disturbance $\boldsymbol{D}_{i2}$, there exists an unknown positive constant $\bar{D}_{i2m}$ such that

$$\dot{\boldsymbol{D}}_{i2}^T\dot{\boldsymbol{D}}_{i2} \leq \bar{D}_{i2m} \tag{8.33}$$

The prediction error under the RBFNN and DO is constructed as

$$\begin{cases} \boldsymbol{\Upsilon}_{i2} = \boldsymbol{x}_{i2} - \hat{\boldsymbol{x}}_{i2} \\ \dot{\hat{\boldsymbol{x}}}_{i2} = \Gamma_2^{-1}\hat{\boldsymbol{W}}_{i2}^T\boldsymbol{\varphi}_{i2} + \boldsymbol{g}_{i2N}\boldsymbol{u}_{i0} + \hat{\boldsymbol{D}}_{i2} + k_8\boldsymbol{\Upsilon}_{i2} \end{cases} \tag{8.34}$$

where $\hat{\boldsymbol{W}}_{i2}$ and $\hat{\boldsymbol{D}}_{i2}$ are the estimations of $\boldsymbol{W}_{i2}^*$ and $\boldsymbol{D}_{i2}$, respectively. k_8 is a positive design parameter.

Then, to estimate $\boldsymbol{D}_{i2}$, the DO is designed as

$$\begin{cases} \hat{\boldsymbol{D}}_{i2} = \boldsymbol{\pi}_{i2} + k_9 \boldsymbol{x}_{i2} \\ \dot{\boldsymbol{\pi}}_{i2} = -k_9 \boldsymbol{\pi}_{i2} - k_9 \Big[\Gamma_2^{-1} \hat{\boldsymbol{W}}_{i2}^T \boldsymbol{\varphi}_{i2} + \boldsymbol{g}_{i2N} \boldsymbol{u}_{i0} + k_9 \boldsymbol{x}_{i2} - k_9^{-1} (k_{10} \boldsymbol{\Upsilon}_{i2} + \boldsymbol{e}_{i2}) \Big] \end{cases} \tag{8.35}$$

where k_9 and k_{10} are positive design parameters.

By taking the time derivative of (8.35), one has

$$\begin{aligned} \dot{\hat{\boldsymbol{D}}}_{i2} &= \dot{\boldsymbol{\pi}}_{i2} + k_9 \dot{\boldsymbol{x}}_{i2} \\ &= k_9 \tilde{\boldsymbol{D}}_{i2} + k_9 \Gamma_2^{-1} \tilde{\boldsymbol{W}}_{i2}^T \boldsymbol{\varphi}_{i2} + k_{10} \boldsymbol{\Upsilon}_{i2} + \boldsymbol{e}_{i2} \end{aligned} \tag{8.36}$$

where $\tilde{\boldsymbol{W}}_{i2} = \boldsymbol{W}_{i2}^* - \hat{\boldsymbol{W}}_{i2}$ and $\tilde{\boldsymbol{D}}_{i2} = \boldsymbol{D}_{i2} - \hat{\boldsymbol{D}}_{i2}$ are the estimation errors.

Taking the time derivative of $\tilde{\boldsymbol{D}}_{i2}$ yields

$$\dot{\tilde{\boldsymbol{D}}}_{i2} = \dot{\boldsymbol{D}}_{i2} - k_9 \tilde{\boldsymbol{D}} - k_9 \Gamma_2^{-1} \tilde{\boldsymbol{W}}_{i2}^T \boldsymbol{\varphi}_{i2} - k_{10} \boldsymbol{\Upsilon}_{i2} - \boldsymbol{e}_{i2} \tag{8.37}$$

Design the control input signal and adaptive law as

$$\bar{\boldsymbol{u}}_{i0} = \boldsymbol{g}_{i2N}^{-1} \Big(-\Gamma_2^{-1} \hat{\boldsymbol{W}}_{i2}^T \boldsymbol{\varphi}_{i2} - \hat{\boldsymbol{D}}_{i2} + \dot{\boldsymbol{x}}_{i2d} - \boldsymbol{K}_{11} \boldsymbol{e}_{i2} + k_{12} \boldsymbol{\mu}_{i2} - \phi_i \boldsymbol{g}_{i1}^T \boldsymbol{\eta}_{i1} \boldsymbol{E}_{i1} \Big) \tag{8.38}$$

$$\dot{\hat{\boldsymbol{W}}}_{i2} = k_{13} \Big[\boldsymbol{\varphi}_{i2} \Gamma_2^{-1} (\boldsymbol{e}_{i2} + k_{10} \boldsymbol{\Upsilon}_{i2})^T - k_{14} \hat{\boldsymbol{W}}_{i2} \Big] \tag{8.39}$$

where $\boldsymbol{K}_{11}$ is a diagonal matrix with positive elements. k_{12}, k_{13}, and k_{14} are positive design parameters.

In practical engineering applications, actuators often encounter input saturation due to physical limitations, which is given by

$$\boldsymbol{u}_{i0} = \begin{cases} \boldsymbol{u}_{i0\max}, & \bar{\boldsymbol{u}}_{i0} \geq \boldsymbol{u}_{i0\max} \\ \bar{\boldsymbol{u}}_{i0}, & \boldsymbol{u}_{i0\min} < \bar{\boldsymbol{u}}_{i0} < \boldsymbol{u}_{i0\max} \\ \boldsymbol{u}_{i0\min}, & \bar{\boldsymbol{u}}_{i0} \leq \boldsymbol{u}_{i0\min} \end{cases} \tag{8.40}$$

where $\boldsymbol{u}_{i0\max}$ and $\boldsymbol{u}_{i0\min}$ are the maximum and minimum allowable values, respectively.

Inspired by the work in [1], to avoid persistent actuator saturation, the following auxiliary system is introduced as

$$\dot{\boldsymbol{\mu}}_{i2} = \begin{cases} -k_{12}\boldsymbol{\mu}_{i2} - \boldsymbol{\mu}_{i2}\Big[(\boldsymbol{\mu}_{i2}^T\boldsymbol{\mu}_{i2})^{-1}\boldsymbol{e}_{i2}^T\boldsymbol{g}_{i2N}\boldsymbol{\zeta}_{i2} \\ \qquad + \frac{1}{2}(\boldsymbol{\mu}_{i2}^T\boldsymbol{\mu}_{i2})^{-1}\boldsymbol{\zeta}_{i2}^T\boldsymbol{\zeta}_{i2}\Big] + \boldsymbol{\zeta}_{i2} & \boldsymbol{\mu}_{i2}^T\boldsymbol{\mu}_{i2} \geq \mu_{i2b} \\ 0 & \boldsymbol{\mu}_{i2}^T\boldsymbol{\mu}_{i2} < \mu_{i2b} \end{cases} \tag{8.41}$$

where $\boldsymbol{\mu}_{i2} = [\mu_{i21}, \mu_{i22}, \mu_{i23}]^T$. μ_{i2b} is a positive constant to be chosen by the designer. $\boldsymbol{\zeta}_{i2} = \boldsymbol{u}_{i0} - \bar{\boldsymbol{u}}_{i0}$.

By utilizing the auxiliary system (8.41) and a similar analytical procedure to the one in [2], regarding $\boldsymbol{\mu}_{i2}^T\boldsymbol{\mu}_{i2} \geq \mu_{i2b}$, one has

$$\boldsymbol{\mu}_{i2}^T\dot{\boldsymbol{\mu}}_{i2} \leq -k_{12}\boldsymbol{\mu}_{i2}^T\boldsymbol{\mu}_{i2} - \boldsymbol{e}_{i2}^T\boldsymbol{g}_{i2N}\boldsymbol{\zeta}_{i2} + \frac{1}{2}\boldsymbol{\mu}_{i2}^T\boldsymbol{\mu}_{i2} \tag{8.42}$$

When $\boldsymbol{\mu}_{i2}^T\boldsymbol{\mu}_{i2} < \mu_{i2b}$, one has

$$\boldsymbol{\mu}_{i2}^T\dot{\boldsymbol{\mu}}_{i2} = 0 \tag{8.43}$$

Remark 8.1 From the auxiliary system (8.29), it is observed that if the filter error $\boldsymbol{\zeta}_{i1} = 0$, one has $\dot{\boldsymbol{\mu}}_{i1} = -k_5\boldsymbol{\mu}_{i1}$, which shows that $\boldsymbol{\mu}_{i1}$ converges into the set $\boldsymbol{\mu}_{i1}^T\boldsymbol{\mu}_{i1} < \mu_{i1b}$. When $\boldsymbol{\mu}_{i1}$ is in the set $\boldsymbol{\mu}_{i1}^T\boldsymbol{\mu}_{i1} < \mu_{i1b}$, one can reset $\boldsymbol{\mu}_{i1}$ to $\boldsymbol{\mu}_{i1}^T\boldsymbol{\mu}_{i1} \geq \mu_{i1b}$ if saturation occurs, i.e., $\boldsymbol{\zeta}_{i1} \neq 0$. Then, the auxiliary system (8.29) will be reactivated to compensate for the error. Such a procedure can also be used to handle the auxiliary system (8.41) if saturation occurs ($\boldsymbol{\zeta}_{i2} \neq 0$) when $\boldsymbol{\mu}_{i2}$ is in the set $\boldsymbol{\mu}_{i2}^T\boldsymbol{\mu}_{i2} < \mu_{i2b}$.

Remark 8.2 In the presented control scheme, the attitude synchronization tracking errors (8.10) are first transformed into a new set (8.13) by using a smooth and strictly increasing function $\Theta(\cdot)$. Then, based on the transformed attitude error dynamics (8.19) and the angular rate tracking error dynamics (8.32), the NNs with updating laws (8.27), (8.39) and DOs (8.23), (8.35) are integrated to approximate the unknown nonlinear functions due to the actuator faults and inherent nonlinearities in the UAVs. To enhance the approximation ability, the prediction errors (8.22), (8.34) are introduced into the approximators. By utilizing the auxiliary systems (8.29) and (8.41) to deal with the filter error $\boldsymbol{\zeta}_{i1}$ and actuator saturation (8.40), respectively, and integrating the approximated nonlinear functions, the overall controller can be developed as (8.26), (8.38).

8.3.3 Stability Analysis

Theorem 8.1 *Consider a group of UAVs described by (8.1)–(8.2) with Assumptions 8.1, 8.2, and 8.3 satisfied, if the control laws are chosen as (8.26), (8.38), the prediction errors are constructed as (8.22), (8.34), the DOs are designed as (8.23), (8.35), the adaptive laws are developed as (8.27), (8.39), and the auxiliary systems are constructed as (8.29), (8.41), then the attitudes of all UAVs can track their attitude references in a synchronized behavior even in the presence of actuator faults (8.3) and the attitude synchronization tracking errors are strictly confined within the prescribed performance bounds (8.12) and all signals in the closed-loop system are bounded.*

Proof Choose the Lyapunov function candidate as

$$L = \sum_{i=1}^{N} L_i = \sum_{i=1}^{N} \left(\sum_{j=1}^{5} L_{i1j} + \sum_{j=1}^{5} L_{i2j} \right) \tag{8.44}$$

where $L_{i11} = \frac{1}{2}\tilde{\boldsymbol{D}}_{i1}^T \tilde{\boldsymbol{D}}_{i1}$, $L_{i12} = \frac{1}{2}\boldsymbol{E}_{i1}^T \boldsymbol{E}_{i1}$, $L_{i13} = \frac{1}{2k_6}\mathrm{tr}(\tilde{\boldsymbol{W}}_{i1}^T \tilde{\boldsymbol{W}}_{i1})$, $L_{i14} = \frac{1}{2}k_3\phi_i \boldsymbol{\Upsilon}_{i1}^T \boldsymbol{\Upsilon}_{i1}$, $L_{i15} = \frac{1}{2}\boldsymbol{\mu}_{i1}^T \boldsymbol{\mu}_{i1}$, $L_{i21} = \frac{1}{2}\tilde{\boldsymbol{D}}_{i2}^T \tilde{\boldsymbol{D}}_{i2}$, $L_{i22} = \frac{1}{2}\boldsymbol{e}_{i2}^T \boldsymbol{e}_{i2}$, $L_{i23} = \frac{1}{2k_{13}}\mathrm{tr}(\tilde{\boldsymbol{W}}_{i2}^T \tilde{\boldsymbol{W}}_{i2})$, $L_{i24} = \frac{1}{2}k_{10}\boldsymbol{\Upsilon}_{i2}^T \boldsymbol{\Upsilon}_{i2}$, $L_{i25} = \frac{1}{2}\boldsymbol{\mu}_{i2}^T \boldsymbol{\mu}_{i2}$.

Taking the time derivative of L_{i11} along with (8.25) gives

$$\begin{aligned} \dot{L}_{i11} =& \tilde{\boldsymbol{D}}_{i1}^T \dot{\boldsymbol{D}}_{i1} - k_2 \tilde{\boldsymbol{D}}_{i1}^T \tilde{\boldsymbol{D}}_{i1} - k_2 \Gamma_1^{-1} \tilde{\boldsymbol{D}}_{i1}^T \tilde{\boldsymbol{W}}_{i1}^T \boldsymbol{\varphi}_{i1} \\ & - k_3\phi_i \tilde{\boldsymbol{D}}_{i1}^T \boldsymbol{\Upsilon}_{i1} - \tilde{\boldsymbol{D}}_{i1}^T \boldsymbol{E}_{i1} - \phi_i \tilde{\boldsymbol{D}}_{i1}^T \boldsymbol{\eta}_{i1} \boldsymbol{E}_{i1} \end{aligned} \tag{8.45}$$

By substituting (8.26) into (8.20), one has

$$\begin{aligned} \dot{L}_{i12} =& \phi_i \Gamma_1^{-1} \boldsymbol{E}_{i1}^T \boldsymbol{\eta}_{i1} \tilde{\boldsymbol{W}}_{i1}^T \boldsymbol{\varphi}_{i1} + \phi_i \boldsymbol{E}_{i1}^T \boldsymbol{\eta}_{i1} \tilde{\boldsymbol{D}}_{i1} - \boldsymbol{E}_{i1}^T \boldsymbol{K}_4 \boldsymbol{E}_{i1} \\ & + k_5 \boldsymbol{E}_{i1}^T \boldsymbol{\mu}_{i1} + \phi_i \boldsymbol{E}_{i1}^T \boldsymbol{\eta}_{i1} \boldsymbol{g}_{i1} \boldsymbol{\zeta}_{i1} + \phi_i \boldsymbol{E}_{i1}^T \boldsymbol{\eta}_{i1} \boldsymbol{g}_{i1} \boldsymbol{e}_{i2} \end{aligned} \tag{8.46}$$

In view of the developed adaptive law (8.27), the time derivative of L_{i13} is given by

$$\begin{aligned} \dot{L}_{i13} =& -\phi_i \Gamma_1^{-1} \boldsymbol{E}_{i1}^T \boldsymbol{\eta}_{i1} \tilde{\boldsymbol{W}}_{i1}^T \boldsymbol{\varphi}_{i1} - \phi_i \Gamma_1^{-1} k_3 \boldsymbol{\Upsilon}_{i1}^T \tilde{\boldsymbol{W}}_{i1}^T \boldsymbol{\varphi}_{i1} \\ & - k_7 \mathrm{tr}(\tilde{\boldsymbol{W}}_{i1}^T \tilde{\boldsymbol{W}}_{i1}) + k_7 \mathrm{tr}\left(\tilde{\boldsymbol{W}}_{i1}^T \boldsymbol{W}_{i1}^*\right) \end{aligned} \tag{8.47}$$

By taking the time derivative of L_{i14}, one has

$$\dot{L}_{i14} = k_3\phi_i \Gamma_1^{-1} \boldsymbol{\Upsilon}_{i1}^T \tilde{\boldsymbol{W}}_{i1}^T \boldsymbol{\varphi}_{i1} + k_3\phi_i \boldsymbol{\Upsilon}_{i1}^T \tilde{\boldsymbol{D}}_{i1} - k_1 k_3 \phi_i \boldsymbol{\Upsilon}_{i1}^T \boldsymbol{\Upsilon}_{i1} \tag{8.48}$$

Therefore, by considering (8.45)–(8.48), the time derivative of $L_{i1} = \sum_{j=1}^{5} L_{i1j}$ has

$$
\begin{aligned}
\dot{L}_{i1} \leq & \frac{\tilde{\boldsymbol{D}}_{i1}^T \tilde{\boldsymbol{D}}_{i1}}{2h_{11}^2} + \frac{h_{11}^2 \bar{D}_{i1m}}{2} - k_2 \tilde{\boldsymbol{D}}_{i1}^T \tilde{\boldsymbol{D}}_{i1} + \frac{k_2 \Gamma_1^{-1} \tilde{\boldsymbol{D}}_{i1}^T \tilde{\boldsymbol{D}}_{i1}}{2h_{12}^2} \\
& + \frac{k_2 \Gamma_1^{-1} h_{12}^2 \varphi_{i1m}}{2} \operatorname{tr}\left(\tilde{\boldsymbol{W}}_{i1}^T \tilde{\boldsymbol{W}}_{i1}\right) + \frac{\tilde{\boldsymbol{D}}_{i1}^T \tilde{\boldsymbol{D}}_{i1}}{2h_{13}^2} + \frac{h_{13}^2 \boldsymbol{E}_{i1}^T \boldsymbol{E}_{i1}}{2} - \boldsymbol{E}_{i1}^T \boldsymbol{K}_4 \boldsymbol{E}_{i1} \\
& + k_5 \boldsymbol{E}_{i1}^T \boldsymbol{\mu}_{i1} + \phi_i \boldsymbol{E}_{i1}^T \boldsymbol{\eta}_{i1} \boldsymbol{g}_{i1} \boldsymbol{\zeta}_{i1} + \phi_i \boldsymbol{E}_{i1}^T \boldsymbol{\eta}_{i1} \boldsymbol{g}_{i1} \boldsymbol{e}_{i2} \\
& - \frac{k_7}{2} \operatorname{tr}\left(\tilde{\boldsymbol{W}}_{i1}^T \tilde{\boldsymbol{W}}_{i1}\right) + \frac{k_7}{2} \operatorname{tr}\left(\boldsymbol{W}_{i1}^{*T} \boldsymbol{W}_{i1}^*\right) - k_1 k_3 \phi_i \boldsymbol{\Upsilon}_{i1}^T \boldsymbol{\Upsilon}_{i1} + \boldsymbol{\mu}_{i1}^T \dot{\boldsymbol{\mu}}_{i1}
\end{aligned}
\tag{8.49}
$$

It should be stressed that the following inequalities are used in (8.49):

$$
\begin{cases}
\operatorname{tr}\left(\tilde{\boldsymbol{W}}_{i1}^T \boldsymbol{W}_{i1}^*\right) \leq \frac{1}{2} \operatorname{tr}\left(\tilde{\boldsymbol{W}}_{i1}^T \tilde{\boldsymbol{W}}_{i1}\right) + \frac{1}{2} \operatorname{tr}\left(\boldsymbol{W}_{i1}^{*T} \boldsymbol{W}_{i1}^*\right) \\
\tilde{\boldsymbol{D}}_{i1}^T \dot{\boldsymbol{D}}_{i1} \leq \frac{\tilde{\boldsymbol{D}}_{i1}^T \tilde{\boldsymbol{D}}_{i1}}{2h_{11}^2} + \frac{h_{11}^2 \bar{D}_{i1m}}{2} \\
-\tilde{\boldsymbol{D}}_{i1}^T \tilde{\boldsymbol{W}}_{i1}^T \boldsymbol{\varphi}_{i1} \leq \frac{\tilde{\boldsymbol{D}}_{i1}^T \tilde{\boldsymbol{D}}_{i1}}{2h_{12}^2} + \frac{h_{12}^2 \varphi_{i1m}}{2} \operatorname{tr}\left(\tilde{\boldsymbol{W}}_{i1}^T \tilde{\boldsymbol{W}}_{i1}\right) \\
-\tilde{\boldsymbol{D}}_{i1}^T \boldsymbol{E}_{i1} \leq \frac{\tilde{\boldsymbol{D}}_{i1}^T \tilde{\boldsymbol{D}}_{i1}}{2h_{13}^2} + \frac{h_{13}^2}{2} \boldsymbol{E}_{i1}^T \boldsymbol{E}_{i1}
\end{cases}
$$

where $||\boldsymbol{\varphi}_{i1}||^2 \leq \varphi_{i1m}$. h_{11}, h_{12}, and h_{13} are positive constants.

When $\boldsymbol{\mu}_{i1}^T \boldsymbol{\mu}_{i1} \geq \mu_{i1b}$, one has

$$
\begin{aligned}
\dot{L}_{i1} \leq & -\left(k_2 - \frac{1}{2h_{11}^2} - \frac{k_2 \Gamma_1^{-1}}{2h_{12}^2} - \frac{1}{2h_{13}^2}\right) \tilde{\boldsymbol{D}}_{i1}^T \tilde{\boldsymbol{D}}_{i1} \\
& - \left[\lambda_{\min}(\boldsymbol{K}_4) - \frac{h_{13}^2}{2} - \frac{k_5}{2}\right] \boldsymbol{E}_{i1}^T \boldsymbol{E}_{i1} + \phi_i \boldsymbol{E}_{i1}^T \boldsymbol{\eta}_{i1} \boldsymbol{g}_{i1} \boldsymbol{e}_{i2} \\
& - \left(\frac{k_7}{2} - \frac{k_2 \Gamma_1^{-1} h_{12}^2 \varphi_{i1m}}{2}\right) \operatorname{tr}\left(\tilde{\boldsymbol{W}}_{i1}^T \tilde{\boldsymbol{W}}_{i1}\right) - k_1 k_3 \phi_i \boldsymbol{\Upsilon}_{i1}^T \boldsymbol{\Upsilon}_{i1} \\
& - \left(\frac{k_5}{2} - \frac{1}{2}\right) \boldsymbol{\mu}_{i1}^T \boldsymbol{\mu}_{i1} + \frac{k_7}{2} \operatorname{tr}\left(\boldsymbol{W}_{i1}^{*T} \boldsymbol{W}_{i1}^*\right) + \frac{h_{11}^2 \bar{D}_{i1m}}{2}
\end{aligned}
\tag{8.50}
$$

When $\boldsymbol{\mu}_{i1}^T\boldsymbol{\mu}_{i1} < \mu_{i1b}$, one has

$$\begin{aligned}\dot{L}_{i1} \leq & -\left(k_2 - \frac{1}{2h_{11}^2} - \frac{k_2\Gamma_1^{-1}}{2h_{12}^2} - \frac{1}{2h_{13}^2}\right)\tilde{\boldsymbol{D}}_{i1}^T\tilde{\boldsymbol{D}}_{i1} \\ & - \left[\lambda_{\min}(\boldsymbol{K}_4) - \frac{h_{13}^2}{2} - \frac{k_5}{2} - \frac{\phi_i}{2}\right]\boldsymbol{E}_{i1}^T\boldsymbol{E}_{i1} \\ & + \phi_i\boldsymbol{E}_{i1}^T\boldsymbol{\eta}_{i1}\boldsymbol{g}_{i1}\boldsymbol{e}_{i2} - \left(\frac{k_7}{2} - \frac{k_2\Gamma_1^{-1}h_{12}^2\varphi_{i1m}}{2}\right)\operatorname{tr}\left(\tilde{\boldsymbol{W}}_{i1}^T\tilde{\boldsymbol{W}}_{i1}\right) \\ & - k_1k_3\phi_i\boldsymbol{\Upsilon}_{i1}^T\boldsymbol{\Upsilon}_{i1} + \frac{k_7}{2}\operatorname{tr}\left(\boldsymbol{W}_{i1}^{*T}\boldsymbol{W}_{i1}^*\right) + \frac{h_{11}^2\bar{D}_{i1m}}{2} \\ & + \frac{1}{2}\phi_i(\boldsymbol{\eta}_{i1}\boldsymbol{g}_{i1}\boldsymbol{\zeta}_{i1})^T(\boldsymbol{\eta}_{i1}\boldsymbol{g}_{i1}\boldsymbol{\zeta}_{i1}) + \frac{k_5}{2}\mu_{i1b}\end{aligned} \tag{8.51}$$

Similar to the analytical procedure in [2], synthesizing (8.50) and (8.51) yields

$$\begin{aligned}\dot{L}_{i1} \leq & -\left(k_2 - \frac{1}{2h_{11}^2} - \frac{k_2\Gamma_1^{-1}}{2h_{12}^2} - \frac{1}{2h_{13}^2}\right)\tilde{\boldsymbol{D}}_{i1}^T\tilde{\boldsymbol{D}}_{i1} \\ & - \left[\lambda_{\min}(\boldsymbol{K}_4) - \frac{h_{13}^2}{2} - \frac{k_5}{2} - \frac{\phi_i}{2}\right]\boldsymbol{E}_{i1}^T\boldsymbol{E}_{i1} \\ & + \phi_i\boldsymbol{E}_{i1}^T\boldsymbol{\eta}_{i1}\boldsymbol{g}_{i1}\boldsymbol{e}_{i2} - \left(\frac{k_7}{2} - \frac{k_2\Gamma_1^{-1}h_{12}^2\varphi_{i1m}}{2}\right)\operatorname{tr}\left(\tilde{\boldsymbol{W}}_{i1}^T\tilde{\boldsymbol{W}}_{i1}\right) \\ & - k_1k_3\phi_i\boldsymbol{\Upsilon}_{i1}^T\boldsymbol{\Upsilon}_{i1} - \left(\frac{k_5}{2} - \frac{1}{2}\right)\boldsymbol{\mu}_{i1}^T\boldsymbol{\mu}_{i1} + \frac{k_7}{2}\operatorname{tr}\left(\boldsymbol{W}_{i1}^{*T}\boldsymbol{W}_{i1}^*\right) \\ & + \frac{h_{11}^2\bar{D}_{i1m}}{2} + \frac{1}{2}\phi_i(\boldsymbol{\eta}_{i1}\boldsymbol{g}_{i1}\boldsymbol{\zeta}_{i1})^T(\boldsymbol{\eta}_{i1}\boldsymbol{g}_{i1}\boldsymbol{\zeta}_{i1}) + \frac{k_5}{2}\mu_{i1b}\end{aligned} \tag{8.52}$$

Next, by recalling (8.37), the time derivative of L_{i21} is given by

$$\dot{L}_{i21} = \tilde{\boldsymbol{D}}_{i2}^T\dot{\boldsymbol{D}}_{i2} - k_9\tilde{\boldsymbol{D}}_{i2}^T\tilde{\boldsymbol{D}}_{i2} - k_9\Gamma_2^{-1}\tilde{\boldsymbol{D}}_{i2}^T\tilde{\boldsymbol{W}}_{i2}^T\boldsymbol{\varphi}_{i2} - k_{10}\tilde{\boldsymbol{D}}_{i2}^T\boldsymbol{\Upsilon}_{i2} - \tilde{\boldsymbol{D}}_{i2}^T\boldsymbol{e}_{i2} \tag{8.53}$$

Taking the time derivative of L_{i22} yields

$$\begin{aligned}\dot{L}_{i22} = & \Gamma_2^{-1}\boldsymbol{e}_{i2}^T\tilde{\boldsymbol{W}}_{i2}^T\boldsymbol{\varphi}_{i2} + \boldsymbol{e}_{i2}^T\tilde{\boldsymbol{D}}_{i2} + \boldsymbol{e}_{i2}^T\boldsymbol{g}_{i2N}\boldsymbol{\zeta}_{i2} \\ & - \boldsymbol{e}_{i2}^T\boldsymbol{K}_{11}\boldsymbol{e}_{i2} + k_{12}\boldsymbol{e}_{i2}^T\boldsymbol{\mu}_{i2} - \phi_i\boldsymbol{e}_{i2}^T\boldsymbol{g}_{i1}^T\boldsymbol{\eta}_{i1}\boldsymbol{E}_{i1}\end{aligned} \tag{8.54}$$

In view of (8.39), the time derivative of L_{i23} is calculated as

$$\begin{aligned}\dot{L}_{i23} = &-\Gamma_2^{-1}\boldsymbol{e}_{i2}^T\tilde{\boldsymbol{W}}_{i2}^T\boldsymbol{\varphi}_{i2} - \Gamma_2^{-1}k_{10}\boldsymbol{\Upsilon}_{i2}^T\tilde{\boldsymbol{W}}_{i2}^T\boldsymbol{\varphi}_{i2} \\ &- k_{14}\mathrm{tr}\left(\tilde{\boldsymbol{W}}_{i2}^T\tilde{\boldsymbol{W}}_{i2}\right) + k_{14}\mathrm{tr}\left(\tilde{\boldsymbol{W}}_{i2}^T\boldsymbol{W}_{i2}^*\right)\end{aligned} \tag{8.55}$$

Taking the time derivative of L_{i24} yields

$$\dot{L}_{i24} = k_{10}\Gamma_2^{-1}\boldsymbol{\Upsilon}_{i2}^T\tilde{\boldsymbol{W}}_{i2}^T\boldsymbol{\varphi}_{i2} + k_{10}\boldsymbol{\Upsilon}_{i2}^T\tilde{\boldsymbol{D}}_{i2} - k_8k_{10}\boldsymbol{\Upsilon}_{i2}^T\boldsymbol{\Upsilon}_{i2} \tag{8.56}$$

In the subsequent analysis, the following inequalities are used:

$$\begin{cases}\mathrm{tr}\left(\tilde{\boldsymbol{W}}_{i2}^T\boldsymbol{W}_{i2}^*\right) \le \frac{1}{2}\mathrm{tr}\left(\tilde{\boldsymbol{W}}_{i2}^T\tilde{\boldsymbol{W}}_{i2}\right) + \frac{1}{2}\mathrm{tr}\left(\boldsymbol{W}_{i2}^{*T}\boldsymbol{W}_{i2}^*\right) \\ \tilde{\boldsymbol{D}}_{i2}^T\dot{\boldsymbol{D}}_{i2} \le \frac{\tilde{\boldsymbol{D}}_{i2}^T\tilde{\boldsymbol{D}}_{i2}}{2h_{21}^2} + \frac{h_{21}^2\bar{D}_{i2m}}{2} \\ -\tilde{\boldsymbol{D}}_{i2}^T\tilde{\boldsymbol{W}}_{i2}^T\boldsymbol{\varphi}_{i2} \le \frac{\tilde{\boldsymbol{D}}_{i2}^T\tilde{\boldsymbol{D}}_{i2}}{2h_{22}^2} + \frac{h_{22}^2\varphi_{i2m}}{2}\mathrm{tr}\left(\tilde{\boldsymbol{W}}_{i2}^T\tilde{\boldsymbol{W}}_{i2}\right)\end{cases}$$

where $||\boldsymbol{\varphi}_{i2}||^2 \le \varphi_{i2m}$. h_{21} and h_{22} are positive constants.

By using the inequalities mentioned above and taking the time derivative of $L_{i2} = \sum_{j=1}^5 L_{i2j}$, one has

$$\begin{aligned}\dot{L}_{i2} \le &\frac{\tilde{\boldsymbol{D}}_{i2}^T\tilde{\boldsymbol{D}}_{i2}}{2h_{21}^2} + \frac{h_{21}^2\bar{D}_{i2m}}{2} - k_9\tilde{\boldsymbol{D}}_{i2}^T\tilde{\boldsymbol{D}}_{i2} + \frac{k_9\Gamma_2^{-1}\tilde{\boldsymbol{D}}_{i2}^T\tilde{\boldsymbol{D}}_{i2}}{2h_{22}^2} \\ &+ \frac{k_9\Gamma_2^{-1}h_{22}^2\varphi_{i2m}}{2}\mathrm{tr}\left(\tilde{\boldsymbol{W}}_{i2}^T\tilde{\boldsymbol{W}}_{i2}\right) + \boldsymbol{e}_{i2}^T\boldsymbol{g}_{i2N}\boldsymbol{\zeta}_{i2} - \boldsymbol{e}_{i2}^T\boldsymbol{K}_{11}\boldsymbol{e}_{i2} \\ &+ k_{12}\boldsymbol{e}_{i2}^T\boldsymbol{\mu}_{i2} - \phi_i\boldsymbol{e}_{i2}^T\boldsymbol{g}_{i1}^T\boldsymbol{\eta}_{i1}\boldsymbol{E}_{i1} - \frac{k_{14}}{2}\mathrm{tr}\left(\tilde{\boldsymbol{W}}_{i2}^T\tilde{\boldsymbol{W}}_{i2}\right) \\ &+ \frac{k_{14}}{2}\mathrm{tr}\left(\boldsymbol{W}_{i2}^{*T}\boldsymbol{W}_{i2}^*\right) - k_8k_{10}\boldsymbol{\Upsilon}_{i2}^T\boldsymbol{\Upsilon}_{i2} + \boldsymbol{\mu}_{i2}^T\dot{\boldsymbol{\mu}}_{i2}\end{aligned} \tag{8.57}$$

By recalling (8.42) and (8.43), and taking the procedure similar to (8.50), (8.51), and (8.52), one has

$$\begin{aligned}\dot{L}_{i2} \le &-\left(k_9 - \frac{1}{2h_{21}^2} - \frac{k_9\Gamma_2^{-1}}{2h_{22}^2}\right)\tilde{\boldsymbol{D}}_{i2}^T\tilde{\boldsymbol{D}}_{i2} - \left[\lambda_{\min}(\boldsymbol{K}_{11}) - \frac{k_{12}}{2} - \frac{1}{2}\right]\boldsymbol{e}_{i2}^T\boldsymbol{e}_{i2} \\ &-\left(\frac{k_{14}}{2} - \frac{h_{22}^2\varphi_{i2m}k_9\Gamma_2^{-1}}{2}\right)\mathrm{tr}\left(\tilde{\boldsymbol{W}}_{i2}^T\tilde{\boldsymbol{W}}_{i2}\right) - \phi_i\boldsymbol{e}_{i2}^T\boldsymbol{g}_{i1}^T\boldsymbol{\eta}_{i1}\boldsymbol{E}_{i1}\end{aligned}$$

$$
- k_8 k_{10} \boldsymbol{\Upsilon}_{i2}^T \boldsymbol{\Upsilon}_{i2} - \left(\frac{k_{12}}{2} - \frac{1}{2}\right) \boldsymbol{\mu}_{i2}^T \boldsymbol{\mu}_{i2} + \frac{1}{2} (\boldsymbol{g}_{i2N} \boldsymbol{\zeta}_{i2})^T (\boldsymbol{g}_{i2N} \boldsymbol{\zeta}_{i2})
$$

$$
+ \frac{k_{12}}{2} \mu_{i2b} + \frac{k_{14}}{2} \mathrm{tr}\left(\boldsymbol{W}_{i2}^{*T} \boldsymbol{W}_{i2}^*\right) + \frac{h_{21}^2 \bar{D}_{i2m}}{2} \tag{8.58}
$$

By combining (8.52) and (8.58), the time derivative of L_i is given by

$$
\dot{L}_i \leq -\ell_i L_i + \Psi_i \tag{8.59}
$$

where $\ell_i = \min\Big\{2\Big(k_2 - \frac{1}{2h_{11}^2} - \frac{k_2 \Gamma_1^{-1}}{2h_{12}^2} - \frac{1}{2h_{13}^2}\Big),\ 2\Big[\lambda_{\min}(\boldsymbol{K}_4) - \frac{h_{13}^2}{2} - \frac{k_5}{2} - \frac{\phi_i}{2}\Big],\ 2k_6\Big(\frac{k_7}{2} - \frac{k_2 \Gamma_1^{-1} h_{12}^2 \varphi_{i1m}}{2}\Big),\ 2k_1,\ k_5 - 1,\ 2\Big(k_9 - \frac{1}{2h_{21}^2} - \frac{k_9 \Gamma_2^{-1}}{2h_{22}^2}\Big),\ 2\Big[\lambda_{\min}(\boldsymbol{K}_{11}) - \frac{k_{12}}{2} - \frac{1}{2}\Big],\ 2k_{13}\Big(\frac{k_{14}}{2} - \frac{h_{22}^2 \varphi_{i2m} k_9 \Gamma_2^{-1}}{2}\Big),\ 2k_8,\ k_{12} - 1\Big\}$. $\Psi_i = \frac{k_7}{2} \mathrm{tr}(\boldsymbol{W}_{i1}^{*T} \boldsymbol{W}_{i1}^*) + \frac{h_{11}^2 \bar{D}_{i1m}}{2} + \frac{1}{2} \phi_i (\boldsymbol{\eta}_{i1} \boldsymbol{g}_{i1} \boldsymbol{\zeta}_{i1})^T (\boldsymbol{\eta}_{i1} \boldsymbol{g}_{i1} \boldsymbol{\zeta}_{i1}) + \frac{k_5}{2} \mu_{i1b} + \frac{1}{2} (\boldsymbol{g}_{i2N} \zeta_{i2})^T (\boldsymbol{g}_{i2N} \zeta_{i2}) + \frac{k_{12}}{2} \mu_{i2b} + \frac{k_{14}}{2} \mathrm{tr}(\boldsymbol{W}_{i2}^{*T} \boldsymbol{W}_{i2}^*) + \frac{h_{21}^2 \bar{D}_{i2m}}{2}$.

Then, taking the time derivative of (8.44) yields

$$
\dot{L} \leq -\ell L + \Psi \tag{8.60}
$$

where $\ell = \min\{\ell_i\}$, $\Psi = \sum_{k=1}^{N} \Psi_k$.

Furthermore, L can be calculated as

$$
0 \leq L \leq \frac{\Psi}{\ell} + \left[L(0) - \frac{\Psi}{\ell}\right] e^{-\ell t} \tag{8.61}
$$

From (8.61), it is known that $L \to \frac{\Psi}{\ell}$ as $t \to \infty$. Therefore, it can be concluded that all signals in the Lyapunov function (8.44) are UUB. According to (8.13), it can be seen that the synchronization tracking error $\boldsymbol{e}_{i1}$ is confined within the prescribed bounds since the error $\boldsymbol{E}_{i1}$ is UUB. By recalling (8.11), it is observed that the individual attitude tracking error $\tilde{\boldsymbol{x}}_{i1}$ is also UUB. This concludes the proof. □

8.4 Simulation Results

The proposed control scheme is verified on a network of four UAVs. The communication topology and the corresponding weights are illustrated in Fig. 8.1. In the simulations, the initial conditions of all UAVs are chosen as $\mu_i(0) = 1.719$ deg, $\alpha_i(0) = 2.865$ deg, $\beta_i(0) = 1.719$ deg, $p_i(0) = q_i(0) = r_i(0) = 0$ deg/s, $i = 1, 2, 3, 4$.

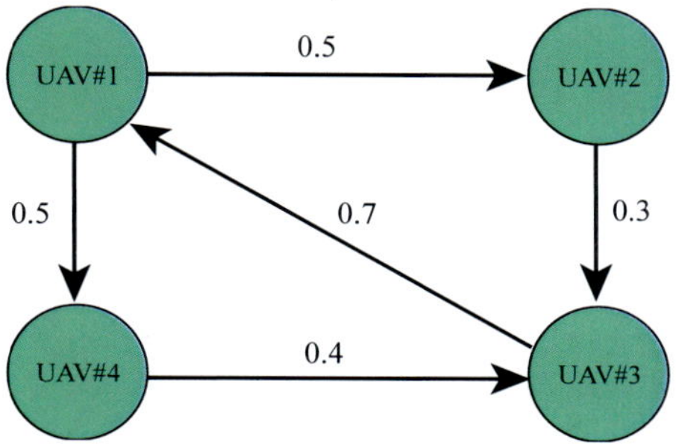

Fig. 8.1 Communication topology

In the simulation, the sideslip angle commands of four UAVs are chosen as 0 deg. The bank angle and angle of attack references μ_{id}, α_{id} can be obtained by shaping the bank angle command μ_{ic} and the angle of attack command α_{ic}, $i = 1, 2, 3, 4$, with a filter to generate the smooth reference commands, respectively. The bank angle and angle of attack commands are chosen as

$$\begin{cases} \mu_{ic} = 0^\circ, \ \alpha_{ic} = 2^\circ, & 0\ \text{s} \le t < 6\ \text{s} \\ \mu_{ic} = 10^\circ, \ \alpha_{ic} = 5^\circ, & 6\ \text{s} \le t \le 20\ \text{s} \end{cases} \tag{8.62}$$

where $i = 1, 2, 3, 4$, and the second-order filter is chosen as

$$\frac{\mu_{id}}{\mu_{ic}} = \frac{\alpha_{id}}{\alpha_{ic}} = \frac{\omega_d^2}{s^2 + 2\xi_d\omega_d s + \omega_d^2} \tag{8.63}$$

where $\omega_d = 0.8$, $\xi_d = 0.9$.

The input saturation upper and lower limits are assumed to be $\boldsymbol{u}_{i0\max} = [60^\circ, 60^\circ, 60^\circ]^T$ and $\boldsymbol{u}_{i0\min} = [-60^\circ, -60^\circ, -60^\circ]^T$, respectively. To verify the fault-tolerant capability of the proposed control scheme, it is assumed that UAV#1, UAV#2, and UAV#3 encounter aileron, elevator, and rudder actuator faults at $t = 3$ s, $t = 8$ s, and $t = 12$ s, respectively. Based on the fault model (8.3), the following faults are adopted in the simulation:

(1) UAV#1 aileron fault ($t = 3$ s):

$$\begin{cases} \rho_{11} = 1, \ u_{1f1} = 0^\circ, & 0\ \text{s} \le t < 3\ \text{s} \\ \rho_{11} = 0.7, \ u_{1f1} = 17.19^\circ, & 3\ \text{s} \le t \le 20\ \text{s} \end{cases} \tag{8.64}$$

(2) UAV#2 elevator fault ($t = 8$ s):

$$\begin{cases} \rho_{22} = 1, \ u_{2f2} = 0^\circ, & 0\ \text{s} \le t < 8\ \text{s} \\ \rho_{22} = 0.7, \ u_{2f2} = 8.595^\circ, & 8\ \text{s} \le t \le 20\ \text{s} \end{cases} \tag{8.65}$$

(3) UAV#3 rudder fault ($t = 12$ s):

$$\begin{cases} \rho_{33} = 1, \ u_{3f3} = 0^\circ, & 0\,\text{s} \le t < 12\,\text{s} \\ \rho_{33} = 0.5, \ u_{3f3} = 17.19^\circ, & 12\,\text{s} \le t \le 20\,\text{s} \end{cases} \tag{8.66}$$

The design parameters are chosen as $\lambda_1 = 2$, $\lambda_2 = 1.5$, $\underline{k}_{i1\nu} = 1$, $\varepsilon_{i1\nu 0} = 22.92^\circ$, $\varepsilon_{i1\nu\infty} = 8.595^\circ$, $\iota_{i1\nu} = 0.3$, $k_1 = 2$, $k_2 = 50$, $k_3 = 5$, $\boldsymbol{K}_4 = \text{diag}\{4,\ 12,\ 3.87\}$, $k_5 = 1.5$, $k_6 = 30$, $k_7 = 0.02$, $\tau_i = 0.05$, $k_8 = 3$, $k_9 = 3$, $k_{10} = 4$, $\boldsymbol{K}_{11} = \text{diag}\{15.7,\ 8,\ 7\}$, $k_{12} = 4$, $k_{13} = 30$, $k_{14} = 0.15$, $\Gamma_1 = 3$, $\Gamma_2 = 2$.

From Fig. 8.2, it is obviously observed that the bank angles, angles of attack, and sideslip angles of UAVs#1–4 can successfully track the references μ_{id}, α_{id}, β_{id}, respectively, even when UAV#1 encounters an aileron fault at t=3 s, UAV#2 encounters an elevator fault at t=8 s, and UAV#3 confronts a rudder fault at t=12 s, sequentially.

Figure 8.3 demonstrates the attitude tracking errors of four UAVs with respect to their individual attitude references μ_{id}, α_{id}, β_{id}, $i = 1,\ 2,\ 3,\ 4$. It can be seen that slightly worse transient performance is induced when the aileron of UAV#1 is subjected to an actuator fault at $t = 3$ s, the elevator of UAV#2 encounters an actuator fault at $t = 8$ s, and the rudder of UAV#3 is subjected to an actuator fault at $t = 12$ s, respectively. Moreover, it can also be observed that the bank angle tracking

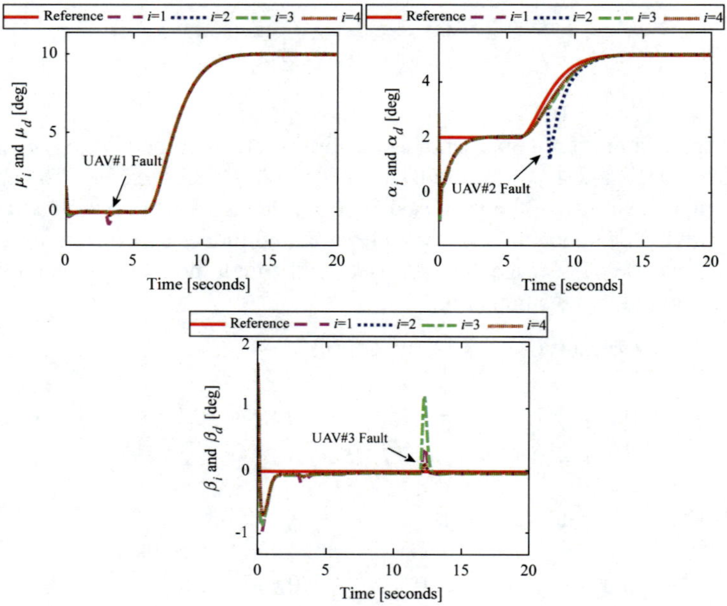

Fig. 8.2 Bank angles μ_i, angles of attack α_i, and sideslip angles β_i of four UAVs, $i = 1, 2, 3, 4$

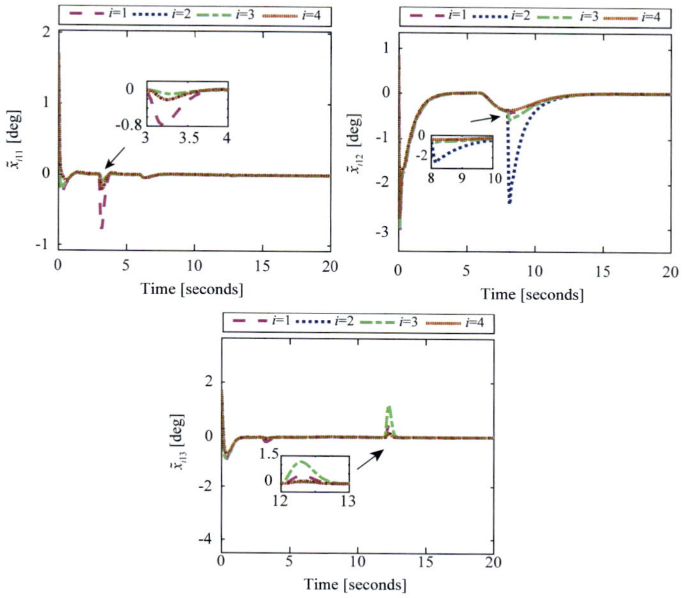

Fig. 8.3 Bank angle, angle of attack, and sideslip angle tracking errors $\tilde{x}_{i11}$, $\tilde{x}_{i12}$, $\tilde{x}_{i13}$ of four UAVs with respect to their attitude references, $i = 1, 2, 3, 4$

errors of UAVs#2–4, the angle of attack tracking errors of UAVs#1, 3, 4, and the sideslip angle tracking errors of UAVs#1, 2, 4 encounter very slight performance degradations at $t = 3\,\text{s}, 8\,\text{s}, 12\,\text{s}$, respectively. This is due to the fact that in the communication network of numerous UAVs, the state variations of faulty UAVs can be sent to other healthy UAVs. Large performance degradations or instability can be caused if prompt actions are not adopted. Fortunately, under the proposed decentralized FTCC scheme, the tracking errors are driven into the region containing zero in a timely manner, and the stability of overall multi-UAVs system is hence guaranteed. Figure 8.4 shows the attitude synchronization tracking errors and the bank angle, angle of attack, and sideslip angle synchronization tracking errors are strictly evolved in the region between the prescribed upper bound $\overline{k}_{i1\nu}\varepsilon_{i1\nu}$ and lower bound $-\underline{k}_{i1\nu}\varepsilon_{i1\nu}$, even in the presence of actuator faults.

The control inputs are shown in Fig. 8.5. To attenuate the adverse effects of aileron, elevator, and rudder faults on the tracking performance, the control inputs are adjusted to achieve satisfactory tracking performance, and these can be observed in Figs. 8.2, 8.3, 8.4. It can be seen that both the aileron control input of UAV#1 and the rudder control input of UAV#3 get saturated at the beginning. The control input signals sent to the UAV system are strictly restricted between $\boldsymbol{u}_{i0\min}$ and $\boldsymbol{u}_{i0\max}$ by imposing the saturation limits (8.40) on the control input signal (8.38). To compensate for the error between the control input signal and the saturated control input signal in the event of input saturation, an auxiliary dynamic system (8.41) is

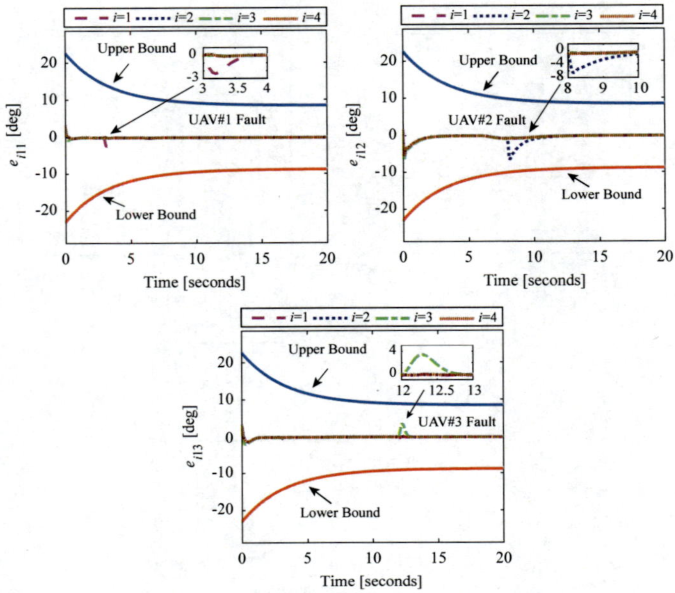

Fig. 8.4 Attitude synchronization tracking errors $e_{i11}, e_{i12}, e_{i13}$ of four UAVs, $i = 1, 2, 3, 4$

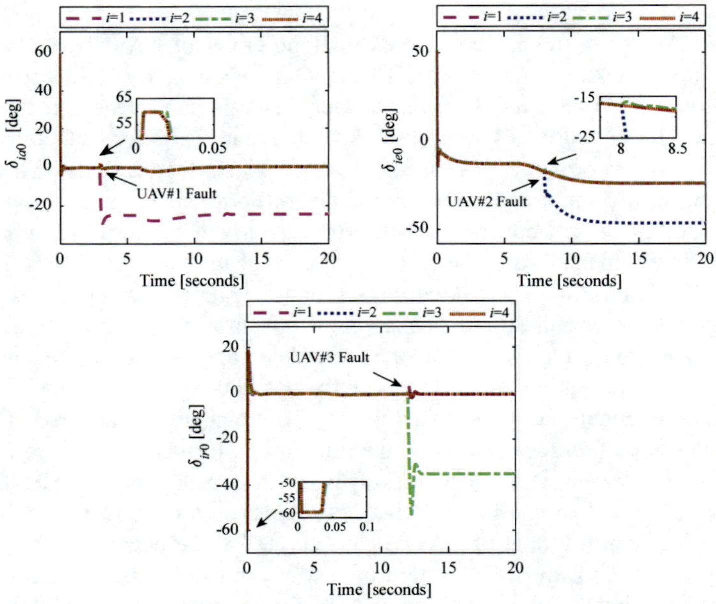

Fig. 8.5 Aileron, elevator, and rudder control inputs of four UAVs

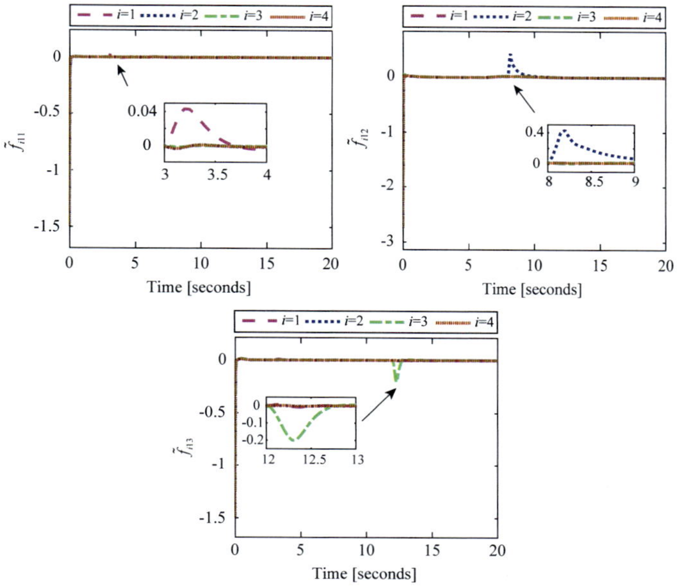

Fig. 8.6 Approximation errors of NN and DO under (8.23) and (8.27)

introduced to pull the saturated control input signal back to the unsaturated region, without persistent actuator saturation.

Figures 8.6 and 8.7 illustrate the approximation errors under the NNs and the DOs. Figure 8.6 displays that the nonlinear function $\boldsymbol{f}_{i1}$ in (8.5) can be well approximated using (8.22), (8.23), and (8.27) even when aileron, elevator, and rudder faults are encountered by UAV#1, UAV#2, UAV#3, respectively. The approximation errors $\tilde{\boldsymbol{\Xi}}_i = [\tilde{\Xi}_{i1}, \tilde{\Xi}_{i2}, \tilde{\Xi}_{i3}]^T = (\boldsymbol{F}_{i2} + \boldsymbol{d}_i) - (\Gamma_2^{-1}\hat{\boldsymbol{W}}_{i2}\boldsymbol{\varphi}_{i2} + \hat{\boldsymbol{D}}_{i2})$ are shown in Fig. 8.7, which rapidly converge into the regions containing zero under (8.34), (8.35), and (8.39). When UAV#1, UAV#2, and UAV#3 encounter the aileron, elevator, and rudder faults, slightly degraded but acceptable approximation performances are induced. Under the proposed NNs and the DOs, the approximation errors $\tilde{\Xi}_{i1}$, $\tilde{\Xi}_{i2}$, and $\tilde{\Xi}_{i3}$ are pulled into the small regions containing zero again. Therefore, the approximation ability can be guaranteed. Furthermore, considering the simulation scenario that UAV#1, UAV#2, and UAV#3 encounter aileron, elevator, and rudder faults, respectively, and UAV#4 is healthy, one can conclude that the proposed decentralized FTCC scheme with prescribed attitude synchronization tracking performance can be applied to UAVs in both faulty and healthy conditions.

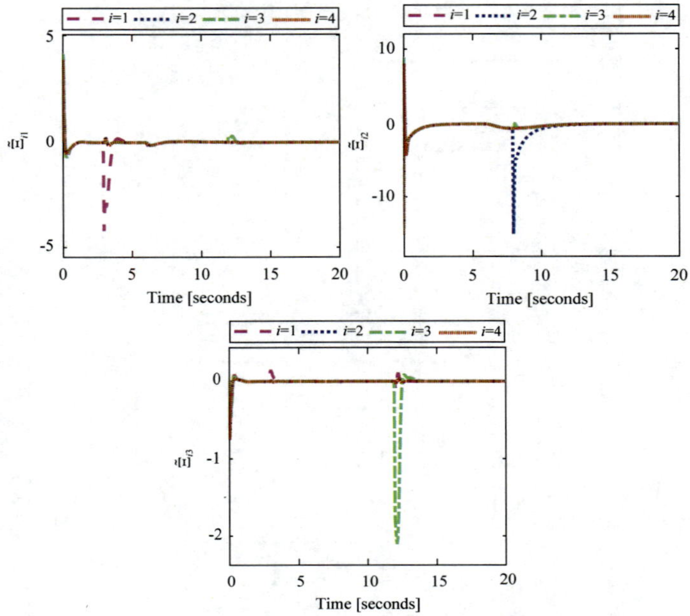

Fig. 8.7 Approximation errors of NN and DO under (8.35) and (8.39)

8.5 Conclusions

In this chapter, a decentralized FTCC scheme has been proposed for a group of UAVs to achieve the prescribed attitude synchronization tracking performance in a directed communication network. NNs and DOs are integrated to approximate unknown nonlinear functions, as well as nonlinear terms due to actuator faults. Moreover, the prediction errors are integrated into the adaptive laws and DOs to enhance the approximation ability. By imposing the prescribed performance functions on the attitude synchronization errors, prescribed synchronization tracking control can be achieved. It has been proven theoretically that the proposed control scheme can make the attitudes of all UAVs synchronously track their attitude references. Simulation results on a network of four UAVs have demonstrated the effectiveness of the proposed control scheme.

References

1. M. Chen, S.Z.S. Ge, B.B. Ren, Adaptive tracking control of uncertain MIMO nonlinear systems with input constraints. Automatica **47**(3), 452–465 (2011)
2. J.L. Du, X. Hu, M. Krstić, Y.Q. Sun, Robust dynamic positioning of ships with disturbances under input saturation. Automatica **73**, 207–214 (2016)

3. L. He, X.X. Sun, Y. Lin, Distributed output-feedback formation tracking control for unmanned aerial vehicles. Int. J. Syst. Sci. **47**(16), 3919–3928 (2016)
4. W. Lin, Distributed UAV formation control using differential game approach. Aerosp. Sci. Technol. **35**, 54–62 (2014)
5. N. Nigam, S. Bieniawski, I. Kroo, J. Vian, Control of multiple UAVs for persistent surveillance: algorithm and flight test results. IEEE Trans. Control Syst. Technol. **20**(5), 1236–1251 (2012)
6. S. Waharte, N. Trigoni, Supporting search and rescue operations with UAVs, in *International Conference on Emerging Security Technologies*, Canterbury (2010)
7. B.L. Wu, D.W. Wang, E.K. Poh, Decentralized robust adaptive control for attitude synchronization under directed communication topology. J. Guid. Control Dyn. **34**(4), 1276–1282 (2011)
8. R.B. Xue, J.M. Song, G.H. Cai, Distributed formation flight control of multi-UAV system with nonuniform time-delays and jointly connected topologies. Proc. Inst. Mech. Eng. Part G. J. Aerosp. Eng. **230**(10), 1871–1881 (2016)
9. M.D. Yan, X. Zhu, X.X. Zhang, Y.H. Qu, Consensus-based three-dimensional multi-UAV formation control strategy with high precision. Front. Inform. Technol. Elect. Eng. **18**(7), 968–977 (2017)
10. Z.Q. Yu, Y.M. Zhang, B. Jiang, C.Y. Su, J. Fu, Y. Jin, T.Y. Chai, Nussbaum-based finite-time fractional-order backstepping fault-tolerant flight control of fixed-wing UAV against input saturation with hardware-in-the-loop validation. Mech. Syst. Signal Proc. **153**, 107406 (2021)
11. Z.Q. Yu, J.X. Li, Y.W. Xu, Y.M. Zhang, B. Jiang, C.Y. Su, Reinforcement learning-based fractional-order adaptive fault-tolerant formation control of networked fixed-wing UAVs with prescribed performance. IEEE Trans. Neural Netw. Learn. Syst. (2023). Published online https://doi.org/10.1109/TNNLS.2023.3281403
12. C. Yuan, Y.M. Zhang, Z.X. Liu, A survey on technologies for automatic forest fire monitoring, detection, and fighting using unmanned aerial vehicles and remote sensing techniques. Can. J. Forest Res. **45**(7), 783–792 (2015)
13. Z.J. Zhao, J. Zhang, Z.J. Liu, W. He, K.S. Hong, Adaptive quantized fault-tolerant control of a 2-DOF helicopter system with actuator fault and unknown dead zone. Automatica **148**, 110792 (2023)
14. A.M. Zou, Z.G. Hou, M. Tan, Adaptive control of a class of nonlinear pure-feedback systems using fuzzy backstepping approach. IEEE Trans. Fuzzy Syst. **16**(4), 886–897 (2008)

Chapter 9
Decentralized FTCC of Multi-UAVs for Cooperative Forest Fire Monitoring

9.1 Introduction

Forest fire has been one of the most dangerous threats to the forest, wildlife, and firefighters, causing disastrous economic problems and unrepairable environmental effects [5, 16, 17, 39]. Nowadays, the traditional human fire-fighting methods are still widely used, and the firefighters usually go to the fire scene to put out the fire, imposing fatal threats to the safety of firefighters. Forest fire fighting is usually based on the visual observations from the fire-fighting experts, which may cause calamitous results if the spread direction of the forest fire, easily changed by the varied wind direction, is not monitored promptly [15]. Since UAV can provide the aerial image of the forest fire, observe the spread direction, and put out the forest fire by spraying water or launching fire extinguishing bombs, it is a desired platform for forest fire monitoring and fighting [5, 23, 24, 28, 39, 40].

If only one UAV is adopted to detect/monitor a large forest fire with a fast spread speed, the fire information may not be duly captured due to the limited coverage area of sensors on-board one UAV. To improve the monitoring efficiency and simultaneously obtain the fire information from different positions along the fire spread perimeter, multi-UAVs need to be utilized to cooperatively monitor the forest fire. In [3], a cooperative surveillance scheme was developed for UAVs to oversee the fires. By incorporating infrared and visual images from the cameras on-board the UAVs, a cooperative fire detection method was investigated in [13] to provide an accurate fire location. Then, the authors further studied a cooperative perception system with infrared, visual cameras, and fire detectors for multiple heterogeneous UAVs [14]. The newly proposed system combined multipurpose low-level image processing functions and information fusion algorithms. In [25], a decision-making method was studied to assist the monitoring of multiple forest fires using UAVs. By exchanging the critical coordination information, a cooperative perimeter surveillance algorithm was developed for multi-UAVs [7]. More recently, in [22], by using a group of UAVs and remote sensing technology, a monitoring

Z. Yu et al., *Fault-Tolerant Cooperative Control of Unmanned Aerial Vehicles*,
https://doi.org/10.1007/978-981-99-7661-4_9

system was developed to increase the fire prediction efficiency. The aforementioned detecting/monitoring methods can achieve good results. However, very few works have considered the unexpected cooperative monitoring problem when multiple fixed-wing UAVs are adopted in the monitoring task and a portion of UAVs is subjected to the actuator faults, which often significantly degrade the monitoring performance if appropriate actions are not adopted promptly.

As the generalized form of integer-order (IO) calculus, FO calculus has been investigated in many fields, such as FO control [4, 10, 34, 37], image processing [8, 30], and bioengineering [11]. Regarding the FO control, there exist two types: (1) FO controller design for FO system and (2) FO controller design for IO system. It is natural to develop FO controllers for FO systems. For the second type, incorporating FO calculus into the controller for IO systems can provide extra degrees of freedom and simultaneously enhance the control performance. Recently, FO PID, FO sliding-mode, and FO backstepping control methods have been developed for various systems [6, 9, 18, 26, 27, 31]. However, the results of introducing FO calculus into the fault-tolerant time-varying formation control of multiple fixed-wing UAVs for cooperative monitoring of forest fire are very rare.

Motivated by the above analysis, by using the elliptical spread information of forest fire, this chapter will develop a decentralized FTCC scheme for multi-UAVs with application to the cooperative monitoring mission of forest fire. With the aid of sliding-mode differentiators (SMDs) and reference systems, SMDOs are designed to estimate the lumped disturbance induced by the external disturbances and actuator faults. Then, by using FO calculus, an FO SMC scheme is developed to achieve the cooperative monitoring of forest fire. The main contributions of this chapter are listed as follows:

(1) Different from the forest fire monitoring strategy using only one UAV [38], and in view of the advantages of rapid response, long endurance, and high maneuverability over rotary-wing UAVs, this chapter further investigates the cooperative monitoring of forest fire using multiple fixed-wing UAVs, which is more practical and challenging than the single UAV case and can provide the forest fire information for large area and different positions along the perimeter of forest fire.
(2) Compared with the forest fire monitoring strategies presented in [2, 3], by utilizing the elliptically spread information of forest fire, an FTCC scheme is proposed for multi-UAVs by simultaneously considering the actuator faults and time-varying displacement of each UAV with respect to the forest fire center.
(3) Different from the extensively studied IO control schemes [32, 33, 35, 36], by introducing SMDOs from the SMDs and reference systems to estimate the lumped disturbances and utilizing FO calculus, the FO SMC scheme is investigated for multi-UAVs to cooperatively monitor the forest fire.

The remainder of this chapter is listed as follows. Section 9.2 outlines the preliminaries and problem statement. The controller design and stability analysis are presented in Sect. 9.3. Simulation results are provided in Sect. 9.4 to show the effectiveness of the proposed monitoring scheme. Section 9.5 concludes the conducted work.

(*Remark: The main control schemes and contents of this chapter are from the published journal paper "Z. Q. Yu, Y. M. Zhang, B. Jiang, and X. Yu. Fault-tolerant time-varying elliptical formation control of multiple fixed-wing UAVs for cooperative forest fire monitoring, Journal of Intelligent & Robotic Systems, 2021, 101: 48." The authors appreciate the permission from the Springer Nature to reuse the results published in the relevant Journal.*)

9.2 Preliminaries and Problem Statement

9.2.1 *Faulty UAV Position Model*

In this chapter, a group of N fixed-wing UAVs is used to cooperatively monitor the forest fire. Different from the longitudinal model used in Chaps. 3, 4, 5, 6 and the inner-loop attitude model used in Chaps. 7 and 8, the outer-loop position model is utilized in this chapter to construct the fault-tolerant time-varying elliptical formation control scheme. By recalling the outer-loop position model (2.15)–(2.16), the following position model is formulated [12]:

$$\begin{cases} \dot{x}_i = V_i \cos\gamma_i \cos\chi_i \\ \dot{y}_i = V_i \cos\gamma_i \sin\chi_i \\ \dot{z}_i = V_i \sin\gamma_i \end{cases} \tag{9.1}$$

$$\begin{cases} \dot{V}_i = (T_i - D_i)/m_i + b_{i1} - g\sin\gamma_i \\ \dot{\chi}_i = (gn_i \sin\phi_i + b_{i2})/(V_i\cos\gamma_i) \\ \dot{\gamma}_i = -g\cos\gamma_i/V_i + (gn_i\cos\phi_i + b_{i3})/V_i \end{cases} \tag{9.2}$$

where $i = 1, 2, \ldots, N$. n_i represents the lift factor and ϕ_i is the bank angle. b_{i1}, b_{i2}, and b_{i3} are the external disturbances. With respect to the point-mass model (9.1), (9.2), the output and control input vectors are chosen as $\boldsymbol{P}_i = [x_i, y_i, z_i]^T$ and $\boldsymbol{u}_i = [T_i, n_i, \phi_i]^T$, respectively.

The loss-of-effectiveness and bias faults are considered, expressed by

$$\boldsymbol{u}_i = \boldsymbol{\rho}_i \boldsymbol{u}_{i0} + \boldsymbol{\varpi}_i \tag{9.3}$$

where $\boldsymbol{u}_i$ is the applied signal, and $\boldsymbol{u}_{i0} = [T_{i0}, n_{i0}, \phi_{i0}]^T$ represents the designed control signal. $\boldsymbol{\rho}_i = \text{diag}\{\rho_{i1}, \rho_{i2}, \rho_{i3}\}$ with $0 < \rho_{i1}, \rho_{i2}, \rho_{i3} \leq 1$ represents the remaining actuation effectiveness matrix. $\boldsymbol{\varpi}_i = [\varpi_{i1}, \varpi_{i2}, \varpi_{i3}]^T$ denotes the bias fault vector. Note that ρ_{i1} and ϖ_{i1} represent the faults encountered by the engine;

ρ_{i2} and ϖ_{i2} reflect the lift variations due to the elevator faults; ρ_{i3} and ϖ_{i3} are the bank angle variations caused by the aileron and rudder faults.

By substituting the fault model (9.3) into (9.2), one can render

$$\begin{cases} \dot{V}_i = (T_{i0} - D_i)/m_i - g \sin \gamma_i + (\rho_{i1} T_i - T_{i0} + \varpi_{i1})/m_i + b_{i1} \\ \dot{\chi}_i = g n_{i0} \sin \phi_{i0}/(V_i \cos \gamma_i) + (b_{i2} - g n_{i0} \sin \phi_{i0})/(V_i \cos \gamma_i) \\ \qquad + g(\rho_{i2} n_i + \varpi_{i2}) \sin(\rho_{i3}\phi_i + \varpi_3)/(V_i \cos \gamma_i) \\ \dot{\gamma}_i = g n_{i0} \cos \phi_{i0}/V_i - g \cos \gamma_i/V_i + (b_{i3} - g n_{i0} \cos \phi_{i0})/V_i \\ \qquad + (\rho_{i2} n_i + \varpi_{i2}) \cos(\rho_{i3}\phi_i + \varpi_{i3})/V_i \end{cases} \tag{9.4}$$

By recalling (9.1) and using (9.4), the following control-oriented model can be obtained:

$$\begin{aligned} \ddot{\boldsymbol{P}}_i &= \boldsymbol{R}_i \boldsymbol{\mu}_{im} + \boldsymbol{G} + \boldsymbol{R}_i \boldsymbol{D}_{i0} \\ &= \boldsymbol{\mu}_i + \boldsymbol{D}_{ir} \end{aligned} \tag{9.5}$$

where $\boldsymbol{\mu}_i = \boldsymbol{R}_i \boldsymbol{\mu}_{im} + \boldsymbol{G}$ is the virtual control signal, $\boldsymbol{D}_{ir} = \boldsymbol{R}_i \boldsymbol{D}_{i0}$, $\boldsymbol{G} = [0, 0, -g]^T$. $\boldsymbol{\mu}_{im}$, $\boldsymbol{D}_{i0}$, and $\boldsymbol{R}_i$ are obtained from (9.1), (9.4), (9.5), and given by

$$\boldsymbol{\mu}_{im} = \begin{bmatrix} (T_{i0} - D_i)/m_i - g \sin \gamma_i \\ g n_{i0} \sin \phi_{i0} \\ g n_{i0} \cos \phi_{i0} \end{bmatrix} \tag{9.6}$$

$$\boldsymbol{D}_{i0} = \begin{bmatrix} (\rho_{i1} T_i - T_{i0} + \varpi_{i1})/m_i + b_{i1} \\ [g(\rho_{i2} n_i + \varpi_{i2}) \sin(\rho_{i3}\phi_i + \varpi_3) - g n_{i0} \sin \phi_{i0} + b_{i2}]/V_i \cos \gamma_i \\ [g(\rho_{i2} n_i + \varpi_{i2}) \cos(\rho_{i3}\phi_i + \varpi_{i3}) - g n_{i0} \cos \phi_{i0} + b_{i3}]/V_i \end{bmatrix} \tag{9.7}$$

$$\boldsymbol{R}_i = \begin{bmatrix} \cos \gamma_i \cos \chi_i & -\sin \chi_i & -\sin \gamma_i \cos \chi_i \\ \cos \gamma_i \sin \chi_i & \cos \chi_i & -\sin \gamma_i \sin \chi_i \\ \sin \gamma_i & 0 & \cos \gamma_i \end{bmatrix} \tag{9.8}$$

According to (9.5)–(9.8), the control input signal can be extracted as

$$\boldsymbol{u}_{i0} = \begin{bmatrix} m[\boldsymbol{\mu}_{im}(1) + g \sin \gamma_i] + D_i \\ \boldsymbol{\mu}_{im}(3)/(g \cos \phi_i) \\ \arctan[\boldsymbol{\mu}_{im}(2)/\boldsymbol{\mu}_{im}(3)] \end{bmatrix} \tag{9.9}$$

where $\boldsymbol{\mu}_{im}$ can be calculated from the virtual control signal $\boldsymbol{\mu}_i$, i.e., $\boldsymbol{\mu}_{im} = \boldsymbol{R}_i^{-1}(\boldsymbol{\mu}_i - \boldsymbol{G})$.

9.2.2 Fire Spread Model

According to the analysis in [20], empirical model can be used to simulate the fire growth and provide a satisfactory estimation of the fire perimeter. In this section, it is assumed that the fire ignites at a point (x_{f0}, y_{f0}) and spreads with the elliptical form. Figure 9.1 illustrates the fire spread model, which is mathematically defined by Perry [20]

$$\begin{cases} x_{fi} = ct_f + at_f \cos\beta_i + x_{f0} \\ y_{fi} = bt_f \sin\beta_i + y_{f0} \end{cases} \tag{9.10}$$

where $0 \le \beta_i < 2\pi$ is the included angle associated with the point i on the fire perimeter. t_f is the spread time of fire, and (x_{f0}, y_{f0}) is the ignition point. a and b denote the spread rates of fire along the X, Y axes, respectively. c represents the moving rate of the center of ellipse in the wind direction. Without loss of generality, the wind direction is assumed to be parallel to the X-axis.

With respect to the cooperative monitoring of forest fire with multi-UAVs, the ith fixed-wing UAV is associated with the point i in Fig. 9.1. Then, all UAVs need to track the central point of fire $\boldsymbol{P}_0 = [x_{f0} + ct_f, y_{f0}, 0]^T$ with time-varying relative displacements $\boldsymbol{T}_{ir} = [at_f \cos\beta_i + \delta_{ix}, bt_f \sin\beta_i + \delta_{iy}, \delta_{iz}]^T$, and $\boldsymbol{\delta}_i = [\delta_{ix}, \delta_{iy}, \delta_{iz}]^T$ is the safe distance vector of the ith UAV with respect to the fire border. Therefore, the desired position signal for the ith fixed-wing UAV is defined as $\boldsymbol{P}_{id} = \boldsymbol{P}_0 + \boldsymbol{T}_{ir}$.

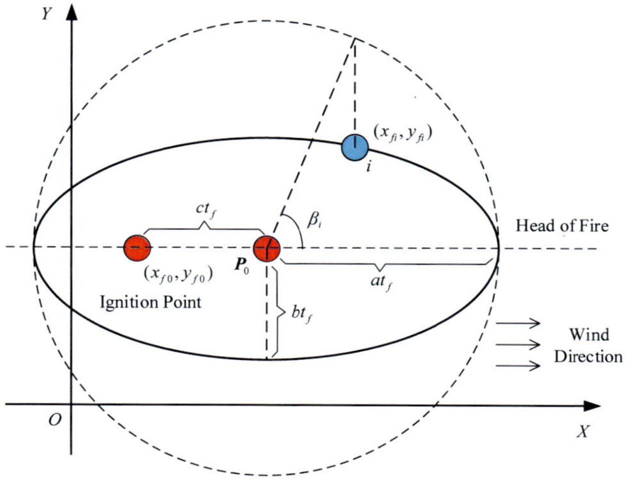

Fig. 9.1 Fire spread model

9.2.3 Fractional-Order Calculus

To develop the FO control scheme for fixed-wing UAVs with application to the cooperative monitoring of forest fire, the definitions and property of FO calculus are first introduced.

Definition 9.1 The Riemann–Liouville (RL) fractional integral of a function $f(t)$ is described as [21]

$$ {}_{t_0}\mathcal{D}_t^{-a} f(t) = \frac{1}{\Gamma(a)} \int_0^t \frac{f(\tau)}{(t-\tau)^{1-a}} d\tau \tag{9.11} $$

where $\Gamma(\cdot)$ is the Gamma function, $0 < a < 1$ is the FO operator, and t_0 is the initial time.

Definition 9.2 The RL fractional derivative of $f(t)$ is defined as [21]

$$ {}_{t_0}\mathcal{D}_t^{a} f(t) = \frac{1}{\Gamma(m-a)} \frac{d^m}{dt^m} \int_{t_0}^t \frac{f(\tau)}{(t-\tau)^{a-m+1}} d\tau \tag{9.12} $$

where $m - 1 < a \leq m$, $m \in N$.

To simplify the expression of FO calculus, the notations $\mathcal{D}^a$ and $\mathcal{D}^{-a}$ are used to represent the fractional derivative and integral, respectively.

Property 9.1 Regarding the RL FO calculus, the following relationship of FO and IO derivatives holds [19]:

$$ \frac{d^n}{dt^n}(D^a f(t)) = D^a \left(\frac{d^n f(t)}{dt^a} \right) = D^{a+n} f(t) \tag{9.13} $$

To move on, the following lemma is introduced to facilitate the SMDO design.

Lemma 9.1 *For a function $f(t)$, there exists the following SMD [1]:*

$$ \begin{cases} \dot{\eta}_1 = \eta_2 - \zeta_1 |\eta_1 - f(t)|^{\alpha_1} \operatorname{sign}(\eta_1 - f(t)) \\ \qquad -\xi_1 |\eta_1 - f(t)|^{\gamma_1} \operatorname{sign}(\eta_1 - f(t)) \\ \quad \vdots \\ \dot{\eta}_i = \eta_{i+1} - \zeta_i |\eta_1 - f(t)|^{\alpha_i} \operatorname{sign}(\eta_1 - f(t)) \\ \qquad -\xi_i |\eta_1 - f(t)|^{\alpha_i} \operatorname{sign}(\eta_1 - f(t)) \\ \quad \vdots \\ \dot{\eta}_n = -\zeta_n |\eta_1 - f(t)|^{\alpha_n} \operatorname{sign}(\eta_1 - f(t)) \\ \qquad -\xi_n |\eta_1 - f(t)|^{\alpha_n} \operatorname{sign}(\eta_1 - f(t)) \end{cases} \tag{9.14} $$

such that the estimation errors $e_{1sm} = \eta_1(t) - f(t)$, $e_{ism} = \eta_i(t) - f^{(i-1)}(t), \ldots,$ *and* $e_{nsm} = \eta_n(t) - f^{(n-1)}(t)$ *converge to the origin in fixed time,* $i = 1, 2, \ldots, n$. *The parameter* α_i *is selected as* $\alpha_i \in (0, 1)$, $i = 1, 2, \ldots, n$, $\alpha_i = i\alpha_1 - (i - 1)$, $\alpha_1 \in (1 - \varepsilon_0, 1)$, $i = 2, \ldots, n$. *The parameter* γ_i *is selected as* $\gamma_i > 1$, $i = 1, 2, \ldots, n$, $\gamma_i = i\gamma_1 - (i - 1)$, $\gamma_1 \in (1, 1 + \varepsilon_1)$, $i = 2, \ldots, n$. ε_0 *and* ε_1 *are sufficiently small values. The parameters* ζ_i, ξ_i, $i = 1, 2, \ldots, n$, *are selected such that the matrices* A_0 *and* A_1 *are Hurwitz, given by*

$$A_0 = \begin{bmatrix} -\zeta_1 & 1\,0 \cdots 0 \\ -\zeta_2 & 0\,1 \cdots 0 \\ \vdots & \\ -\zeta_{n-1} & 0\,0 \cdots 1 \\ -\zeta_n & 0\,0 \cdots 0 \end{bmatrix}, A_1 = \begin{bmatrix} -\xi_1 & 1\,0 \cdots 0 \\ -\xi_2 & 0\,1 \cdots 0 \\ \vdots & \\ -\xi_{n-1} & 0\,0 \cdots 1 \\ -\xi_n & 0\,0 \cdots 0 \end{bmatrix} \tag{9.15}$$

9.2.4 Control Objective

The control objective of this chapter is to design an FTCC scheme for N fixed-wing UAVs (9.1)–(9.2), such that all UAVs can cooperatively monitor the elliptically spread forest fire with time-varying relative displacements, even when a portion of UAVs is encountered by actuator faults and all UAVs are subjected to external disturbances.

9.3 Main Results

In this section, the FO FTCC scheme is designed for multiple fixed-wing UAVs. To effectively estimate the lumped disturbances caused by the external disturbances and actuator faults, SMDOs are designed based on the SMD in Lemma 9.1.

9.3.1 Controller Design

By defining the individual position tracking error as $\tilde{\boldsymbol{P}}_i = \boldsymbol{P}_i - \boldsymbol{P}_{id}$, then the cooperative tracking error is defined as

$$\boldsymbol{e}_{ip} = \lambda_1 \tilde{\boldsymbol{P}}_i + \lambda_2 \sum_{j \in N_i} a_{ij} (\tilde{\boldsymbol{P}}_i - \tilde{\boldsymbol{P}}_j) \tag{9.16}$$

where $\boldsymbol{e}_{ip} = [e_{ip1}, e_{ip2}, e_{ip3}]^T$, λ_1 and λ_2 are positive parameters, and $\tilde{\boldsymbol{P}}_j$ is the individual position tracking error associated with the jth fixed-wing UAV. Note that the parameters λ_1 and λ_2 are used to regulate the position tracking and cooperative monitoring performances, respectively.

By recalling the basic graph theory in Sect. 7.2.2 of Chap. 7 and rewriting the cooperative tracking errors of all fixed-wing UAVs into one expression, one can render

$$\boldsymbol{e}_p = [(\lambda_1 \boldsymbol{I}_N + \lambda_2 \mathcal{L}) \otimes \boldsymbol{I}_3]\tilde{\boldsymbol{P}} \tag{9.17}$$

where $\boldsymbol{e}_p = [\boldsymbol{e}_{1p}^T, \boldsymbol{e}_{2p}^T, \ldots, \boldsymbol{e}_{Np}^T]^T$, $\tilde{\boldsymbol{P}} = [\tilde{P}_1^T, \tilde{P}_2^T, \ldots, \tilde{P}_N^T]^T$.

According to the analysis in Sect. 7.2.2 of Chap. 7 and using (9.17), the position tracking error $\tilde{\boldsymbol{P}} = [(\lambda_1 \boldsymbol{I}_N + \lambda_2 \mathcal{L}) \otimes \boldsymbol{I}_3]^{-1}\boldsymbol{e}_p$ converges to zero once the cooperative tracking error $\boldsymbol{e}_p$ converges to zero successfully. To this end, the following cooperative monitoring controller will be designed based on the error $\boldsymbol{e}_{ip}$.

To estimate the lumped disturbance $\boldsymbol{D}_{ir} = [D_{i1}, D_{i2}, D_{i3}]^T$ in (9.5), a reference system is introduced as

$$\dot{\hat{\boldsymbol{\Psi}}}_i = \boldsymbol{\mu}_i \tag{9.18}$$

By defining $\boldsymbol{\varepsilon}_i = [\varepsilon_{i1}, \varepsilon_{i2}, \varepsilon_{i3}]^T = \boldsymbol{\Psi}_i - \hat{\boldsymbol{\Psi}}_i$, where $\boldsymbol{\Psi}_i = \dot{\boldsymbol{P}}_i$, one has

$$\dot{\boldsymbol{\varepsilon}}_i = \boldsymbol{D}_{ir} \tag{9.19}$$

Then, by using the SMD in Lemma 9.1 and the newly constructed reference system (9.18), we design the following SMDO to estimate $\boldsymbol{D}_{ir}$:

$$\begin{cases} \dot{\eta}_{i1\varrho} = \eta_{i2\varrho} - \zeta_1|\eta_{i1\varrho} - \varepsilon_{i\varrho}|^{\alpha_1}\operatorname{sign}(\eta_{i1\varrho} - \varepsilon_{i\varrho}) \\ \qquad -\xi_1|\eta_{i1\varrho} - \varepsilon_{i\varrho}|^{\gamma_1}\operatorname{sign}(\eta_{i1\varrho} - \varepsilon_{i\varrho}) \\ \dot{\eta}_{i2\varrho} = \eta_{i3\varrho} - \zeta_2|\eta_{i1\varrho} - \varepsilon_{i\varrho}|^{\alpha_2}\operatorname{sign}(\eta_{i1\varrho} - \varepsilon_{i\varrho}) \\ \qquad -\xi_2|\eta_{i1\varrho} - \varepsilon_{i\varrho}|^{\gamma_2}\operatorname{sign}(\eta_{i1\varrho} - \varepsilon_{i\varrho}) \\ \dot{\eta}_{i3\varrho} = -\zeta_3|\eta_{i1\varrho} - \varepsilon_{i\varrho}|^{\alpha_3}\operatorname{sign}(\eta_{i1\varrho} - \varepsilon_{i\varrho}) \\ \qquad -\xi_3|\eta_{i1\varrho} - \varepsilon_{i\varrho}|^{\gamma_3}\operatorname{sign}(\eta_{i1\varrho} - \varepsilon_{i\varrho}) \end{cases} \tag{9.20}$$

where $\varrho = 1, 2, 3$. $\eta_{i1\varrho}$, $\eta_{i2\varrho}$, and $\eta_{i3\varrho}$ are the estimations of $\varepsilon_{i\varrho}$, $D_{i\varrho}$, and $\dot{D}_{i\varrho}$, respectively. ζ_1, ζ_2, ζ_3, ξ_1, ξ_2, ξ_3, α_1, α_2, α_3, γ_1, γ_2, and γ_3 are positive design parameters. The values of these parameters should be chosen according to the parameter selection criterion mentioned in Lemma 9.1. $\boldsymbol{\eta}_{i1} = [\eta_{i11}, \eta_{i12}, \eta_{i13}]^T$, $\boldsymbol{\eta}_{i2} = [\eta_{i21}, \eta_{i22}, \eta_{i23}]^T$, $\boldsymbol{\eta}_{i3} = [\eta_{i31}, \eta_{i32}, \eta_{i33}]^T$.

According to Lemma 9.1, $\boldsymbol{\eta}_{i1}$, $\boldsymbol{\eta}_{i2}$, and $\boldsymbol{\eta}_{i3}$ can exactly converge to $\boldsymbol{\varepsilon}_i$, $\dot{\boldsymbol{\varepsilon}}_i$, and $\ddot{\boldsymbol{\varepsilon}}_i$ in fixed time T_0, respectively. By using (9.19), one can conclude that the estimation errors $\boldsymbol{e}_{iD} = \dot{\boldsymbol{\varepsilon}}_i - \boldsymbol{\eta}_{i2} = \boldsymbol{D}_{ir} - \boldsymbol{\eta}_{i2}$ and $\boldsymbol{e}_{iDd} = \ddot{\boldsymbol{\varepsilon}}_i - \boldsymbol{\eta}_{i3} = \dot{\boldsymbol{D}}_{ir} - \boldsymbol{\eta}_{i3}$ are bounded.

Since IO derivative is the special case of FO derivative when the FO operator is chosen as a positive integer, it can be assumed that $D^a \boldsymbol{e}_{iD}$ is bounded by $|D^a \boldsymbol{e}_{iD}| \leq \bar{\boldsymbol{D}}_{ia}$ as $t < T_0$ and $D^a \boldsymbol{e}_{iD} = 0$ as $t \geq T_0$.

By combining (9.5) and (9.16), the following error dynamics can be obtained:

$$\begin{aligned} \ddot{\boldsymbol{e}}_{ip} &= \left(\lambda_1 + \lambda_2 \sum_{j \in N_i} a_{ij}\right) \ddot{\bar{\boldsymbol{P}}}_i - \lambda_2 \sum_{j \in N_i} a_{ij} \ddot{\bar{\boldsymbol{P}}}_j \\ &= \Phi_i (\boldsymbol{\mu}_i + \boldsymbol{D}_{ir} - \ddot{\boldsymbol{P}}_{id}) - \lambda_2 \sum_{j \in N_i} a_{ij} \ddot{\bar{\boldsymbol{P}}}_j \end{aligned} \tag{9.21}$$

where $\Phi_i = \lambda_1 + \lambda_2 \sum_{j \in N_i} a_{ij}$.

Introduce the following FO sliding-mode surface:

$$\boldsymbol{S}_i = D^{a_1+1} \boldsymbol{e}_{ip} + \kappa_1 \boldsymbol{e}_{ip} + \kappa_2 \boldsymbol{\alpha}_1 (\boldsymbol{e}_{ip}) \tag{9.22}$$

where $0 < a_1 < 1$ is an FO operator, and κ_1 and κ_2 are positive design parameters. The nonlinear function $\boldsymbol{\alpha}_1(\boldsymbol{e}_{ip})$ is defined as

$$\boldsymbol{\alpha}_1(\boldsymbol{e}_{ip}) = \begin{cases} \boldsymbol{e}_{ip}^{\frac{p_1}{q_1}}, & \text{if } C_{i1} \text{ is satisfied} \\ \boldsymbol{e}_{ip}, & \text{if } C_{i2} \text{ is satisfied} \end{cases} \tag{9.23}$$

and C_{i1} represents the condition $\bar{\boldsymbol{S}}_i = 0$ or $\bar{\boldsymbol{S}}_i \neq 0,\ |\boldsymbol{e}_{ip}| > \boldsymbol{\epsilon}_{ip}$, C_{i2} denotes the condition $\bar{\boldsymbol{S}}_i \neq 0,\ |\boldsymbol{e}_{ip}| \leq \boldsymbol{\epsilon}_{ip}$. $\boldsymbol{\epsilon}_{ip}$ is a positive constant vector. The definitions of $\bar{\boldsymbol{S}}_i$, $D^{a_1+1}\boldsymbol{e}_{ip}$, and $\boldsymbol{e}_{ip}^{\frac{p_1}{q_1}}$ are given by

$$\begin{cases} \bar{\boldsymbol{S}}_i = D^{a_1+1} \boldsymbol{e}_{ip} + \kappa_1 \boldsymbol{e}_{ip} + \kappa_2 \boldsymbol{e}_{ip}^{\frac{p_1}{q_1}} \\ D^{a_1+1} \boldsymbol{e}_{ip} = \left[D^{a_1+1} e_{ip1},\ D^{a_1+1} e_{ip2},\ D^{a_1+1} e_{ip3} \right]^T \\ \boldsymbol{e}_{ip}^{\frac{p_1}{q_1}} = \left[e_{ip1}^{\frac{p_1}{q_1}},\ e_{ip2}^{\frac{p_1}{q_1}},\ e_{ip3}^{\frac{p_1}{q_1}} \right]^T \end{cases} \tag{9.24}$$

By differentiating (9.22) with respect to time and recalling the Property 9.1 of FO calculus, the following expression can be obtained:

$$\dot{\boldsymbol{S}}_i = D^{a_1} \ddot{\boldsymbol{e}}_{ip} + \kappa_1 \dot{\boldsymbol{e}}_{ip} + \kappa_2 \dot{\boldsymbol{\alpha}}_1 (\boldsymbol{e}_{ip}) \tag{9.25}$$

where $\dot{\boldsymbol{\alpha}}_1(\boldsymbol{e}_{ip})$ is given by

$$\dot{\boldsymbol{\alpha}}_1(\boldsymbol{e}_{ip}) = \begin{cases} \frac{p_1}{q_1}\boldsymbol{e}_{ip}^{\frac{p_1-q_1}{q_1}}\dot{\boldsymbol{e}}_{ip}, & \text{if } C_{i1} \text{ is satisfied} \\ \dot{\boldsymbol{e}}_{ip}, & \text{if } C_{i2} \text{ is satisfied} \end{cases} \tag{9.26}$$

Substituting (9.21) into (9.25) yields

$$\dot{\boldsymbol{S}}_i = D^{a_1}\left[\Phi_i(\boldsymbol{\mu}_i + \boldsymbol{D}_{ir} - \ddot{\boldsymbol{P}}_{id}) - \lambda_2 \sum_{j\in N_i} a_{ij}\ddot{\tilde{\boldsymbol{P}}}_j\right] + \kappa_1\dot{\boldsymbol{e}}_{ip} + \kappa_2\dot{\boldsymbol{\alpha}}_1(\boldsymbol{e}_{ip}) \tag{9.27}$$

Based on the error dynamics (9.27), the formation control signal is designed as

$$\begin{aligned} \boldsymbol{\mu}_i = & -\boldsymbol{\eta}_{i2} + \ddot{\boldsymbol{P}}_{id} - \Phi_i^{-1}D^{-a_1}(\kappa_1\dot{\boldsymbol{e}}_{ip}) - \Phi_i^{-1}D^{-a_1}(\boldsymbol{K}_1\boldsymbol{S}_i) \\ & + \Phi_i^{-1}\lambda_2\sum_{j\in N_i} a_{ij}\ddot{\tilde{\boldsymbol{P}}}_j - \Phi_i^{-1}D^{-a_1}[\kappa_2\dot{\boldsymbol{\alpha}}(\boldsymbol{e}_{ip})] \end{aligned} \tag{9.28}$$

where $\boldsymbol{K}_1$ is a positive diagonal matrix.

9.3.2 Stability Analysis

Theorem 9.1 *Consider N fixed-wing UAVs described by (9.1)–(9.2) with the actuator fault (9.3) and the forest fire spread model (9.10). If each fixed-wing UAV is controlled by the protocol (9.28) with the FO sliding-mode surface (9.22), the reference system (9.18), and the SMDO (9.20), then all fixed-wing UAVs can cooperatively monitor the forest fire in an elliptical time-varying formation form against actuator faults and external disturbances.*

Proof Choose the Lyapunov function as

$$L_i = \frac{1}{2}\boldsymbol{S}_i^T\boldsymbol{S}_i \tag{9.29}$$

By combining (9.27) and (9.28), one has

$$\dot{\boldsymbol{S}}_i = \Phi_i D^{a_1}\boldsymbol{e}_{iD} - \boldsymbol{K}_1\boldsymbol{S}_i \tag{9.30}$$

Then, in view of (9.29) and (9.30), one can render

$$
\begin{aligned}
\dot{L}_i =& \boldsymbol{S}_i^T \Bigg[D^{a_1} \Bigg[\Phi_i(\boldsymbol{\mu}_i + \boldsymbol{D}_{ir} - \ddot{\boldsymbol{P}}_{id}) - \lambda_2 \sum_{j \in N_i} a_{ij} \ddot{\bar{\boldsymbol{P}}}_j \Bigg] + \kappa_1 \dot{\boldsymbol{e}}_{ip} + \kappa_2 \dot{\boldsymbol{\alpha}}_1(\boldsymbol{e}_{ip}) \Bigg] \\
=& \boldsymbol{S}_i^T (\Phi_i D^{a_1} \boldsymbol{e}_{iD} - \boldsymbol{K}_1 \boldsymbol{S}_i) \\
\leq& - [2\lambda_{\min}(\boldsymbol{K}_1) - 1] L_i + \frac{\Phi_i^2 (D^{a_1} \boldsymbol{e}_{iD})^T (D^{a_1} \boldsymbol{e}_{iD})}{2} \\
\leq& - [2\lambda_{\min}(\boldsymbol{K}_1) - 1] L_i + \Xi_i
\end{aligned}
\tag{9.31}
$$

According to the analysis of SMDO (9.20), when $t < T_0$, one can conclude that the term Ξ_i in (9.31) is bounded by $|\Xi_i| \leq \Xi_{im}$, where Ξ_{im} is an unknown constant; when $t \geq T_0$, one has $\Xi_i = 0$. Therefore, it can be further concluded that the error $\boldsymbol{S}_i$ is bounded and eventually converges to zero.

When the system trajectory reaches to the sliding-mode surface, one has $D^{a_1+1}\boldsymbol{e}_{ip} + \kappa_1 \boldsymbol{e}_{ip} + \kappa_2 \boldsymbol{\alpha}_1(\boldsymbol{e}_{ip}) = \boldsymbol{0}$. Then, with respect to the case $|\boldsymbol{e}_{ip}| > \boldsymbol{\epsilon}_{ip}$, one has $\bar{\boldsymbol{S}}_i = D^{a_1+1}\boldsymbol{e}_{ip} + \kappa_1 \boldsymbol{e}_{ip} + \kappa_2 \boldsymbol{e}_{ip}^{\frac{p_1}{q_1}} = \boldsymbol{0}$. According to the analyses in [6, 19], the FO sliding-mode surface in this case is stable. Regarding the case $|\boldsymbol{e}_{ip}| \leq \boldsymbol{\epsilon}_{ip}$, one has $\bar{\boldsymbol{S}}_i = D^{a_1+1}\boldsymbol{e}_{ip} + \kappa_1 \boldsymbol{e}_{ip} + \kappa_2 \boldsymbol{e}_{ip}$. By borrowing the FO stability theorem from [41], which proved the stability of fractional differential systems with order lying in $(1, 2)$, the stability of FO sliding-mode surface in this case can be guaranteed. Therefore, the cooperative tracking errors are UUB, which means that the errors successfully converge into the very small region containing zero. This ends the proof. □

9.4 Simulation Results

To illustrate the effectiveness of cooperative forest fire monitoring strategy, four fixed-wing UAVs are adopted in the simulation. Figure 9.2 shows the communication network.

The mass of the ith UAV is $m_i = 20\,\text{kg}$, and the drag force D_i is derived as [29]

$$
D_i = 0.5\rho(V_i - V_{iw})^2 S C_{D0} + \frac{2k_d k_n^2 L^2/g^2}{\rho(V_i - V_{iw})^2 S} \tag{9.32}
$$

where $\rho = 1.225\,\text{kg/m}^3$ is the atmospheric density, $S = 1.37\ \text{m}^2$ is the wing area, $C_{D0} = 0.02$ is the zero-lift coefficient, $k_n = 1$ is the load-factor effectiveness, $k_d = 0.1$ is the induced drag coefficient, and V_{iw} is gust, which can be chosen as

$$
V_{iw} = \underbrace{0.215 V_m \log_{10}(z_i) + 0.285 V_m}_{\text{term 1}} + \underbrace{\Delta V_{iw}}_{\text{term 2}} \tag{9.33}
$$

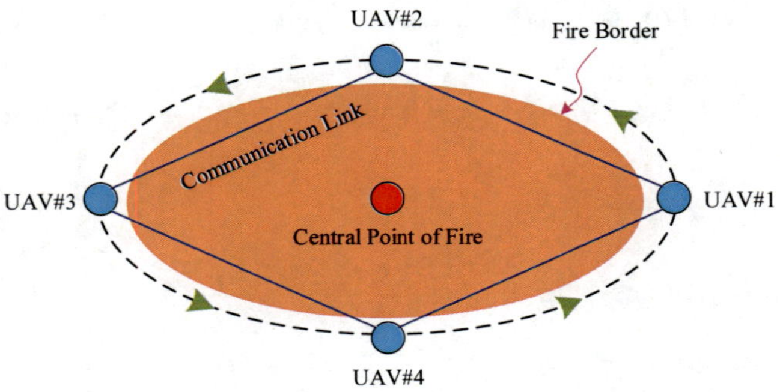

Fig. 9.2 Communication network of four UAVs for monitoring the forest fire

where terms 1 and 2, respectively, represent the normal wind shear and the wind gust turbulence, which is assumed to a Gaussian random variable with zero mean and a standard deviation of $0.09V_m$, and V_m is the mean wind speed with a value of 10 m/s.

In the simulation, it is assumed that the fire has been spread for 1800 s when the four UAVs are distributed along the perimeter to cooperatively monitor the forest fire. The parameters of fire spread model are assumed as $a = 1.2$ m/s, $b = 0.8$ m/s, $c = 0.3$ m/s, $x_{f0} = 3$ m, and $y_{f0} = 4$ m. The safe distance vectors of four UAVs are set as $\boldsymbol{\delta}_1 = [0.5, 0.5, 100]^T$ m, $\boldsymbol{\delta}_2 = [1, 1, 105]^T$ m, $\boldsymbol{\delta}_3 = [1.5, 1.5, 110]^T$ m, $\boldsymbol{\delta}_4 = [2, 2, 115]^T$ m, respectively. Then, the desired position references for four UAVs are chosen as

$$\boldsymbol{P}_{1d} = \begin{bmatrix} 3 + 0.3(t + 1800) + 1.2(t + 1800)\cos(0.0327t) + 0.5 \\ 4 + 0.8(t + 1800)\sin(0.0327t) + 0.5 \\ 100 \end{bmatrix}$$

$$\boldsymbol{P}_{2d} = \begin{bmatrix} 3 + 0.3(t + 1800) + 1.2(t + 1800)\cos(0.0327t + 0.5\pi) + 1 \\ 4 + 0.8(t + 1800)\sin(0.0327t + 0.5\pi) + 1 \\ 105 \end{bmatrix}$$

$$\boldsymbol{P}_{3d} = \begin{bmatrix} 3 + 0.3(t + 1800) + 1.2(t + 1800)\cos(0.0327t + \pi) + 1.5 \\ 4 + 0.8(t + 1800)\sin(0.0327t + \pi) + 1.5 \\ 110 \end{bmatrix}$$

$$\boldsymbol{P}_{4d} = \begin{bmatrix} 3 + 0.3(t + 1800) + 1.2(t + 1800)\cos(0.0327t + 1.5\pi) + 2 \\ 4 + 0.8(t + 1800)\sin(0.0327t + 1.5\pi) + 2 \\ 115 \end{bmatrix}$$

The initial states of four UAVs are set as $x_1(0) = 2703.5$ m, $y_1(0) = 4.5$ m, $z_1(0) = 100$ m, $V_1(0) = 47$ m/s, $\chi_1(0) = \pi/2$ rad, $\gamma_1(0) = 0$ rad; $x_2(0) = 544$ m,

$y_2(0) = 1445\,\text{m}$, $z_2(0) = 105\,\text{m}$, $V_2(0) = 70\,\text{m/s}$, $\chi_2(0) = \pi$ rad, $\gamma_2(0) = 0$ rad; $x_3(0) = -1615.5\,\text{m}$, $y_3(0) = 5.5\,\text{m}$, $z_3(0) = 110\,\text{m}$, $V_3(0) = 47\,\text{m/s}$, $\chi_3(0) = 3\pi/2$ rad, $\gamma_3(0) = 0$ rad; $x_4(0) = 545\,\text{m}$, $y_4(0) = -1434\,\text{m}$, $z_4(0) = 115\,\text{m}$, $V_4(0) = 71\,\text{m/s}$, $\chi_4(0) = 0$ rad, $\gamma_4(0) = 0$ rad. In the simulation, four UAVs encounter external disturbances $b_{i1} = -2$, $b_{i2} = 1.2$, $b_{i3} = 0.6$ at $t = 20$ s. Based on the fault model (9.3), when all UAVs are healthy, one has $\rho_{i1} = \rho_{i2} = \rho_{i3} = 1$, $\varpi_{i1} = \varpi_{i2} = \varpi_{i3} = 0$, $i = 1, 2, 3, 4$. It is assumed that UAV#4 is healthy in the simulation, and UAVs#1, 2, 3 are subjected to the following fault signals at 50 s, 100 s, and 150 s, respectively:

(1) UAV#1 faults:

$$\begin{cases} \rho_{11} = 0.4e^{-10(t-50)} + 0.6, & t < 50\,\text{s} \\ \rho_{12} = 0.4e^{-10(t-50)} + 0.6, & t < 50\,\text{s} \\ \rho_{13} = 0.4e^{-10(t-50)} + 0.6, & t < 50\,\text{s} \\ \varpi_{11} = -3e^{-10(t-50)} + 3, & t \geq 50\,\text{s} \\ \varpi_{12} = -0.1e^{-10(t-50)} + 0.1, & t \geq 50\,\text{s} \\ \varpi_{13} = -0.15e^{-10(t-50)} + 0.15, & t \geq 50\,\text{s} \end{cases} \tag{9.34}$$

(2) UAV#2 faults:

$$\begin{cases} \rho_{21} = 0.5e^{-10(t-100)} + 0.5, & t < 100\,\text{s} \\ \rho_{22} = 0.35e^{-10(t-100)} + 0.65, & t < 100\,\text{s} \\ \rho_{23} = 0.35e^{-10(t-100)} + 0.65, & t < 100\,\text{s} \\ \varpi_{21} = -3.5e^{-10(t-100)} + 3.5, & t \geq 100\,\text{s} \\ \varpi_{22} = -0.12e^{-10(t-100)} + 0.12, & t \geq 100\,\text{s} \\ \varpi_{23} = -0.18e^{-10(t-100)} + 0.18, & t \geq 100\,\text{s} \end{cases} \tag{9.35}$$

(3) UAV#3 faults:

$$\begin{cases} \rho_{31} = 0.35e^{-10(t-150)} + 0.65, & t < 150\,\text{s} \\ \rho_{32} = 0.3e^{-10(t-150)} + 0.7, & t < 150\,\text{s} \\ \rho_{33} = 0.3e^{-10(t-150)} + 0.7, & t < 150\,\text{s} \\ \varpi_{31} = -2e^{-10(t-150)} + 2, & t \geq 150\,\text{s} \\ \varpi_{32} = -0.1e^{-10(t-150)} + 0.1, & t \geq 150\,\text{s} \\ \varpi_{33} = -0.2e^{-10(t-150)} + 0.2, & t \geq 150\,\text{s} \end{cases} \tag{9.36}$$

Figure 9.3 illustrates the trajectories of four fixed-wing UAVs and the propagation of forest fire. It can be seen from Fig. 9.3 that these four UAVs can cooperatively monitor the forest fire with safe displacements. Figures 9.4, 9.5,

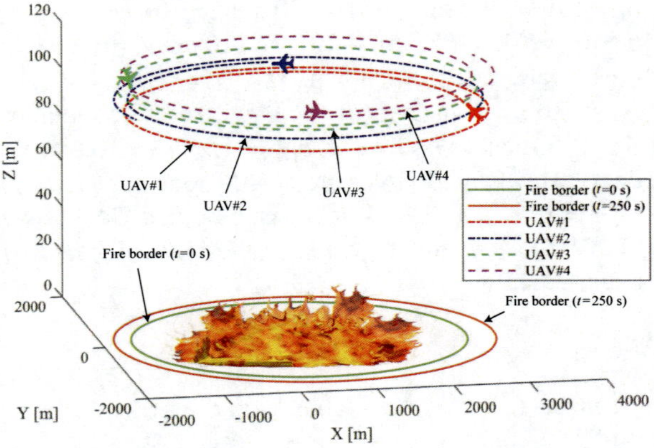

Fig. 9.3 The trajectories of four UAVs and the border variation of fire

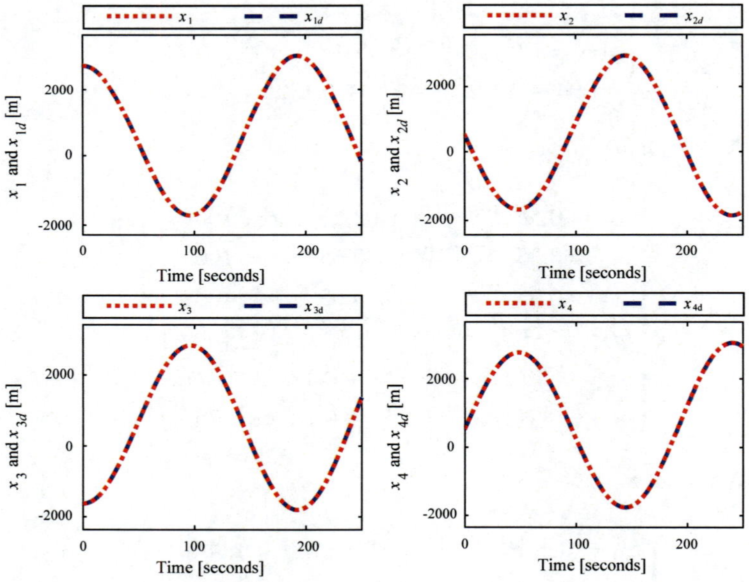

Fig. 9.4 Time histories of $x_i, i = 1, 2, 3, 4$

and 9.6 show the time histories of forward, lateral, and vertical positions of four UAVs, and it is observed that the position signals can track the desired references even if four UAVs are subjected to the external disturbances, gust, and UAVs#1, 2, 3 encounter actuator faults.

The position tracking errors of four UAVs are demonstrated in Fig. 9.7. From Fig. 9.7, it is illustrated that all tracking errors encounter very slight performance

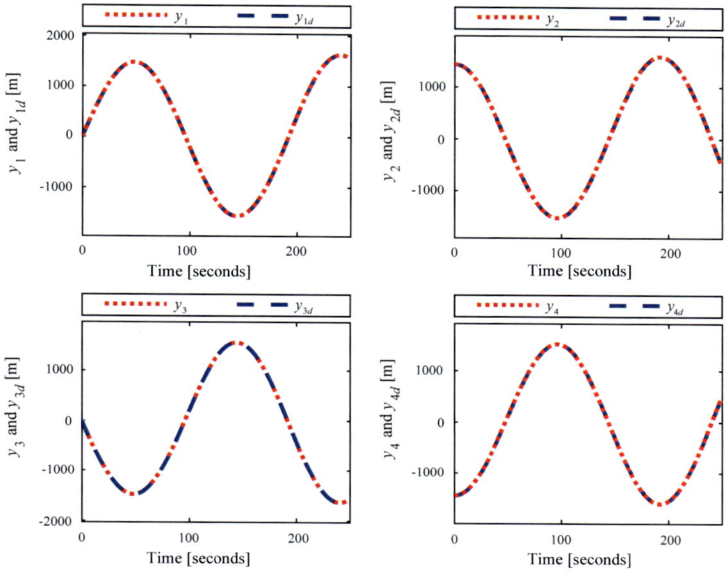

Fig. 9.5 Time histories of y_i, $i = 1, 2, 3, 4$

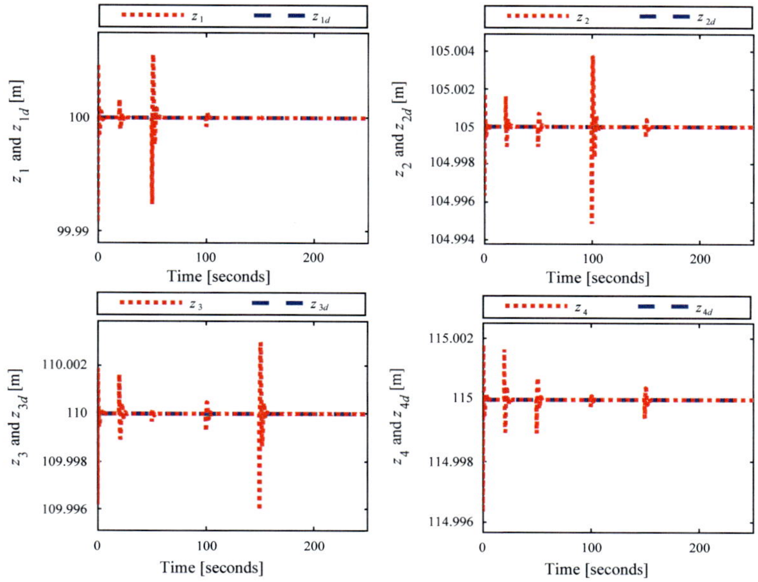

Fig. 9.6 Time histories of z_i, $i = 1, 2, 3, 4$

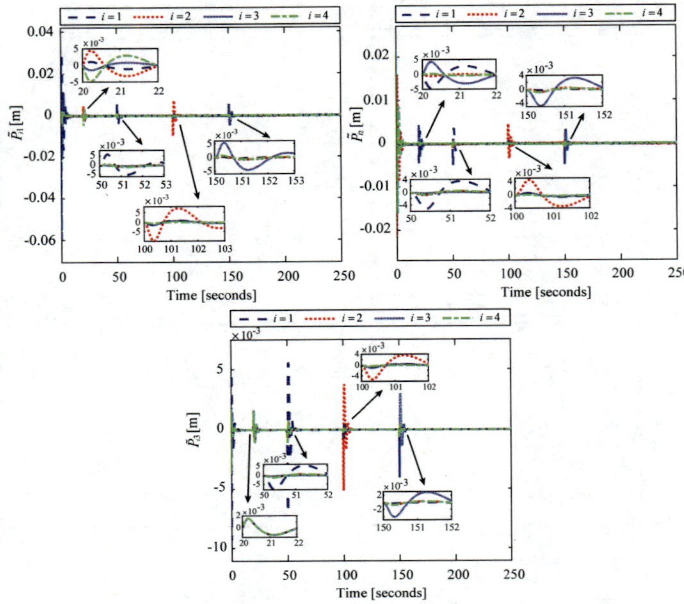

Fig. 9.7 Position tracking errors of four UAVs

degradations when all UAVs are subjected to the external disturbances at $t = 20$ s. The position tracking errors $\tilde{P}_{11}$, $\tilde{P}_{12}$, and $\tilde{P}_{13}$ also encounter perturbations when UAV#1 becomes faulty at $t = 50$ s. Similar performance degradations are also observed when UAV#2 and UAV#3 suffer from actuator faults at $t = 100$ s and $t = 150$ s, respectively. Then, under the developed control scheme, the degraded performances are improved, and the stability of the UAV team is regained. Figure 9.8 illustrates the cooperative position tracking errors of four UAVs, and it can be seen that the cooperative tracking errors of four UAVs are UUB and the cooperative monitoring of forest fire is achieved.

Figure 9.9 gives the velocities, heading angles, and flight path angles of four fixed-wing UAVs. It can be seen that all these states are bounded. Note that regarding the heading angle, the values of 0 rad and 2π rad represent the same heading angle. Therefore, the heading angles in Fig. 9.9 are continuous and bounded. Figure 9.10 illustrates the control inputs, i.e., thrusts, lift factors, and bank angles, which are bounded even when UAVs are subjected by external disturbances, gust, and UAVs#1, 2, 3 encounter actuator faults. It is also observed in Fig. 9.10 that the control inputs adjust their signals promptly to attenuate the adverse effects of disturbances and actuator faults. Figure 9.11 illustrates the lumped disturbances estimated by the SMDOs with the combination of SMDs and reference systems. With the help of the corresponding estimations of lumped disturbances induced by external disturbances and actuator faults under the proposed DOs, the cooperative monitoring performance of forest fire using a group of fixed-wing UAVs can be

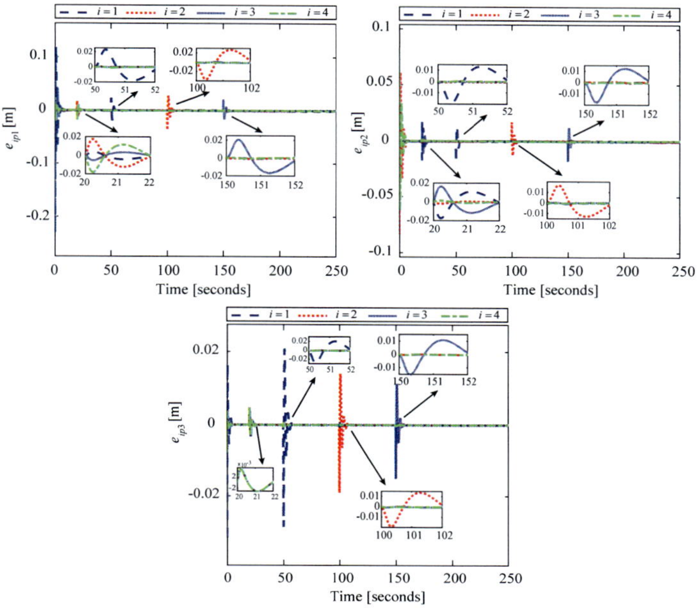

Fig. 9.8 Cooperative position tracking errors of four UAVs

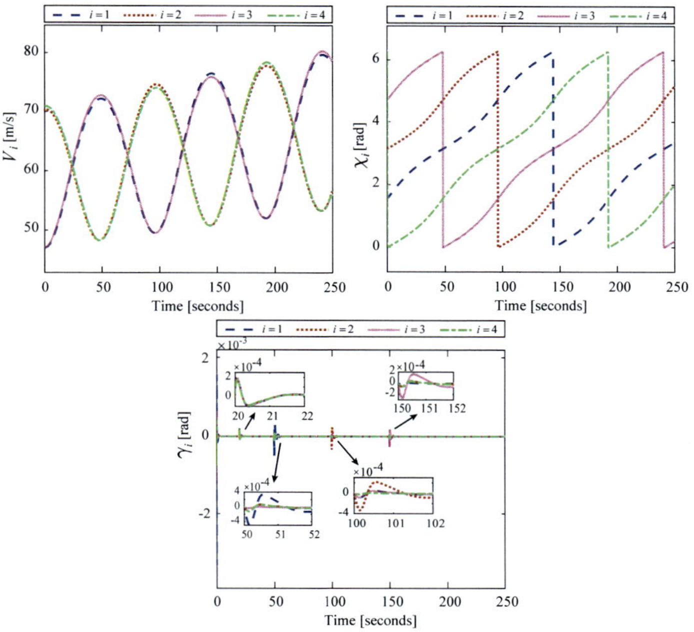

Fig. 9.9 Velocities, heading angles, and flight path angles of four UAVs

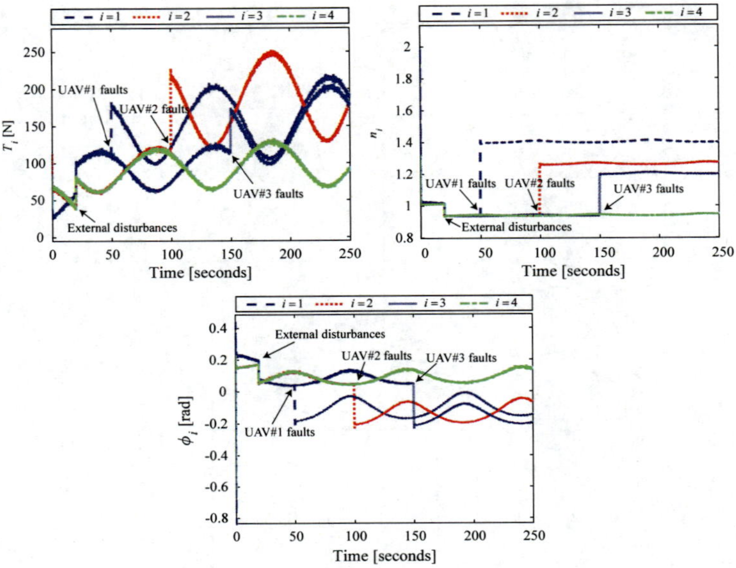

Fig. 9.10 Control inputs of four UAVs

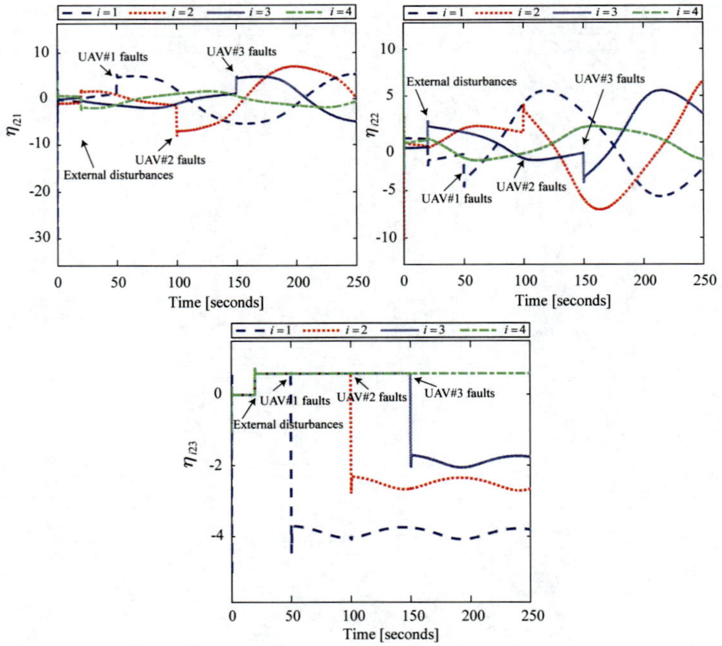

Fig. 9.11 Estimated lumped disturbances

guaranteed even when all UAVs encounter external disturbances and a portion of UAVs is subjected to the actuator faults.

9.5 Conclusions

This chapter has proposed a cooperative forest fire monitoring method by using the time-varying elliptical formation control strategy with multiple fixed-wing UAVs even when a subset of UAVs encounter actuator faults and all UAVs are subjected to external disturbances. By using SMDs and reference systems to construct SMDOs, lumped disturbances have been expeditiously estimated. Then, by incorporating the FO calculus into the cooperative monitoring method, FO SMC scheme has been developed for a group of fixed-wing UAVs. It has been proven theoretically that the presented control scheme can make the positions of all UAVs cooperatively monitor the border of forest fire with safe distances, and the cooperative tracking errors are UUB. Simulation results have shown the effectiveness of the proposed method.

References

1. M. Basin, P. Yu, Y. Shtessel, Finite- and fixed-time differentiators utilising HOSM techniques. IET Contr. Theory Appl. **11**(8), 1144–1152 (2016)
2. D.W. Casbeer, R.W. Beard, T.W. McLain, S.M. Li, R.K. Mehra, Forest fire monitoring with multiple small UAVs, in *American Control Conference*, Portland, OR, USA (2005)
3. D.W. Casbeer, D. Kingston, R.W. Beard, T.W. Mclain, Cooperative forest fire surveillance using a team of small unmanned air vehicles. Int. J. Syst. Sci. **37**(6), 351–360 (2006)
4. J.L. Han, L. Di, C. Coopmans, Y.Q. Chen, Pitch loop control of a VTOL UAV using fractional order controller. J. Intell. Robot. Syst. **73**(1–4), 187–195 (2014)
5. F.M.A. Hossain, Y.M. Zhang, M.A. Tonima, Forest fire flame and smoke detection from UAV-captured images using fire-specific color features and multi-color space local binary pattern. J. Unmanned Veh. Syst. **8**, 285–309 (2020)
6. C.C. Hua, J.N. Chen, X.P. Guan, Fractional-order sliding mode control of uncertain QUAVs with time-varying state constraints. Nonlinear Dyn. **95**(2), 1347–1360 (2019)
7. D. Kingston, R.W. Beard, R.S. Holt, Decentralized perimeter surveillance using a team of UAVs. IEEE Trans. Robot. **24**(6), 1394–1404 (2008)
8. B. Li, W. Xie, Adaptive fractional differential approach and its application to medical image enhancement. Comput. Electr. Eng. **45**, 324–335 (2015)
9. H.S. Li, Y. Luo, Y.Q. Chen, A fractional order proportional and derivative (FOPD) motion controller: tuning rule and experiments. IEEE Trans. Control Syst. Technol. **18**(2), 516–520 (2009)
10. Z. Li, L. Liu, S. Dehghan, Y.Q. Chen, D.Y. Xue, A review and evaluation of numerical tools for fractional calculus and fractional order controls. Int. J. Control **90**(6), 1165–1181 (2017)
11. R.L. Magin, *Fractional Calculus in Bioengineering* (Begell House Redding, Danbury, 2006)
12. P.K. Menon, G.D. Sweriduk, B. Sridhar, Optimal strategies for free-flight air traffic conflict resolution. J. Guid. Control Dyn. **22**(2), 202–211 (1999)

13. L. Merino, F. Caballero, J.R.M.D. Dios, A. Ollero, Cooperative fire detection using unmanned aerial vehicles, in *International Conference on Robotics and Automation*, Barcelona, Spain (2005)
14. L. Merino, F. Caballero, J.R.M.D. Dios, J. Ferruz, A. Ollero, A cooperative perception system for multiple UAVs: application to automatic detection of forest fires. J. Field Robot. **23**(3–4), 165–184 (2006)
15. L. Merino, F. Caballero, J.R.M.D. Dios, I. Maza, A. Ollero, Automatic forest fire monitoring and measurement using unmanned aerial vehicles, in *International Congress on Forest Fire Research*, Coimbra, Portugal (2010)
16. L. Merino, F. Caballero, J.R. Martínez-De-Dios, I. Maza, A. Ollero, An unmanned aircraft system for automatic forest fire monitoring and measurement. J. Intell. Robot. Syst. **65**(1–4), 533–548 (2012)
17. M.F. Mysorewala, D.O. Popa, F.L. Lewis, Multi-scale adaptive sampling with mobile agents for mapping of forest fires. J. Intell. Robot. Syst. **54**(4), 535–565 (2009)
18. N. Nikdel, M. Badamchizadeh, V. Azimirad, M.A. Nazari, Fractional-order adaptive backstepping control of robotic manipulators in the presence of model uncertainties and external disturbances. IEEE Trans. Ind. Electron. **63**(10), 6249–6256 (2016)
19. D. Nojavanzadeh, M. Badamchizadeh, Adaptive fractional-order non-singular fast terminal sliding mode control for robot manipulators. IET Contr. Theory Appl. **10**(13), 1565–1572 (2016)
20. G.L.W. Perry, Current approaches to modelling the spread of wildland fire: a review. Prog. Phys. Geog. **22**(2), 222–245 (1998)
21. I. Podlubny, *Fractional Differential Equations*, vol. 198 (Academic Press, Cambridge, 1999)
22. V. Sherstjuk, M. Zharikova, I. Sokol, Forest fire-fighting monitoring system based on UAV team and remote sensing, in *International Conference on Electronics and Nanotechnology*, Kiev, Ukraine (2018)
23. P. Skorput, S. Mandzuka, H. Vojvodic, The use of unmanned aerial vehicles for forest fire monitoring, in *International Symposium on Electronics in Marine*, Zadar, Croatia (2016)
24. S. Sudhakar, V. Vijayakumar, C.S. Kumar, V. Priya, L. Ravi, V. Subramaniyaswamy, Unmanned aerial vehicle (UAV) based forest fire detection and monitoring for reducing false alarms in forest-fires. Comput. Commun. **149**, 1–16 (2020)
25. P.B. Sujit, D. Kingston, R. Beard, Cooperative forest fire monitoring using multiple UAVs, in *IEEE Conference on Decision and Control*, New Orleans, LA, USA (2007)
26. G.H. Sun, L.G. Wu, Z.A. Kuang, Z.Q. Ma, J.X. Liu, Practical tracking control of linear motor via fractional-order sliding mode. Automatica **94**, 221–235 (2018)
27. Y.Y. Wang, L.Y. Gu, Y.H. Xu, X.X. Cao, Practical tracking control of robot manipulators with continuous fractional-order nonsingular terminal sliding mode. IEEE Trans. Ind. Electron. **63**(10), 6194–6204 (2016)
28. E. Wardihani, M. Ramdhani, A. Suharjono, T.A. Setyawan, S.S. Hidayat, S.W. Helmy, E. Triyono, F. Saifullah, Real-time forest fire monitoring system using unmanned aerial vehicle. J. Eng. Technol. **13**(6), 1587–1594 (2018)
29. Y.J. Xu, Nonlinear robust stochastic control for unmanned aerial vehicles. J. Guid. Control Dynam. **32**(4), 1308–1319 (2009)
30. Q. Yang, D.L. Chen, T.B. Zhao, Y.Q. Chen, Fractional calculus in image processing: a review. Fract. Calc. Appl. Anal. **19**(5), 1222–1249 (2016)
31. C. Yin, Y.Q. Chen, S.M. Zhong, Fractional-order sliding mode based extremum seeking control of a class of nonlinear systems. Automatica **50**(12), 3173–3181 (2014)
32. Z.Q. Yu, Y.H. Qu, Y.M. Zhang, Safe control of trailing UAV in close formation flight against actuator fault and wake vortex effect. Aerosp. Sci. Technol. **77**, 189–205 (2018)
33. Z.Q. Yu, Y.H. Qu, Y.M. Zhang, Fault-tolerant containment control of multiple unmanned aerial vehicles based on distributed sliding-mode observer. J. Intell. Robot. Syst. **93**(1–2), 163–177 (2019)

34. Z.Q. Yu, Y.M. Zhang, B. Jiang, C.Y. Su, J. Fu, Y. Jin, T.Y. Chai, Decentralized fractional-order backstepping fault-tolerant control of multi-UAVs against actuator faults and wind effects. Aerosp. Sci. Technol. **104**, 105939 (2020)
35. Z.Q. Yu, Y.M. Zhang, B. Jiang, X. Yu, J. Fu, Y. Jin, T.Y. Chai, Distributed adaptive fault-tolerant close formation flight control of multiple trailing fixed-wing UAVs. ISA Trans. **106**, 181–199 (2020)
36. Z.Q. Yu, Y.M. Zhang, Z.X. Liu, Y.H. Qu, C.Y. Su, B. Jiang, Decentralized finite-time adaptive fault-tolerant synchronization tracking control for multiple UAVs with prescribed performance. J. Frankl. Inst. **357**(16), 11830–11862 (2020)
37. Z.Q. Yu, Y.M. Zhang, B. Jiang, J. Fu, Y. Jin, T.Y. Chai, Composite adaptive disturbance observer-based decentralized fractional-order fault-tolerant control of networked UAVs. IEEE Trans. Syst. Man Cybern. -Syst. **52**(2), 799–813 (2022)
38. C. Yuan, Z.X. Liu, Y.M. Zhang, UAV-based forest fire detection and tracking using image processing techniques, in *International Conference on Unmanned Aircraft Systems (ICUAS)*, Denver, CO, USA (2015)
39. C. Yuan, Y.M. Zhang, Z.X. Liu, A survey on technologies for automatic forest fire monitoring, detection, and fighting using unmanned aerial vehicles and remote sensing techniques. Can. J. Forest Res. **45**(7), 783–792 (2015)
40. C. Yuan, Z.X. Liu, Y.M. Zhang, Aerial images-based forest fire detection for firefighting using optical remote sensing techniques and unmanned aerial vehicles. J. Intell. Robot. Syst. **88**(2–4), 635–654 (2017)
41. F.R. Zhang, C.P. Li, Stability analysis of fractional differential systems with order lying in (1, 2). Adv. Differ. Equ. **2011**(1), 213485 (2011)

Chapter 10
Conclusions and Future Directions

10.1 Conclusions

For multi-UAVs system, FTCC techniques are generally utilized to ensure the stability, reliability, autonomy, and intelligence of the formation team, especially in the presence of actuator/sensor/communication/system component faults. This monograph focuses on the FTCC design and the stability analysis for multi-UAVs. In view of the proposed FTCC schemes in the previous chapters, the contributions and achievements of this monograph are summarized as the following aspects:

(1) A prescribed performance-based distributed neural adaptive FTCC scheme is presented to achieve the longitudinal motions of multi-UAVs. A distributed SMO is first utilized to estimate the leader UAV' s reference. Then, by transforming the tracking errors of follower UAVs with respect to the estimated references into a new set of errors, a distributed neural adaptive FTCC protocol is developed based on the combination of DSC and MLPNN. Moreover, auxiliary dynamic systems are exploited to deal with input saturation.
(2) A distributed FTCC scheme is developed to achieve the longitudinal motions of multi-UAVs in the presence of actuator faults and input saturation. To eliminate the "differential explosion" in the traditional backstepping architecture, the DSC architecture is utilized to construct the distributed FTCC scheme. Moreover, the DO technique is used to handle the actuator faults and external disturbances. Furthermore, to reduce the adverse influence caused by the control input saturation of UAVs, auxiliary dynamic systems are introduced to regulate the control signals when the input signals are saturated for a long time.
(3) By considering the formation team consisting of multiple leader UAVs and multiple follower UAVs in the presence of faults and input saturation, a distributed FTCC scheme is developed to regulate the longitudinal motions of multi-UAVs by using the distributed SMO, DO, and HGO, such that all follower UAV can be steered into the convex hull spanned by the leader UAVs.

Z. Yu et al., *Fault-Tolerant Cooperative Control of Unmanned Aerial Vehicles*,
https://doi.org/10.1007/978-981-99-7661-4_10

(4) By incorporating the finite-time feature into the FTCC design, a distributed finite-time FTCC scheme is further presented to adjust the longitudinal motions of multi-UAVs against faults and input saturation, such that the adverse effects of faults can be attenuated in a timely manner. The distributed finite-time SMO, NN, MLPNN, and FOSMD are integrated to facilitate the FTCC design.
(5) By simultaneously considering the finite-time and prescribed performance requirements, a decentralized FTCC scheme is developed to achieve the attitude synchronization tracking control of multi-UAVs against faults. By integrating the prescribed performance functions into the synchronization tracking errors, a finite-time attitude synchronization tracking control scheme is developed by using NNs and finite-time differentiators. To reduce the computational burden caused by estimating the weight vectors, the norms of weight vectors are used for the estimation, such that the number of adaptive parameters is significantly reduced and independent from the number of neurons.
(6) By considering the directed communications, a decentralized FTCC scheme is developed to achieve the attitude synchronization tracking control of multi-UAVs. NNs are used to approximate unknown nonlinear terms due to the inherent nonlinearities in UAV models and the loss-of-effectiveness actuator faults. To further compensate for NN approximation errors and actuator bias faults, the DO technique is incorporated into the control scheme to increase the composite approximation capability. Moreover, the prediction errors, which represent the approximation qualities of the states induced by NNs and DOs to the measured states, are integrated into the developed FTCC scheme.
(7) A practical cooperative forest fire monitoring example is implemented on multi-UAVs to achieve the time-varying elliptical formation flight against faults. By using the FO SMC strategy, an FTCC is developed for multi-UAVs to monitor the elliptical spread of forest fire. To estimate the lumped disturbances induced by the external disturbances and actuator faults, SMDOs are developed by introducing reference systems and SMDs.

10.2 Future Research Directions

This monograph presents systematical and comprehensive descriptions of the FTCC schemes for multi-UAVs against actuator/sensor/communication/system component faults. It provides some interesting and effective solutions for the FTCC design. However, there still exist some open and challenging problems to be addressed for the advanced FTCC design, such as the integrated design of FTCC and FDD, simultaneous consideration of outer-loop kinematics and inner-loop dynamics, finite/fixed-time convergence, various physical constraints, simultaneous consideration of faults, and attacks. Some possible future research directions are listed as follows:

(1) In practical formation flight, the formation pattern and FTCC performance are highly coupled, and it is necessary to extend existing research outcomes in formation flight to address the FTCC problem by simultaneously considering the outer-loop kinematics and inner-loop dynamics.
(2) With respect to the FTCC of multi-UAVs, the existing results are mainly focused on the passive FTCC design without the FDD module, which cannot handle untested severe faults. To increase the fault-tolerant capability of the formation team, it is necessary to investigate the active FTCC design for multi-UAVs by incorporating the FDD module.
(3) Although several finite-time FTCC results have been obtained for multi-UAVs, such a finite-time feature should be further explored for multi-UAVs to react to the faults in a timely manner by simultaneously considering the avoidance of the second damage to the faulty actuators or system components. Moreover, fixed-time FTCC results are very rare for multi-UAVs, which should be further investigated.
(4) In practical formation flight, various constraints, such as input saturation, state constraint, error constraint, and inner/outer collision avoidance, inevitably exist in the formation flight team. However, the existing FTCC results for multi-UAVs usually are considered one of the above constraints. Therefore, how to design the FTCC scheme for multi-UAVs with the consideration of the aforementioned practical constraints in a unified framework needs further research.
(5) When multi-UAVs perform tasks in a complex environment, inner faults and outer attacks may be simultaneously encountered by the formation team. Therefore, how to ensure the formation flight safety of multi-UAVs in the presence of faults and attacks should be explored to provide effective and safe cooperative control protocols.

Printed in the United States
by Baker & Taylor Publisher Services